Analytical Calorimetry

Volume 5

A Continuation Order Plan is available for this series. A continuation order will bring delivery of each new volume immediately upon publication. Volumes are billed only upon actual shipment. For further information please contact the publisher.

Analytical Calorimetry

Volume 5

Edited by

Julian F. Johnson

Institute of Materials Science
University of Connecticut
Storrs, Connecticut

and

Philip S. Gill

Du Pont Company
Wilmington, Delaware

PLENUM PRESS • NEW YORK AND LONDON

The Library of Congress cataloged the first volume of this title as follows:

Analytical calorimetry. v. 1
  1968–
  New York. Plenum Press.

  v. ill. 26 cm.
  Editors: 1968–   R. S. Porter and J. F. Johnson.
  Vols for 1968–   contain the proceedings of the American Chemical Society
Symposium on Analytical Calorimetry.
  ISSN 0066-1538

  1. Thermal analysis—Congresses—Collected works. 2. Calorimeters and calorimetry—Congress—Collected works. I. Porter, Roger Stephen. 1928–   ed. II.
Johnson, Julian Frank, 1923–   ed. III. American Chemical Society Symposium
on Analytical Calorimetry.
QD79.T38A5   547′.308′6   68-8862

Library of Congress Catalog Card Number 68-8862

ISBN-13: 978-1-4612-9677-5   e-ISBN-13: 978-1-4613-2699-1
DOI: 10.1007/978-1-4613-2699-1

Proceedings of an International Symposium on Analytical Calorimetry, a part of the
185th National Meeting of the American Chemical Society, held March 20–25, 1983,
in Seattle, Washington

© 1984 Plenum Press, New York
Softcover reprint of the hardcover 1st edition 1984

A Division of Plenum Publishing Corporation
233 Spring Street, New York, N.Y. 10013

PREFACE

This Volume 5 in a continuing series represents the compilation of papers presented at the International Symposium on Analytical Calorimetry as part of the 185th National Meeting of the American Chemical Society, Seattle, Washington, March 20-25th, 1983.

A much broader variety of topics are covered than in previous volumes, due to the growth in the field of Thermal Analysis. Specific topics covering such techniques as differential scanning calorimetry, combined thermogravimetric procedures, dynamic mechanical analysis and a variety of novel kinetic analyses are covered.

A wide range of material types are included in this volume such as polymers (alloys, blends and composites), fossil fuels, biological products, liquid crystals and inorganic materials.

The co-editors of this volume would like to thank all the contributors for their efforts in conforming to the manuscript requirements, and for being prompt in the preparation. We would also like to thank those who presided over sessions during the course of the symposium; Professor Anselm C. Griffin, Professor Roger S. Porter and Dr. Edith A. Turi.

Philip S. Gill

Julian F. Johnson

CONTENTS

COUPLING OF DSC AND DYNAMIC MECHANICAL EXPERIMENTS FOR PROBING PROCESSING-STRUCTURE-PROPERTY RELATIONSHIPS OF CATALYST MODIFIED HIGH PERFORMANCE EPOXY MATRICES

Thomas E. Munns and James C. Seferis[*]

Polymeric Composites Laboratory
Department of Chemical Engineering
University of Washington
Seattle, Washington

INTRODUCTION

Structures fabricated from advanced composite materials are being included in critical, load bearing design applications due to their outstanding mechanical properties and light weight. The structural characterization of the composite matrix materials is important because the matrix controls physical properties such as shear strength, toughness, chemical resistance, and load transfer, as well as the processing requirements of the composite. The most common advanced composite matrix materials are high performance epoxy systems. Therefore, it is necessary to understand the processing-structure-property relationships for epoxy resins.

An important epoxy matrix material for composite applications is the system consisting of tetraglycidyl 4,4'-diaminodiphenyl methane (TGDDM) and polyglycidyl ether of Bisphenol-A Novalac cured with 4,4'-diaminodiphenyl sulfone (DDS). This TGDDM-Novalac-DDS system is similar to commercially available systems including the Narmco 5208 [1]. Earlier work performed in our laboratory has focused on characterizing the processing-structure-property relationships for the TGDDM-Novalac-DDS system, using a battery of experiments including dynamic mechanical, differential scanning calorimetry, density and titration measurements [1,2].

Boron trifluoride complexes are added to epoxy matrix materials to accelerate the cure of the polymer in order to reduce

---

[*]To whom correspondence should be addressed

the processing time for the composite part. Commercial systems
which contain BF_3 catalysts include Fiberite 934 and Hercules
3501 (3,4). The present study on catalyzed systems coupled with
our earlier investigation (5) was undertaken to elucidate further
the effect of adding BF_3 catalyst to our well-characterized
TGDDM-Novalac-DDS epoxy system as a function of processing
conditions.

Films were produced for experimentation by melt (2) or
solution (6) processes developed in our laboratory. The melt
process is similar to commercial processes for producing composite
parts from unidirectional pre-impregnated tape. Woven fabric
reinforced composites are impregnated by submerging the fabric in
an epoxy solution. Thus, the solution process which is primarily
emphasized in this study provides additional information to
characterize and compare matrices produced by the two processes.

Dynamic mechanical and differential scanning calorimetry
experiments were used to analyze the BF_3 catalyzed
TGDDM-Novalac-DDS system. Dynamic mechanical experiments have
been utilized in the past to characterize this epoxy resin
(1,2,5,6,7). They have provided information concerning the
viscoelastic response of the polymer as well as important
morphological information. On the other hand, Differential
Scanning Calorimetry (DSC) experiments provide a means for
measuring extent of reaction, reaction kinetics and mechanisms of
the epoxy system curing process (2).

The purpose of this work is to combine the results from
dynamical mechanical and DSC experiments to understand the effect
of BF_3 catalysis on the TGDDM-Novalac-DDS epoxy system. The
dynamic mechanical and DSC results for solution processed epoxy
films are compared to melt processed results to determine the
effect of processing conditions on the properties of the catalyzed
TGDDM-Novalac-DDS system.

EXPERIMENTAL

Resin batches were prepared containing 88.5% by weight TGDDM
(MY720), 11.5% by weight Novalac (Celanese EPI-REZ-SU-8), and 25
PHR DDS (Ciba-Geigy Eporal) with 0, 1, 2, and 3 PHR boron
trifluoride monoethylamine (BF_3:MEA). PHR is based on 100 total
parts by weight of both epoxy resins. The resin constituents used
in these systems were chosen due to their widespread usage in
commercial formulations and are not the pure monomers.

The experimental resin batches were mixed according to the
procedures outlined in our earlier work (5). Melt processed films
were cured at 177°C for two hours as has been previously described

in detail (2,5). Similarly, solution processed films were made
using a method developed in our laboratory but modified
appropriately to accommodate the addition of the BF_3:MEA
catalyst (6). In summary, the mixed epoxy resin was dissolved in
dry N,N-dimethyl acetamide (DMAc). The solution was brought to
135-140°C and reacted for six hours with constant stirring. The
epoxy solution was slow cooled to 120°C and poured over preheated
(120°C) mercury. The solvent was driven off under reduced
pressure at 115-120°C for sixteen hours.

Samples measuring 0.4cm x 5cm x 0.01cm thick were cut from
the melt and solution processed films using a scratch and break
technique. Sample edges were sanded smooth to produce an edge of
high quality for dynamic mechanical experiments. The dynamic
mechanical experiments and error analysis were conducted on a
modified Rheovibron DDVII at a frequency of 11 Htz and a heating
rate of 1°C/min according to our well-established procedures (8).
Corrected data are reported here in the traditional tan δ and
dynamic modulus forms.

Differential scanning calorimetry experiments were performed
using a Dupont 910 DSC module and 1090 thermal analyzer. DSC
experiments were conducted on uncured resin, melt processed, and
solution processed samples at a heating rate of 2°C/min.

RESULTS AND DISCUSSION

Four principle reactions have been identified for the cure of
TGDDM-DDS and TGDDM-Novalac-DDS epoxy systems (3,5). These
include primary amine, secondary amine, hydroxyl and
homopolymerization reactions. The primary amine reaction involves
reaction between primary amino hydrogens and epoxide groups. The
secondary amine reaction involves reaction between secondary amino
hydrogens and epoxide groups (2). The hydroxyl reaction involves
reaction between epoxide groups and hydroxyl groups that are a
result of impurities in the TGDDM or previous amine-epoxy
reactions (3). Reaction mechanisms involving epoxy
homopolymerization have been described for epoxy systems catalyzed
by BF_3 compounds (9,10). If stepwise amine-epoxy reaction is
assumed, crosslinks result from all secondary amine reactions or
whenever three or more epoxide groups from a monomer molecule have
participated in amine, hydroxyl, or homopolymerization reactions
(2).

For these catalyzed systems, two main viscoelastic
transitions were observed in the dynamic mechanical data as
described in our previous publication (5) and shown in Figure 1(a)
for tan δ. The low temperature, β, transition provides
information concerning the amine-epoxide reactions. Since the

magnitude and breadth of the β transition increase with increasing amine-epoxy reaction, the relative extent of amine-epoxy reaction can be determined (2,7).

The high temperature, α transition, splits into two transitions, α_1 and α_2, for a partially cured sample that undergoes further reaction during the dynamic mechanical experiment (1,2,5). From the temperature of the α_1 peak, the relative extent of crosslinking before additional curing reaction can be determined. The temperature of the α_1 transition will increase as crosslinking increases due to the lower mobility of a more highly crosslinked network. The magnitude of the α_1 transition decreases as the amount of cure increases for similarly crosslinked networks. Therefore, the relative extent of reaction for the original, partially cured, network can be determined from the α_1 peak magnitude. The α_2 transition is attributed to the glass transition of the fully cured network. The α_2 transition temperature can be related to the extent of crosslinking in the final network.

Dynamic mechanical results for melt processed films, as reported in our earlier work, reflect the change in networks due to BF_3:MEA catalysis (5). A small increase in amine-epoxy reaction with increasing BF_3:MEA content was indicated by a slight increase in the magnitude and breadth of the β transition in tan δ. This was attributed to the heat input to the system by the exothermic catalyzed reaction. A significant decrease in α_1 transition magnitude, with only a slight increase in β transition, indicated that as BF_3:MEA concentration was increased, the extent of homopolymerization and total epoxide group reaction increased. The temperature of the α_2 peak maximum increased with increasing BF_3:MEA content, indicating an increase in the crosslink density of the resulting network. The glass transition temperature (T_g), defined from dynamic modulus data as the temperature where the sample's modulus is one half its room temperature value (1,11), increased with increasing BF_3:MEA concentration due to the higher extent of crosslinking for the catalyzed network.

The dynamic mechanical results for the solution processed samples shown in Figure 1 provide information concerning the changes in network structure as catalyst concentration was varied. The magnitude and breadth of the β-transition decreased slightly with increasing BF_3:MEA concentration indicating a lower extent of amine-epoxy reaction with increasing catalyst content. It can be safely assumed for this case that the heat from the exothermic catalyzed reaction can be quickly dissipated in solution and should not affect the cure reactions. Thus, the amount of amine-epoxy reaction may have decreased due to the competition with catalyzed reactions.

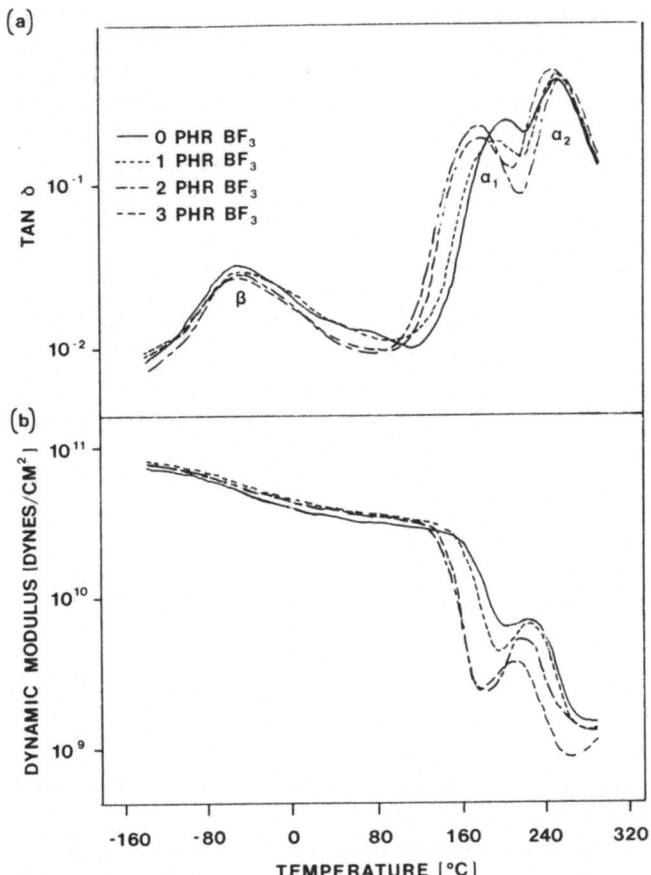

Fig. 1.   Dynamic mechanical properties for solution processed
          BF$_3$:MEA catalyzed TGDDM-Novalac-DDS.
          a) Tan δ as a function of temperature
          b) Dynamic modulus as a function of temperature

     The temperature of the α$_1$ transition maximum, for solution
processed films, decreased significantly with increasing BF$_3$:MEA
concentration. This indicates a lower degree of crosslinking as
catalyst is added to the resin. The decrease in crosslinking can
be attributed to the increased importance of the catalyzed
homopolymerization reaction, which is primarily a chain extension
reaction due to the high diffusive mobility of molecules in
solution.

     The dynamic modulus, shown in Figure 1(b), shows no
significant variation attributable to BF$_3$:MEA catalysis below

120°C. However, above 120°C, the dynamic modulus varies
significantly due to the lower extent of crosslinking in the
catalyzed networks. The decrease in both magnitude and
temperature of the final dynamic modulus maxima with increasing
catalyst content indicates that the network structure of the
catalyzed systems remains less crosslinked after the additional

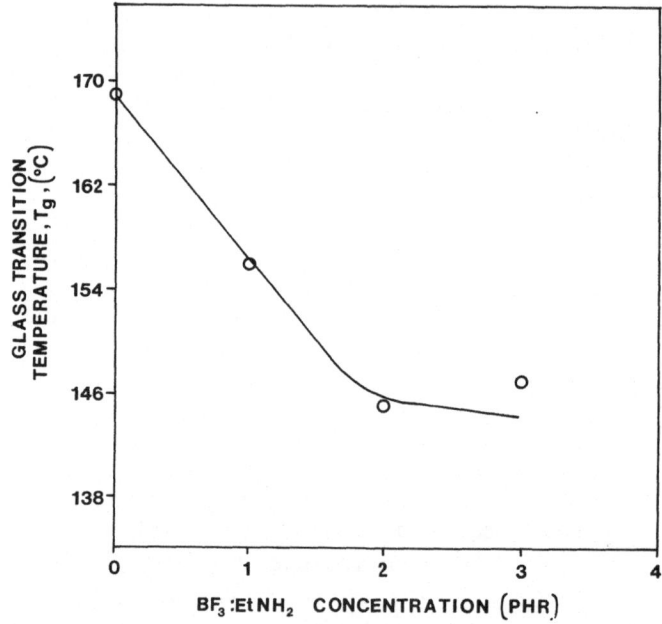

Fig. 2   Glass transition temperature (T$_g$) of the solution
           processed TGDDM-Novalac-DDS epoxy as a function of
           BF$_3$:MEA concentration. Tg is defined as the
           temperature where the sample's dynamic modulus equals
           half of its value at room temperatre (E$ = E_{RT}$/2)

reaction during the dynamic mechanical experiment. The glass
transition temperature defined on modulus basis (Tg=T$_{E_{rt}/2}$)
decreases as the BF$_3$:MEA is increased and becomes asymptotic
between 2 and 3 PHR BF$_3$:MEA as shown in Figure 2. This is
indicative of the higher extent of chain extension and lower
crosslinking as the BF$_3$:MEA concentration is increased.

DSC thermograms for the uncured epoxies are shown in
Figure 3. The uncatalyzed system exhibits two reaction peaks, as
observed in our earlier work (12). The lower temperature shoulder
is attributable to amine-epoxy reaction and the higher temperature
peak is due to hydroxyl and homopolymerization reactions. The

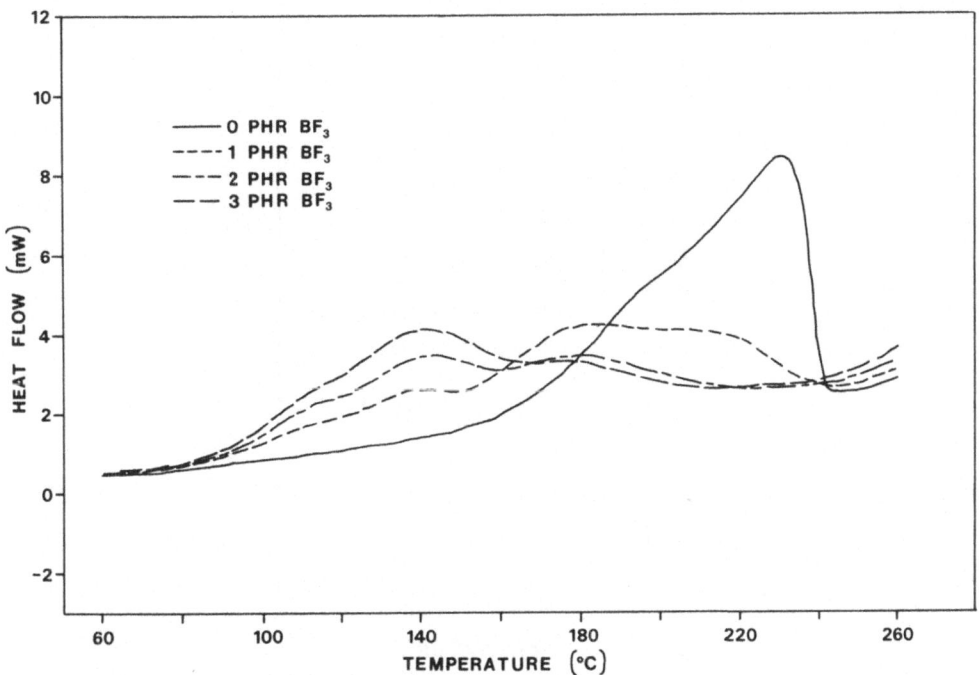

Fig. 3.   DSC thermograms for uncured TGDDM-Novalac-DDS BF_3:MEA
systems

catalyzed systems exhibit two additional low temperature reaction
peaks. The additional peaks may be due to BF_3:MEA catalyzed
homopolymerization and catalytically aided amine-epoxy or hydroxyl
reactions. As the BF_3:MEA concentration is increased, the low
temperature reactions become more important, at the expense of the
hydroxyl and, later, of the amine-epoxy reactions.

Kinetic analyses were performed on the initial reaction peaks for the uncured epoxy systems according to the method of Borchardt and Daniels (13). The calculated kinetic parameters are shown in Table 1. When BF_3:MEA is added to the uncatalyzed system, the reaction order increases and the activation energy and

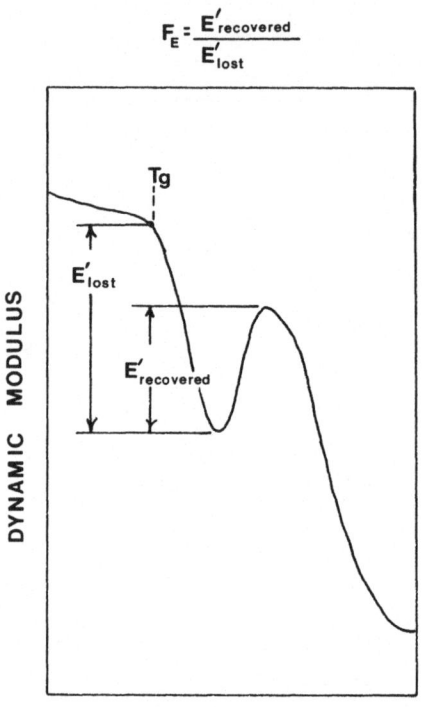

Fig. 4. Method used to calculate the modulus recovery ratio, F_E.

pre-exponential factor decrease. As the concentration of catalyst is increased, the reaction order decreases and the activation energy and pre-exponential increase.

DSC experiments were also conducted on the melt and solution processed cured films to determine the heat of additional reaction, ΔH_A. The magnitude in tan δ of the α_1 transition provides an indication of the relative extent of additional reaction for similarly crosslinked networks. The α_1 transition has been shown to be a composite peak, consisting of contributions of enhanced network molecular mobility due to the onset of the glass transition for the partially cured network, and the decrease in molecular mobility attributed to further cure reaction (1). Highly crosslinked networks would receive a smaller contribution to the α_1 peak from the enhanced molecular mobility than less crosslinked networks. The modulus recovery ratio, F_E, of the increase in dynamic modulus due to additional cure to the decrease in dynamic modulus due to the glass transition of the partially cured network was calculated, as shown schematically in Figure 4, as a measure of additional reaction that is insensitive to network molecular mobility.

For melt processed films, ΔH_A decreases with increasing BF_3:MEA concentration. As shown in Table 2, values of ΔH_A correlate well with indicators of relative extents of further reaction from dynamic mechanical experiments, the magnitude of tan δ at the α_1 peak and F_E. No value for α_1 peak height is included for the melt processed system containing 3 PHR BF_3:MEA because no α_1 transition was observed. For melt processed batches containing 2 and 3 PHR BF_3:MEA, no values for F_E are included because there was no measurable change in dynamic modulus attributable to the additional reaction. These results support the conclusion that the extent of reaction increases with BF_3:MEA concentration for melt processed samples.

As shown in Table 3, for solution processed samples, ΔH_A increases with increasing BF_3:MEA concentration until 3 PHR $BF3$:MEA where ΔH_A decreases significantly. The increase in ΔH_A, for the cured films containing 1 and 2 PHR BF_3:MEA, may be due to the higher molecular mobility in the less crosslinked catalyzed systems, which will allow more interaction and more reaction. The decrease in ΔH_A for the system containing 3 PHR BF_3:MEA may be attributable to a depletion of unreacted epoxides available for additional reaction. The magnitudes of the α_1 peak maximums do not correlate well with ΔH_A results because of the sensitivity of the α_1 transition to changes in molecular mobility. The large decreases in α_1 temperatures with increasing BF_3:MEA concentration indicates large increases in molecular mobility with increasing BF_3:MEA content. In contrast, the ratio F_E correlates well with ΔH_A results because F_E is insensitive to network mobility.

TABLE 1. Kinetic Parameters for TGDDM-Novalac-DDS-BF_3:MEA Uncured Resin Systems

BF_3:MEA Concentration (phr)	Reaction Order, n	Activation Energy, E, (KJ/mole	Pre-Exponential Factor, Log A, (1/min)
0	1.92	112.0	11.1
1	3.50	69.0	7.2
2	3.29	91.2	10.6
3	2.90	97.0	11.5

TABLE 2. Heat of Additional Cure, α_1 Peak Magnitude, and Modulus Recovery Ratio for Melt Processed Films

BF_3:MEA Concentration (PHR)	ΔH_A (J/g)	α_1 Peak Magnitude	F_E
0	130.0	0.095	0.638
1	82.9	0.055	0.252
2	63.2	0.033	-
3	61.6	-	-

TABLE 3. Heat of Additional Cure, α_1 Peak Magnitude, and Modulus Recovery Ratio for Solution Processed Films

BF_3:MEA Concentration (PHR)	ΔH_A (J/g)	α_1 peak Magnitude	F_E
0	58.3	0.268	0.194
1	70.7	0.195	0.292
2	77.9	0.248	0.352
3	29.8	0.220	0.120

CONCLUSIONS

Epoxy film samples were made from BF_3:MEA catalyzed TGDDM-Novalac-DDS resin systems using melt and solution processing techniques. The samples were analyzed using dynamic mechanical and DSC experiments. The results from dynamic mechanical and DSC experiments indicate that the presence of BF_3:MEA catalyst in the TGDDM-Novalac-DDS epoxy system accelerates epoxide homopolymerization reactions. For melt processed films, the magnitude of the low temperature β transition in tan δ increased slightly and the high temperature $α_1$ transition magnitude decreased significantly with increasing BF_3:MEA concentration. These results indicate that the extent of epoxide reaction and degree of crosslinking increased with increasing BF_3:MEA concentration. The catalyzed reactions led to a higher extent of crosslinking for the melt processed films because of the higher cure temperature and the heterogeneous nature of the reactions due to diffusion limitations in the melt. The clustering of reactants leads to more crosslinks than if the monomers were more free to diffuse.

For solution processed films, the β transition magnitude decreased slightly and the $α_1$ transition temperature decreased significantly with increasing BF_3:MEA concentration. These results indicate that the degree of crosslinking decreased as the concentration of BF_3:MEA was increased. The catalyst led to a lower extent of crosslinking due to lower cure temperatures and the homogeneous nature of the reactions due to the easy diffusion of reactants in solution, which favors the catalyzed chain extension reactions.

The DSC results for the unreacted systems exhibited two reaction peaks for the uncatalyzed system and two additional lower temperature peaks for the catalyzed systems. DSC experiments indicate that BF_3:MEA increases the rate of polymerization and decreases the reaction temperatures for the TGDDM-Novalac-DDS epoxy system. As the catalyst concentration is increased, the importance of the low temperature catalyzed reactions increases.

In summary, dynamic mechanical and DSC experiments have proven effective in determining the effect of processing conditions and the addition of BF_3:MEA catalyst on the network structure and properties of the commercially important TGDDM-Novalac-DDS epoxy system.

ACKNOWLEDGMENTS

The authors wish to express their appreciation to the Boeing Commercial Airplane Company for providing financial assistance for this work. We also thank Dr. J.Chen of Boeing and Dr. M. Ibrahim, post-doctoral fellow of the Polymeric Composites Laboratory, for helpful discussions.

REFERENCES

1. J. D. Keenan, J. C. Seferis, and J. T. Quinlivan, J. Appl.
        Polym. Sci., 24, 2375 (1979).
2. H. S. Chu, M. S. Thesis, Department of Chemical Engineering,
        University of Washington (1980); H. S. Chu and
        J. C. Seferis, in "The Role of the Polymeric Matrix on
        the Processing and Properties of Composite Materials,"
        J. C. Seferis and L. Nicolais, Eds., Plenum, in press
        (1983).
3. E. T. Mones, C. M. Walkup, J. A. Happe, and R. J. Morgan,
        Proceedings of the 14th National SAMPE Technical
        Conference, 89 (1982).
4. J. F. Carpenter, N00019-77-C-0155, May 1978.
5. T. E. Munns and J. C. Seferis, J. Appl. Polym. Sci., in
        press (1983).
6. A. M. Ibrahim and J. C. Seferis, in "Interrelations between
        Processing, Structure, and Properties of Polymeric
        Materials," IUPAC Symposium Proceedings, J. C. Seferis
        and P. S. Theocaris, Eds., Elsevier Press, in press
        (1983).
7. M. Ochi, M. Okazaki, and M. Shimbo, J. Polym. Sci:Polym.
        Physics Ed., 20, 689 (1982).
8. A. R. Wedgewood and J. C. Seferis, Polymer, 22, 966 (1981).
9. J. J. Harris and S. C. Temin, J. Appl. Polym. Sci., 10, 523
        (1966).
10. W. G. Potter, "Epoxy Resins," Iliffe, 1970, pp. 50-51.
11. R. J. Morgan and J. E. O'Neal, Polym. Plast. Technol. Eng.,
        10 (1), 49 (1978).
12. E. B. Stark, A. M. Ibrahim, and J. C. Seferis, 28th National
        SAMPE Symp. Exhib., [Proc.], in press (1983).
13. H. J. Borchardt and F. Daniels, JACS, 79, 41-46, 1957.

ON THE GLASS TRANSITION OF POLYACRYLAMIDE

H. K. Yuen, E. P. Tam and J. W. Bulock

Monsanto Company
800 N. Lindbergh Blvd.
St. Louis, MO 63167

ABSTRACT

The effect of a plasticizer on the glass transition tempera-ture (Tg) of an amorphous polymer is well documented. However, similar behavior of water in synthetic polymers has only recently been recognized. The effect of water content (or thermal history) on the Tg's of acrylamide polymers has been studied by means of Differential Scanning Calorimetry (DSC), Thermogravimetry (TG), Thermomechanical Analysis (TMA) and Simultaneous Thermogravimetry - Mass Spectrometry (TG-MS). Commercial samples of polyacrylamide and poly-(acrylamide-co-acrylic acid) from several sources were investigated and were found to contain measurable amounts of water. A pronounced plasticizing capability was detected for a relatively low concentration of water. A lowering of the glass transition temperature of the homopolymer by about 70°C was observed with a concentration of less than 8%. A similar effect was also observed for the β transition and a ratio of 0.66 was obtained for T_β (°K)/ Tg (°K).

INTRODUCTION

Polyacrylamides (PA) are some of the most widely used water soluble polymers. Their long chain-linearity and their versa-tility have led to many successful applications over the past two decades. The biggest market sector for PA's today is municipal and industrial waste water treatment, where they are used as flocculants[1]. Other applications range from enhanced oil recovery to textile sizing.

One of the important physical characteristics which determines the end use of amorphous polymers such as PA's is the glass transition temperature (Tg). From a fundamental standpoint, Tg is a second-order transition in which the polymer changes from a glassy state to a rubbery state. At Tg, long segments of each polymer chain move in random micro-Brownian motion. Below the Tg, the only molecular motions that take place are short range movements of several short chain segments and/or pendant groups. These processes are referred to as "secondary transitions" or "relaxations". The glass transition is usually designated as the α-transition, and any subsequent transitions and relaxations lying below the α-transition are referred to as the β-transition, the γ-transition, etc.

There are numerous references in the literature to the Tg of homopolymers and copolymers. Boyer[2] and McGrum et al.[3] have reviewed this subject thoroughly. There have been a number of studies on the Tg of PA homopolymer, but there is little agreement among them. For instance: a) Chiantore and co-workers[4,5] reported a Tg of 175°C to 179°C using Differential Thermal Analysis; b) Klein and Heitzman[6] obtained a somewhat higher result of 188°C; c) to confuse the picture even further, Illers[7] found the Tg as low as 153°C.

Although water is known as a natural plasticizer for many polar polymers such as nylon[8], polyester resins[9], and cellulosic polymers[10], similar behavior for polyacrylamide and poly(acrylamide-co-acrylic acid) has not been investigated. In this study, the effect of water content (and/or thermal history) on the Tg's of acrylamide-based polymers was studied by Differential Scanning Calorimetry (DSC), Thermogravimetry (TG), Thermomechanical Analysis (TMA), and Simultaneous Thermogravimetry - Mass Spectrometry (TG/MS).

EXPERIMENTAL

Sample

The acrylamide-based polymers used in the study were obtained commercially in powder form. The homopolymers designated as PS-2806, SPP-34 and ALD-18127-7 were supplied by Polysciences, Scientific Polymer Products and Aldrich respectively. Supplied also by Polysciences were our two poly(acrylamide-co-acrylic acid) samples of high and low carboxyl content, PS-2220 and PS-4652. Table I illustrates the molecular weight distribution of these samples as determined by aqueous GPC/Laser light scattering. The acrylic acid contents of PS-2220 and PS-4652 copolymers measured by 90 MHz ^{13}C NMR were approximately 63% and 15% respectively. Both the GPC and the NMR analyses indicated PS-2220 contained appreciable amounts of impurities.

Table I. Molecular Weight Distributions for polyacrylamide and
poly(acrylamide-co-acrylic acid) determined by GPC/Laser Light
Scattering

Sample	Apparent Molecular Weight values				
	$\bar{M}n$	$\bar{M}w$	$\bar{M}z$	$\bar{M}w/\bar{M}n$	$\bar{M}z/\bar{M}n$
SPP-34	1.6×10^6	3.9×10^6	2.7×10^7	2.41	17
ALD-18127-7	1.0×10^6	1.9×10^6	4.8×10^6	1.85	4.8
PS-2806	2.9×10^5	6.7×10^5	2.4×10^6	2.34	8.2
PS-4652	2.1×10^5	5.0×10^5	1.9×10^6	2.45	9.0
PS-2220	Qualitatively, the bulk of the material had $\bar{M}w$ <5000. Impurities of very high molecular weight as well as some with $\bar{M}w$ <1000 were also observed.				

The polymers as received from the suppliers contained about
8-10% water. Samples with water contents ranging between 0-23%
were prepared as follows. For wet samples, a thin layer of the
supplied polymer was put into a 10 ml beaker. The assembly was then
exposed for various amounts of time to saturated vapor inside a 150
ml parafilm covered beaker containing about 20 ml of water. The
polymer readily picked up water in this fashion and in approximately
1 hour, a sample with about 20% water was achieved. Samples drier
than the supplied ones were accomplished by heating in a vacuum
oven at 30°C, 50°C and 80°C for about 15 hours.

Thermal Measurements

The glass transition temperature (Tg) and the β-transition (T_β)
of the polymers were measured with a DuPont 1090 DSC and a 943
thermomechanical analyzer (TMA) respectively. The DSC was equipped
with a 902 PDSC cell and a 901 cell base. Prior to the analyses,
the DSC was calibrated for temperature accuracy with decane,
napthalene and tin standards obtained from Aldrich, NBS and Perkin-
Elmer respectively. The TMA was likewise calibrated using the
recommended procedure with an Indium standard. For the DSC
experiments, the sample was contained in a hermetically sealed gold
pan and was scanned under a nitrogen atmosphere (@ 0.06 SCFH) at
10°C/min. To insure against water leakage, the sample was weighed
before and after the test. The 1090 DSC software determined Tg at
the inflection point of the discontinuity in the heat flow vs
temperature curve. To obtain T_β, the sample was chilled with liquid

nitrogen after it was placed onto the instrument. The expansion
profile of the polymer was monitored at 5°C/min under a nitrogen
purge at 45 cc/min using the regular expansion probe with a load
of 5 gm.

Concurrent with the DSC and the TMA measurements, the water
content of the polymer under study was determined thermogravimetri-
cally (TG) with a Mettler TA-1 thermoanalyzer. The sample was
contained in a quartz cylindrical crucible and was heated to
decomposition at 10°C/min under a nitrogen flow of about 8 ℓ/hr.
As a control, a blank run with the empty crucible was also conducted.
To assist the TG determination for water, simultaneous thermogravimetric - mass spectral analyses (TG/MS) were also performed under
a helium atmosphere on PS-2806, PS-2220 and PS-4652 using a tandem
instrument consisting of a Mettler TA-1 thermoanalyzer and a
Hewlett-Packard 5992 quadrupole mass spectrometer.[11]

RESULTS AND DISCUSSION

As previously stated, the purpose of the TG/MS experiments was
to guide in the determination of the water content of the polymers
by TG analysis. Shown in Figures 1, 2 and 3 are the TG, DTG
(derivative thermogravimetry) and the simultaneously obtained mass
spectral results on PS-2806 homopolymer and PS-4652, PS-2220
copolymers respectively. Water as represented by mass ions 18 and 17
(also m/e=16 for PS-2220 in Figure 3) was observed below about 190°C,
193°C and 238°C in these cases. Above the specified temperatures,
the intensity ratio of the mass ions suggested the presence of NH_3,
probably due to decomposition. In addition, for PS-4652 (Figure 2),
acrylonitrile (m/e=53) was also identified above 193°C, whereas a
possible amine (m/e=30) was found for PS-2220 above 238°C (Figure 3).
Apparent from the figures is the close resemblance betweem the
mass ion 18 and 17 profiles and the DTG curve. The decomposition
temperature cited above is indicated by a maximum in the latter.
This DTG method was hence used to assess the upper temperature limit
to which water content can be measured from the TG for all the
samples under study. Among the homopolymers and copolymers, the
decomposition temperature varied only slightly with water content.

In Figure 4, a typical DSC curve for the homopolymer (ALD-
18127-7 with 20.9% water) is shown. Near 0°C, the endotherm due to
water melting is not observed. This behavior is also common with
the "wet" copolymers and indicates that the water in these polymeric
systems is bound in nature with no freezable clusters present.
The effect of water on the glass transition temperature is
tremendous, as can be seen in Table 2. Water is known to be a
plasticizer for a number of synthetic polar polymers. The most

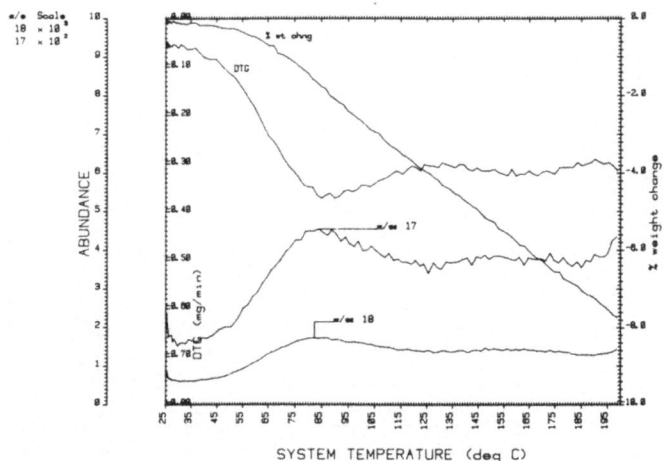

Fig. 1   TG, DTG and MS profiles for polyacrylamide (PS-2806)
        studied under a helium atmosphere.

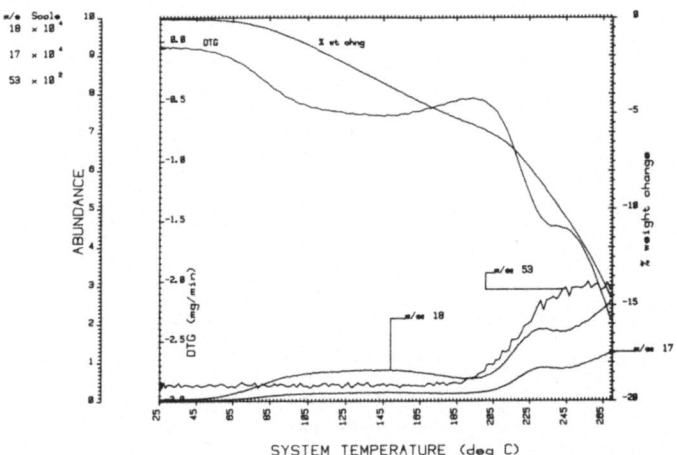

Fig. 2   TG/MS analyses of poly(acrylamide-co-acrylic acid), PS-
        4652, showing the release of water, NH₃ and acrylonitrile.

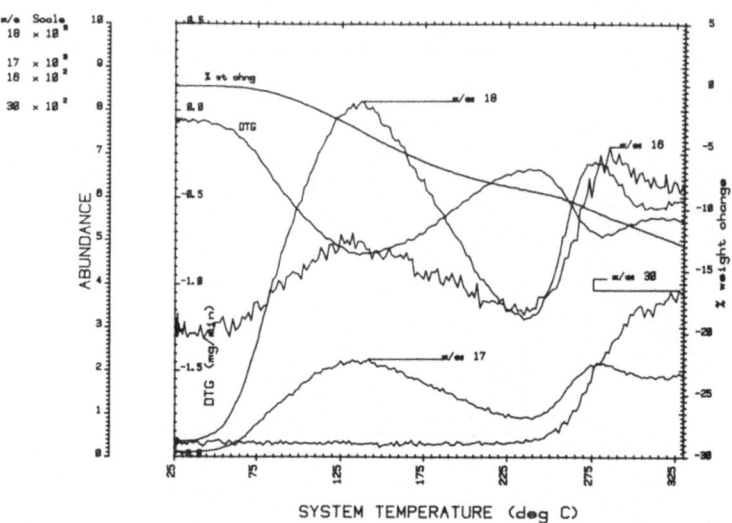

Fig. 3   TG, DTG thermograms plotted together with MS profiles for
          poly(acrylamide-co-acrylic acid), PS-2220.

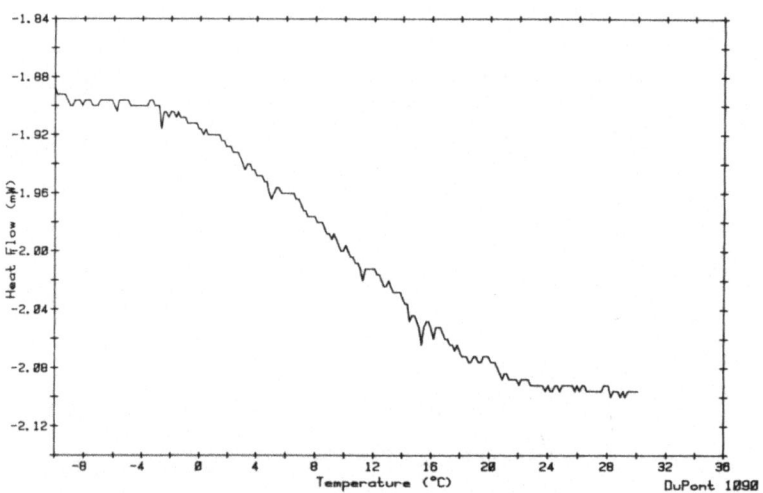

Fig. 4   DSC scan of polyacrylamide (ALD-18127-7) containing 20.9%
          water showing the glass transition and the absence of
          endotherm due to fusion of water.

common equation which describes Tg as a function of solvent concentration is that by Fox and Flory[12]:

$$\frac{1}{Tg} = \frac{W_p}{Tg_p} + \frac{W_s}{Tg_s}$$

where W is the weight fraction, Tg is in °K and the suscripts p and s represent the polymer and the solvent respectively. The equation predicts that a plot of $\frac{1}{Tg}$ vs. W_s is linear for a small amount of the plasticizer. In Figures 5 and 6 we show such plots for the poly-acrylamide homopolymers and its copolymers with acrylic acid respectively. The linearity is well observed in each case to a water content of more than 20%. For the homopolymers (Figure 5), it is seen that for water contents less than 10%, the glass transition temperature is lowest with PS-2806, followed by ALD-18127-7 and SPP-34 respectively. This order corresponds to their molecular weights as shown in Table 1 and thus reflects the effect of molecular weight on Tg. Although the Fox-Flory equation also holds for PS-2220, the copolymer with a high carboxyl content (Fig. 6), its Tg at practically any water content between 6-18% is higher than that of its low carboxyl counter-part, PS-4652; a fact which is in direct contrast to their molecular weight difference (Table 1). We were not able to determine Tg of PS-2220 at water contents below 6% with either DSC or TMA. In consideration of the impurities present in this sample as shown by NMR and GPC (see Sample), and its relatively higher decomposition temperature as previously discussed, it is reasonable to conclude that its high Tg's are probably indicative of the presence of cross-linking.

The plasticizing effect of water on the Tg of polyacrylamide and its copolymer with acrylic acid is greater than that of most other polymers. The comparison is illustrated in Table 3, where ΔTg represents the suppression of the glass transition by the indicated amount of water. It is also interesting to note that polyacrylamide in its glassy bulk form contains a much higher equilibrium amount of water at room temperature conditions, since a glass usually has fewer adsorption sites than its crystalline counterpart.

In addition to the glass transition, which is also called the α transition, β relaxation is common to most polymers and copolymers. We attempted to obtain T_β in a similar fashion as with Tg using DSC, but our efforts were in vain due to a high noise/signal ratio. Thermomechanical analysis (TMA) offered an alternative in this case, although the method was not suitable for measuring the Tg of the sample in a state other than bone-dry. Illustrated in Figure 7 is

Table 2. Glass Transition Temperatures of polyacrylamide and
poly(acrylamide-co-acrylic acid) at various water contents as
determined by DSC

	Sample	Tg(°C)	% Water
	SPP-34	173	1.0
		105.7	5.1
		82	8.6
		33.2	17.7
	ALD-18127-7	167.9	0.3
		107.5	4.5
Homopolymers		83.3	7.9
		10.6	20.9
	PS-2806	168.3	0.0
		145.5	0.7
		87.6	6.6
		83.5	7.2
		19	19.9
	PS-4652	149.9	.0.7
		83.2	7.8
Copolymers		68.1	9.2
		8	17.6
	PS-2220	118.1	5.9
		98.3	8.1
		69.1	11.0
		1.0	18.5

the TMA results obtained in penetration mode on PS-2806 homopolymer
(water content 6.8%) with a 5 gm load. Two breaks corresponding to
transition temperatures of 102°C and 155°C are apparent. While the
first transition is similar to the glass transition temperature
predicted by DSC (Figure 5), the second one corresponds to the Tg of
a sample containing about 0.5% of water. Obviously, the origin of
this "moving" Tg phenomenon is a result of progressive drying of the
sample during the analysis. This problem is of no concern for the
T_β determination, however, as T_β lies below ambient temperature and
decreases with higher water content. Since our sample was in powder
form, T_β was more conveniently measured by expansion than penetration. β relaxation from PS-2806 homopolymer was obtained at three
different water levels and the results are summarized in Table 4
together with their corresponding Tg's from Figure 5.

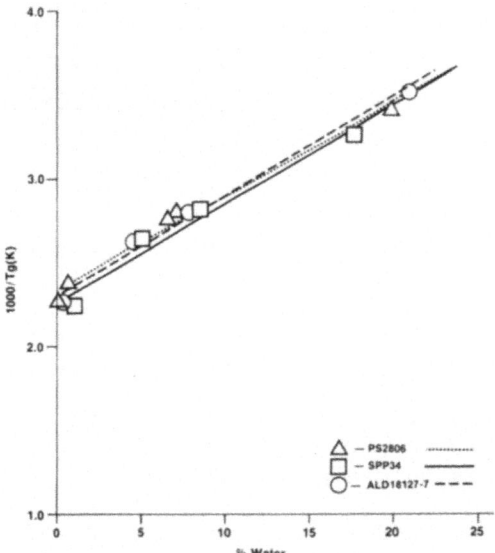

Fig. 5   A plot of Tg(°K)⁻¹ vs. water content for PS-2806, SPP-34
and ALD-18127-7 acrylamide homopolymers showing the
plasticizing effect of water.

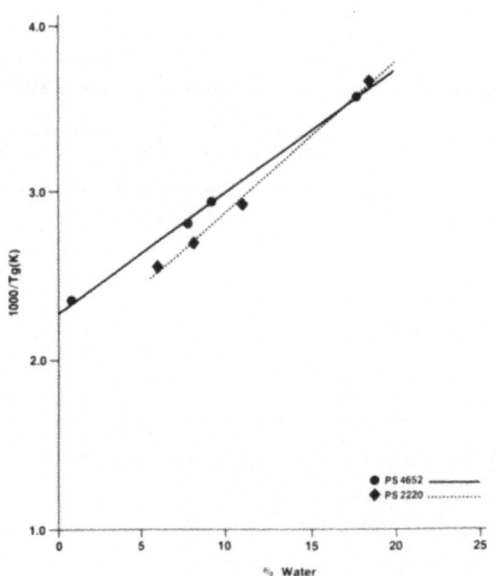

Fig. 6   A plot of Tg(°K)⁻¹ vs. water content for PS-4652 and
PS-2220 acrylamide-acrylic acid copolymers.

Table 3. The Tg Suppression by Water on Various Polymers

Polymer	Form	ΔTg (Dry-wet)	% Water for ΔTg	% Water equil.	Ref.
PET polyester	Fiber	16°	6.8	.4	13
Nylon 6.6	Fiber	30°	6.1	4.2	13
Acetate	Fiber	88°	15.3	5.0	13
"Qiana" nylon	Fiber	85°	10.1	2.5	13
Poly(vinyl alcohol)	Film	64°	6	∿5	14
Poly(vinyl acetate)	Molded part	22°	4.2	<4	15
Polyacrylamide	Powder	72°	8	7.9	present study
Poly(acrylamide-co-acrylic acid): PS-4652	Powder	87°	8	7.8	present study

Table 4. The Effect of Water Content on T_β of Acrylamide homopolymer (PS-2806)

T_β (K)	Calculated Tg (K)*	% Water
283.9	423.5	0.2
243.8	368.9	6.5
187.4	316.4	14.6

*Tg was based on the linear regression of the data on PS-2806 as shown in Figure 5.

A linear regression analysis of the data gives Tg=0.66 T_β. This T_β/Tg ratio was lower than the general correlation of 0.75 suggested by Matsuoka and Ishida.[16] Boyer[17] noted a similar failure with the correlation for aliphatic nylon and other polymers.

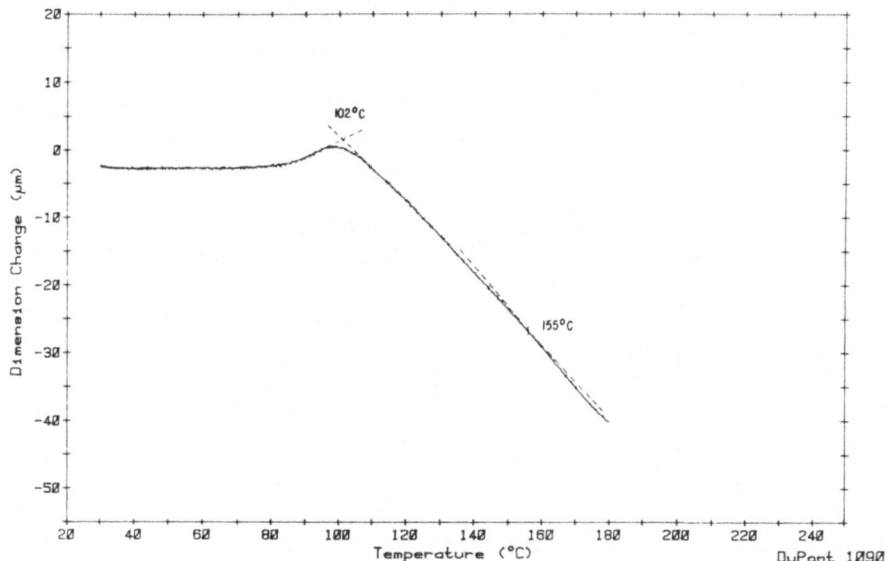

Fig. 7   TMA scan by penetration on PS-2806 homopolymer (6.8% water content) showing two glass transition temperatures.

In fact, our T_β/Tg ratio is quite similar to that for Nylon 6[3] (low crystallinity) which showed a Tg at 80°C and a T_β at -40°C also giving 0.66 for the ratio.

ACKNOWLEDGEMENT

The authors are grateful to Mr. P. D. Gillham and Dr. P. A. Berger of the Physical Sciences Center, Monsanto Company, St. Louis, Missouri for their assistance in determining molecular weights and acid contents respectively.

REFERENCES

1. Yen, Y. C., Acrylamide and Polyacrylamide, Report No. 99,
    Stanford Research Institute, Menlo Park, California, 1976.
2. Boyer, R. F., in Encyclopedia of Polymer Sciences Technology,
    Suppl. Vol II, Wiley, New York, 1977. p. 745.
3. McCrum, N. G., Read, B. E., and Williams, G., Anelastic and
    Dielectric Effects in Polymeric Solids, Wiley, New Yrok, 1967.
4. Chiantore, O., Costa, L., and Guaita, M., Makromol. Chem.,
    Rapid Commun., 1982, 3, 303.
5. Camino G., Casorati, E., Chiantore, O., Costa, L., Guaita, M.,
    and Trossarelli, L., Conv. Ital. Sci. Macromol., 1977, 3, 134.
6. Klein, J., and Heitzmann, R., Makromol. Chem. 1978, 179, 1895.
7. Illers, K. H., Kolloid-Z, 1963, 190, 16.
8. Woodward, A. E., Sauer, J. A., and Wall, R. A., J. Polym. Sci.,
    1961, 50, 117.
9. Boyer, R. F., Polym. Eng. Sci., 1968, 8, 161.
10. Woodward, A. E., and Sauer, J. A., Adv. Polym. Sci., 1958, 1,
    114.
11. Yuen, H. K., Mappes, G. W. and Grote, W. A., Thermochimica
    Acta., 1982, 52, 143.
12. Fox, T. G. and Flory P. J., J. Polym. Sci., 1954, 14, 315.
13. Fuzek, J. F., in Rowland, S. P. (Ed.), Water in Polymers,
    American Chemical Society, Washington, D.C., 1980, p. 515.
14. Pritchard, J. G., in Poly(vinyl alcohol) Basic Properties
    and Uses, Gordon and Breach Science, New York, 1970, p. 60.
15. Johnson, G. E., Bair, H. E., Matsuoka, S., Anderson, E. W.,
    and Scott, J. E., in Rowland, S.P. (Ed.), Water in Polymers,
    American Chemical Society, Washington, D.C., 1980, p. 451.
16. Matsuoka, S. and Ishida, Y., J. Polym. Sci., Polym. Symp.,
    1966, C14, 247.
17. Boyer, R. F., J. Polym. Symp., 1975, 50, 189.

THE BINARY BLENDS, TRANSREACTION AND COPOLYMERS OF

BISPHENOL-A POLYCARBONATE AND A POLYARYLATE

Masao Kimura and Roger S. Porter

Materials Research Laboratory
Polymer Science and Engineering Department
University of Massachusetts
Amherst, Massachusetts   01003

ABSTRACT

Physical blends of bisphenol-A polycarbonate (PC) and a poly-arylate (PAr) exhibit by thermal analysis two amorphous phases: a pure PC phase and a PAr-rich miscible mixed phase. On controlled thermal treatment, transreaction between PC and PAr takes place mainly in the mixed phase, producing a new copolymer. Reaction progression from block to random copolymers has been traced by DSC, ^{13}C NMR and GPC. The final product of transreaction is an amorphous copolymer showing a single T_g depending on the original binary composition.

INTRODUCTION

During the course of compatibility studies on binary polymer blends, the potential of chemical reactions between components can also be of fundamental and practical importance[1]. In the case of polyester pairs, transesterification is readily facilitated, opening a new route for preparation of novel copolymers with various degrees of randomness and composition[2,3].

The concept of transesterification of polyesters has been pre-viously discussed in detail[1,4]. Application to commercial polymers, however, has only begun recently, partly because of limited indus-trial development of new polymers. Moreover, as most pairs of polymer blends are incompatible, partially or completely reacted polymer pairs, i.e., block or random copolymers, could well exhibit better properties than the corresponding polymer blends because of the homogenizing effect of new copolymers made by coreaction.

25

This paper is devoted to a study of the physical blends and transreacted copolymer products of a polyarylate (PAr) and bis-phenol-A polycarbonate (PC). Both polymers are amorphous thermoplastics of relatively high glass transition temperatures.

EXPERIMENTAL

The polyarylate (PAr) used is composed of bisphenol-A (BPA) units with a mixture of terephthaloyl (TP) and isophthaloyl (IP) units (molar ratio BPA:TP:IP = 2:1:1). PAr is a commercial product of Union Carbide Co., with the trade name of ARDEL-D100. Its reported molecular weights are $\bar{M}_n = 1.74 \times 10^4$, $\bar{M}_w = 4.49 \times 10^4$, $\bar{M}_z = 7.31 \times 10^4$ giving a $\bar{M}_w/\bar{M}_n = 2.58$ by GPC. By thermal analysis, T_g was 460K (187°C); no melting or crystallization was observed, as expected.

The bisphenol-A polycarbonate (PC) is a commercial product from the General Electric Co. designated as a Lexan grade, $\bar{M}_n = 1.33 \times 10^4$, $\bar{M}_w = 3.42 \times 10^4$ and $\bar{M}_w/\bar{M}_n = 2.57$. This PC has a glass transition of 419K (146°C) and shows no melting endotherm by DSC, as quenched from 523K (250°C) to room temperature. Solvent-casted PC is known to exhibit crystallinity and melting at various temperatures, depending on prior treatment.

Polymer blends were prepared by a casting method from a mutual chloroform solution. Transreaction was readily avoided in this blending process. After the blend films were dried at 200°C overnight, they were compression-molded at 250°C for 1 min., followed by quenching room temperature.

The glass transition temperatures (T_g) were measured on compression-molded films by using a Perkin-Elmer DSC-2 differential scanning calorimeter with a Data Station at a heating rate of 10K/min. under a nitrogen atmosphere. Transreaction of the binary polymer blends was also carried out directly in the DSC-2, by holding blends at 250°C for up to 16 hours. The transreaction was monitored by DSC, GPC and ^{13}C NMR.

The ^{13}C NMR spectra were obtained on a Varian CFT-20 NMR spectrometer at room temperature and at a concentration of blends at ~10% (w/v) in $CDCl_3$, employing a tetramethylsilane standard, assigned a resonance position of zero PPM (δ scale), as an internal standard.

Molecular weights were measured on a 150 ALC/GPC, Waters Associates, Inc., using polystyrenes as standards for molecular weight and separation. No corrections were made so that the distributions may be slightly more narrow than reported.

RESULTS

I. Physical Binary Blends

Glass transition temperatures (T_g's) and changes of specific heat through the glass transition (ΔC_p's) of blends and of the corresponding pure polymers are summarized in Table 1. From thermal behavior, this PC-PAr pair is clearly incompatible; that is, two T_g's were observed in the physical blends. One at 422°K, a few degrees higher than the T_g of the pure PC (419°K), and another at 451°K, ~10°C lower than the T_g of the pure PAr (460°K). Both T_g's increased slightly as the fraction of PAr was increased. The lower of the two T_g's for a blend could thus be assigned to that of a near pure PC phase. The higher T_g was assigned to a PAr-rich mixed miscible phase.

TABLE 1

Glass Transition Temperatures and Change in Specific Heat Through the Glass Transition Binary Polymer Blends of PAr in Polycarbonate

Weight Ratio	Lower Glass Transition			Higher Glass Transition		
of PAr %	T_g K	ΔC_p cal/g.K $\times 10^{-2}$	ΔC_p/WR cal/K.g(PC) $\times 10^{-2}$	T_g K	ΔC_p cal/g.K $\times 10^{-2}$	ΔC_p/WR cal/g(PAr)K $\times 10^{-2}$
100	--	--	--	460.2	3.98	3.98
75	422.5	0.95	3.80	451.5	3.61	4.80
50	422.0	2.05	4.10	451.0	2.39	4.78
25	421.5	3.36	4.48	450.0	1.32	5.28
0	419.2	4.83	4.83	--	--	--

As shown in Table 1, changes of specific heat on heating through T_g (ΔC_p) were also evidence of partial compatibility in this PC-PAr system. The values of ΔC_p, corrected for the fractional weight of each polymer, are tabulated as ΔC_p/WR in Table 1. The ΔC_p of the lower T_g was much smaller than that of the pure PC. In contrast, the ΔC_p change for the higher T_g was larger than that of the pure PAr. That is, the amount of pure PC phase is apparently smaller than the original fraction while PAr-rich phase appears greater than the initial portion of PAr.

II. Transreaction

Transreaction between the two polymers was carried out at 250°C from one minute to 16 hours using the 50/50 weight fraction blend of PAr/PC. Figure 1 shows the changes of T_g's during transreaction. In the first stage of reaction - up to 1 hour, T_g's of the PC phase and PAr-rich phase remained near constant. A new T_g did appear between them at 444°K, considered to be a new phase of block copolymer. As the transreaction proceeded (1 to 4

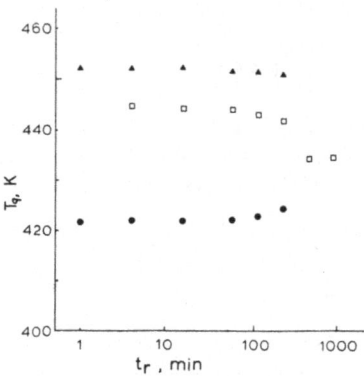

Figure 1: Changes of T_g on transreaction
        Δ : PAr, O : PC, □ transreacted copolymer.

hours), the T_g of PAr-rich phase and block copolymer decreased and T_g of PC phase increased. At the final step of transreaction (after 4 hours), the block copolymers were connected to a random copolymer.

The corresponding changes of molecular weights were also used to monitor the transreaction, as shown in Figure 2. The resultant ratio of weight-average ($\bar{M}_w$) to number-average ($\bar{M}_n$) is consistent with a randomness in the copolymers (Figure 2). Perhaps partly because the molecular weights of the original two polymers in the blend were very close, a decrease of molecular weight was noted only after 4 hours. After 4 hours, with proceeding of randomness, the average molecular weights decreased. $\bar{M}_w$ to $\bar{M}_n$ approached 2.0, which is consistent with a random process[5].

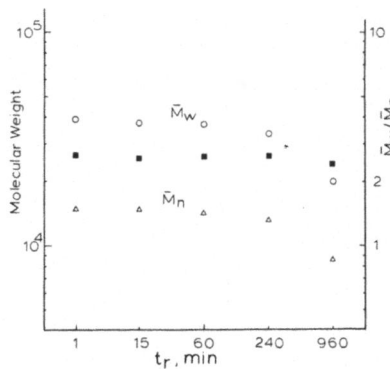

Figure 2: Changes of molecular average weights and their ratio, $\bar{M}_w/\bar{M}_n$. O: $\bar{M}_w$, △: $\bar{M}_n$ □: $\bar{M}_w/\bar{M}_n$

The observed decrease in molecular weight was not due dominantly to degradation, as confirmed by IR and NMR spectra. For ^{13}C NMR it is noted that both polymers contain bisphenol units. The 4 and 4' carbons are thus sensitive to the extent of their reaction as shown in Figure 3. The original peaks regularly decreased in area and the two new peaks appeared and increased systematically in area.

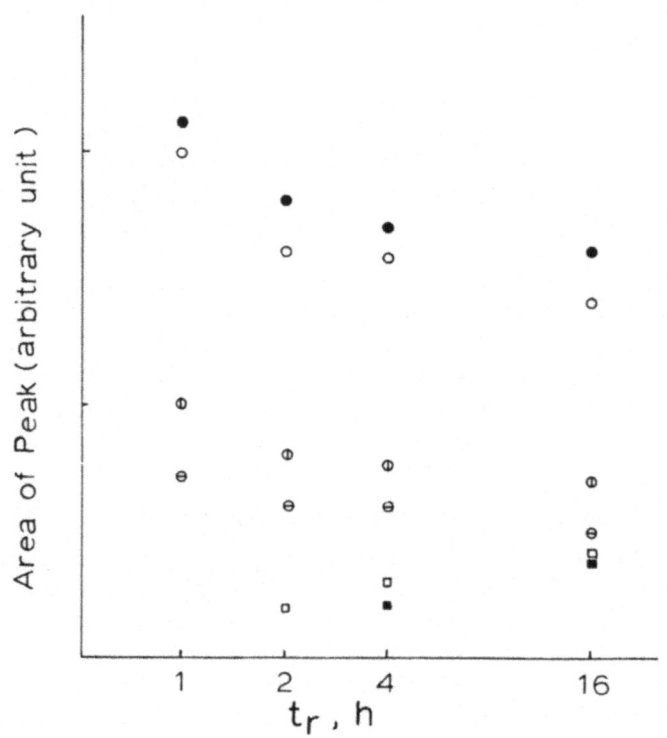

Figure 3: Area changes of ^{13}C NMR spectral peaks by field position in PPM
0: 149.03; 0: 148.77, 0: 133.95,
θ: 134.85; : 126.44, : 126.92

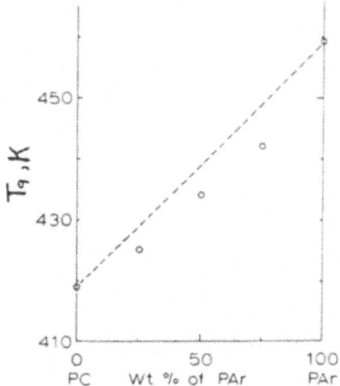

<u>Figure 4</u>: Glass transition temperature transreacted copolymers.

III. Transreacted Blends

Completely transreacted blends showed only single T_g's. It is a few degrees lower than the values calculated (dotted line) by the Fox equation (Figure 4). These new random copolymers are also amorphous, showing no indications by thermal analysis to crystallize or exhibit melting.

CONCLUSIONS

Physical blends of PAr and PC show two T_g's and thus two amorphous phases: a near pure PC phase and a PAr-rich mixed phase. The composition of this miscible mixed phase is likely dependent on the molecular weight for each polymer. The induced transreaction likely occurs in this latter phase because of the intimate contact between chain units of PAr and PC. Therefore, in the first stage of transreaction, the block copolymers produced would likely be PAr rich, with three phases coexisting as suggested by the three T_g's. As the reaction proceeded, the T_g's of the initial two phases are affected because the original polymers would be, at least partially, compatible with the new copolymers. In the final stage of inter-reaction the block copolymers would change to random copolymers, and indeed approach an $\bar{M}_w/\bar{M}_n$ to 2. The final composition was a single-phase amorphous random copolymer having T_g's depending on the original binary composition and always between the T_g's of original PAr and PC polymers.

ACKNOWLEDGEMENTS

The authors wish to give thanks to Mr. H. Stuart for measuring the GPC.

REFERENCES

1. A.M. Kotliar, J. Polym. Sci., Macromol. Rev., 16, 367 (1981).
2. J. Devaux, P. Godard and J.P. Mercier, Polym. Eng. Sci., 22, 229 (1982).
3. D.C. Wahrmund, D.R. Paul and J.W. Barlow, J. Appl. Polym. Sci., 22, 2155 (1978).
4. R. Sredharan and I.M. Mathai, J. Sci. Ind. Res., 33, 178 (1974).
5. A.M. Katliar, J. Polym. Sci., Polym. Chem. Ed., 11, 1157 (1973).

CURIE TEMPERATURES OF THE ICTA CERTIFIED REFERENCE MATERIALS

Paul D. Garn and Ali-Asghar Alamolhoda

Chemistry Department
The University of Akron
Akron, OH   44325   USA

INTRODUCTION

The thermogravimetry reference materials certified by the International Confederation for Thermal Analysis (and distributed by the U.S. National Bureau of Standards as GM-761)[1] provide an indication of the temperature at the sample position in a thermobalance through an apparent weight change as ferromagnetism disappears. The goal at the time of certification was to provide reference points for intercomparison of data. The certification, then, provided assurance that the reference materials were homogeneous within the entire lot, tabular information concerning the values obtained by the several participants in the test program, and a discussion of the reasons for the differences. There was no attempt to establish the Curie temperature from these data. Indeed, the diversity of reported temperatures would render such a value suspect.

Nevertheless, a "true value" is obtainable by an independent calibration. This work was undertaken to provide magnetic transition temperature values that can be used as calibration points for thermobalances--in particular those in which the sample temperature is not measured directly in the sample or the sample holder.

The Curie temperature is defined as that temperature at which ferromagnetism disappears. This disappearance is not really abrupt, although the major effect disappears over a small temperature range. After the major change, there is a tailing-off to the paramagnetic level. The cessation of ferromagnetism is not clear-cut; the Curie temperature is normally found by extrapolation.

Two *ab initio* requirements were set forth in planning this work:
1. The determination must be independent of any thermobalance.
2. The temperature of the sample must be measured directly.

The first condition enables the use of a simple instrument, spe-
cifically designed for this use and having no features added solely
for operator convenience. The second condition avoids the unclear and
variable relation of the temperature of the sample to the temperature
of a measuring point nearby. That relation can be expected to vary
with atmosphere and pressure, with heating rate and with sample size.
Further, reactions that lead to loss or gain of weight will also absorb or evolve heat, changing the relation of temperatures. The form
of measurement chosen was horizontal deflection.

APPARATUS

The furnace assembly (Figure 1) comprised two 30 cm (nominal 12 in.)
half-cylinders mounted so that they were touching one edge and separated by ca. 1 cm along the other (bottom) edge. A bushing for a magnet rod was mounted on one end and a control and a recording thermocouple were inserted through the other. The furnace end supports
were mounted on a base, which also supported the sample assembly. The
sample assembly comprised a mounting bracket, a short length of spring
steel to which two strain gauges were cemented, an upper support piece,
a 1.6 mm (1/16" nominal) two-hole ceramic tube, a Pt vs. 90%Pt-10%Rh
thermocouple 0.25 mm (0.010" nominal) in diameter, and the sample.
The sample pieces were 3-5 mm high and 5-8 mm wide weighing 20-70 mg.
In preparation, the thermocouple wires were welded by capacity dis-

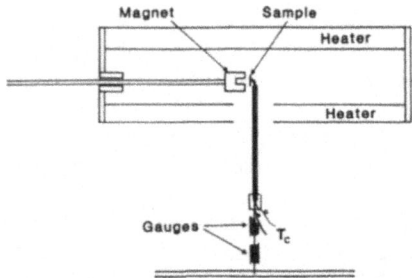

Figure 1. Apparatus for Curie Point Measurements.

charge, 10-13 watt-seconds in a medium-length pulse being supplied from a Hughes Model HRW 50B welder.

The sample was welded to the thermocouple bead, using 13-15 watt-seconds. The ceramic tube could then be replaced upon the support piece and adjusted to the correct height. The protocol given in the ICTA procedure discourages physical contact between the sample and platinum or its alloys; for this purpose, however, the diffusional alloying is of no importance because (a) the thermocouple is renewed for each sample; (b) the thermocouple is formed separately, then welded to the sample; and (c) the time during which the welding is taking place is but a few thousands of a second.

The temperature control apparatus was a Micricon two-channel programmer, by which any desired linear-segment temperature profile could be applied. The power unit was a Leeds & Northrup Zero Voltage Power Pack. A Chromel-Alumel shielded thermocouple provided the control signal.

The sample temperature was measured using a $4\frac{1}{2}$ digit millivolt-meter, DORIC Model 410A, the last digit being 1 μv; for room temperature compensation an Omega-CJ cold junction compensator was used. The movement of the sample was recorded on a strip chart along with a furnace temperature signal from a second type K thermocouple. The sample temperatures in emf were recorded manually at major intervals along the chart paper; the least significant digit represents ca 0.1°C.

A mechanical stop was adjusted to allow ca 1.5 mm movement of the sample, so that the sample was never in contact with the magnet during a measurement. This was adjusted so that the ceramic tube was almost touching one side at rest with the magnet pulled away, the ceramic tube was held against the other side when the sample was being pulled up by the magnet. The magnet was mounted on a 6 mm rod so that it could be inserted to known distances from the at-rest position of the sample.

OPERATING PROCEDURE

Three general temperature-control procedures were used--a fully programmed run with a rapid rise in temperature to near the Curie temperature followed by a slow linear rise through the Curie temperature region, manual heating control by imposing a high voltage on the furnace until the Curie temperature was neared, followed by a selected power input that would bring the sample through the Curie point slowly, and manual cooling control by bringing the sample slightly above the

Curie temperature followed by a power input that ensured a slow cooling through the Curie temperature region. Typically, the Curie temperature region was traversed at less than 1°/min.

To prepare for a run, the previous sample was taken from the furnace by removing the ceramic tube from the support, snipping the
thermocouple close to the sample, pulling through enough wire to form
a new thermal junction, welding the junction, then welding on a
weighed piece of the selected material. The ceramic tube was re-
inserted into the furnace and adjusted to the correct position, that
is, the sample was at the same level as and parallel to the magnet
face, with the thermocouple on the opposite side. The thermocouple
wires to the millivoltmeter were adjusted to minimize force on the
ceramic tube so that the rest position of the sample was virtually unchanged.

The stop was then put in place so that the ceramic tube was just
barely out of contact with it; the gap was approximately 0.1 mm. The
magnet was set at the desired position, the electronic gear turned on
and the program begun.

The bridge circuit containing the strain gauges was adjusted to
give a near-zero signal and the position of that pen adjusted on the
strip chart recorder. The furnace temperature thermocouple, which
was connected through a voltage-offset circuit, was connected to a
second pen and offset so that only a 5 mv. range was displayed.

As the sample went through the Curie temperature region, the output from the sample thermocouple was transcribed onto the chart at
regular intervals and at any unique points.

Table 1.   Observed Magnetic Transition Compared
           with Means of ICTA Points T_2 and T_3

Material	T_2	T_3	T_{obs}
Permanorm 3	259	266	267
Nickel	353	355	364
Mumetal	382	386	388
Permanorm 5	455	459	457

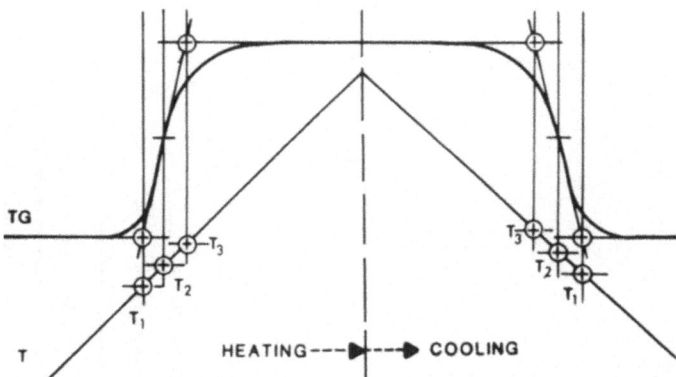

Figure 2.   The defined points on the thermogravimetric temperature
calibration curve [1].

RESULTS AND DISCUSSION

     The data shown in Table 1 must be examined with regard to the
thermobalance data obtained in the ICTA certification program.   The
prima facie expectation is that the measured Curie temperature would
be near T_3 of Figure 2.   This is approximately true.   The degree to
which it is descriptive is in part dependent upon the form of the
actual curve obtained on the thermobalance; the slopes of the magnetic force vs. temperature vary for the several materials.

     On heating, no indication is seen until the attraction between
the sample and the magnet has decreased enough to allow the spring to
pull the sample away from the stop.   This is a function of several
factors--the magnetic susceptibility of the sample material, the
strength of the magnet, and the distance between the sample and magnet.   This movement occurs at some ill-defined point in the curve of
Figure 2.   The actual movement event is less important than the form
of the movement, as seen in Figures 3a and 3b.

     The first movement occurs while there is still attraction, so
the sample position depends upon several influences.   First consider
that the sample is fairly reflective and that it is not in contact
with any heat sink.   Next consider that the magnet--gray after the
first few heatings--is a good absorber of heat radiated from the furnace windings.   The sample is warmed principally from the heated air,

NICKEL

| | Release | 2.986 mV | 2.992 mV |
| | Pull-Up | 2.982 mV | 2.991 mV |

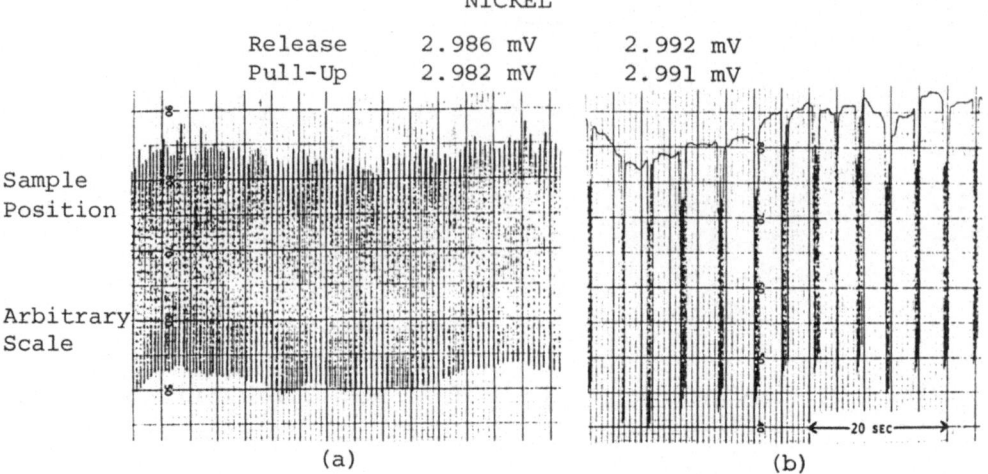

Figure 3.   The sample position as a function of time as the sample
            is heated slowly toward the Curie point.

by conduction from the winding *and* from the magnet.   At some temperature below the Curie temperature, the attraction is less than the
force exerted by the spring and the sample pulls away--not slowly,
because the magnetic force drops off with the second power of the distance.   As soon as the gap is increased, two effects begin--adiabatic
demagnetization and convective cooling.   The sample's susceptibility
is increased and it is pulled back toward the magnet; now the heating
by remagnetization and conduction decreases the susceptibility enough
that it will be released again.   At constant temperature this oscillation, illustrated in Figure 3a, could be maintained indefinitely.

As the temperature is increased, though, the plot changes form.
When the sample is pulled away by the spring, the rapid adiabatic demagnetization does not cool the sample enough so the slower dissipation of heat to the furnace atmosphere is needed to cool the sample
and raise its susceptibility enough to be pulled back.   This is illustrated in Figure 3b.   At some temperature it will no longer pull back
to the stop but will still function as a driven oscillator.   This behavior dies away as the Curie temperature is reached.

The Curie temperature can be defined reasonably well by this cessation of oscillation but the inverse behavior--the inception of ferromagnetism on cooling--appears to be better.   As a sample is cooled
from slightly above the Curie temperature, there is a quite clear
discontinuity between the first very slow movement--presumably due to
paramagnetism--and the substantially more rapid movement as the sample
is pulled to the stop.   The extrapolation of these segments (Figure 4)

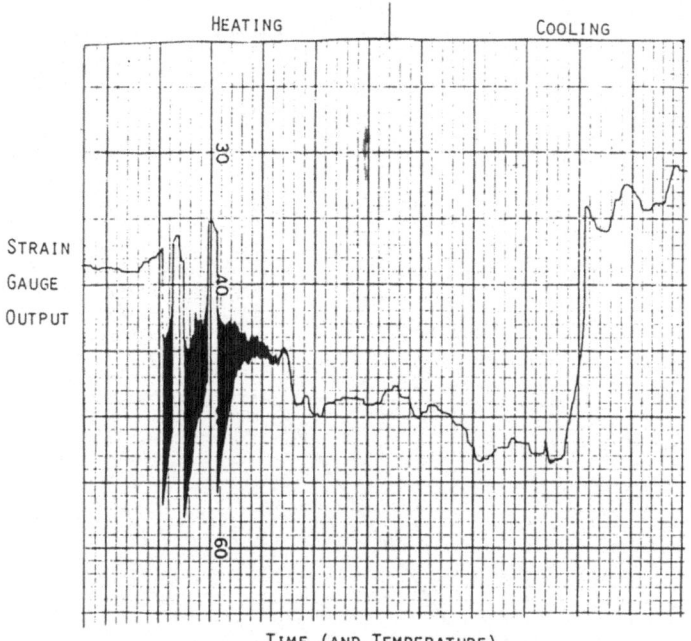

Figure 4. The sample position as a function of time as the sample is
heated and cooled through the Curie point.

is clear and reproducible. Subjectively, this extrapolated value is
in good agreement with the cessation of oscillation on heating.

     The sensitivity of the apparatus leads to a problem, sometimes
poor agreement with the values found during the certification [1,2].
The system sees only the low-level magnetic properties and can give
temperatures well out in the tailing-off region, somewhat above the
T_3 of Figure 2.

     Mumetal has a rather distinctive behavior, two identifiable "ex-
trapolated onset" points (Figure 5). The higher-temperature point is
quite seriously above the range seen by the thermobalances, the other
--which marks the start of the pull-up to the stop--is just above the
mean T_3. It is this lower value that would be found by extrapolation
of the principal slope with the later near-zero slope. This would occur because the apparent weight change between these points is a very
small fraction of the total apparent weight.

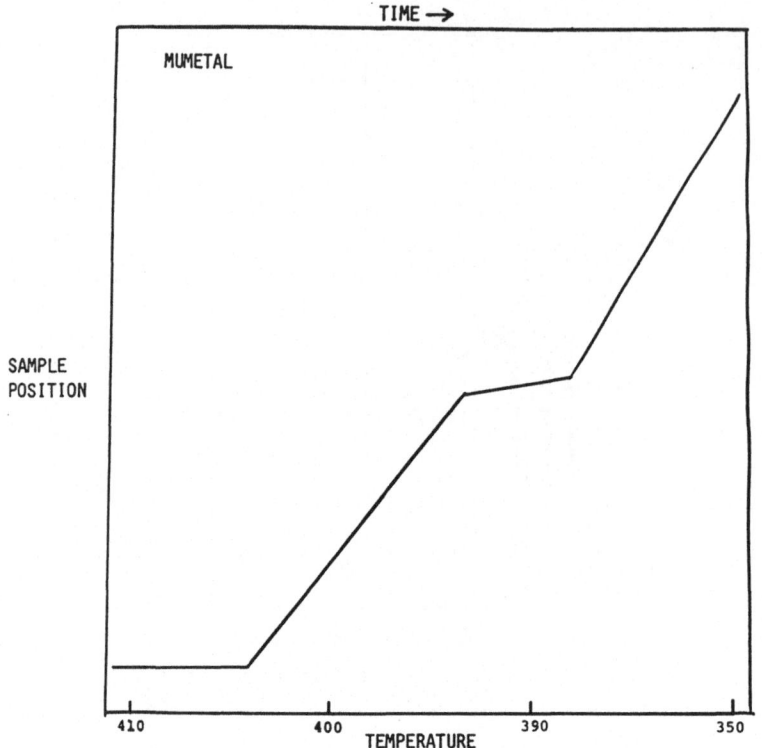

Figure 5.   The sample position as a function temperature on cooling
            for Mumetal showing two discontinuities.

     The Curie point measured for nickel is noticeably higher than
the mean T_3.   This is because of the form of the loss of apparent
weight.   Figure 6 shows the several curves from a single participant
in the ICTA test program.   It is clear that the tailing off for nickel
would yield a much greater difference between the Curie temperature
and T_3 than for any of the three alloys tested.

     In summary, the temperatures of disappearance of ferromagnetism
for four of the ICTA reference materials have been determined.   These
values can be used for calibration of the temperature scale of the
thermobalance by finding the <u>disappearance</u> of the ferromagnetic effect
as compared to the extrapolated end, T_3.   Measurement of this point
may require a sensitivity or range setting that cannot include the
whole apparent weight change.   This calibration does not replace the
reporting of the values T_2 and T_3 as part of a document giving values
for thermal processes; these should still be provided to enable valid

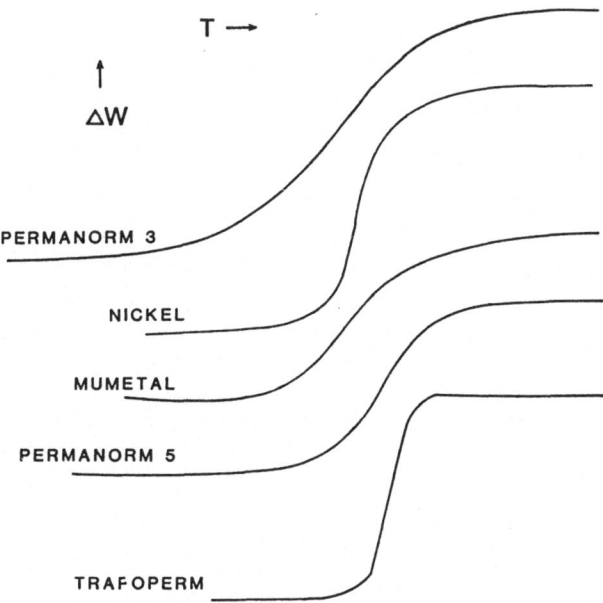

Figure 6.   Apparent-weight curves for the ICTA Certified Reference
            Materials for Thermogravimetry.

comparisons. There are two significant temperatures for Mumetal; the
first is the discontinuity between two levels of ferromagnetism and
is the one that corresponds nearly to the T value, whereas the second
is the end of all ferromagnetism. The second--at 404°C--may not be
discernible in all thermobalances.

REFERENCES

1. Certificate, ICTA Reference Materials for Thermogravimetry,
    issued with GM761 by the U.S. National Bureau of Standards,
    Washington, D.C. 20234. See also Reference 2.

2. P. D. Garn, O. Menis, H.-G. Wiedemann, J. Thermal Analysis, 20,
    185-204 (1981).

This work was supported in part by Faculty Research Grant RG 750 from
The University of Akron.

THERMAL ANALYSIS OF ACRYLAMIDE-BASED POLYMERS

J.J. Maurer, D.N. Schulz, D.B. Siano, and J. Bock

Corporate Research
Science Laboratories
Exxon Research and Engineering Company
Linden, New Jersey
07036·

INTRODUCTION

Examination of the literature regarding the determination of the glass transition temperature (T_g) of polyacrylamide (PAM) reveals several complications. One complication is that PAM is reported to strongly bind about 2-9% water in spite of intensive drying[1]. If this water is not removed, it might be expected to act like a plasticizer and lower the T_g of the polymer. Another complication is that removal of this water requires extensive drying. The use of elevated temperatures for this purpose[2-9] may lead to polymer degradation problems including imidization[1] and crosslinking[5]. Alternatively, use of low temperture drying conditions[4-6] leads to uncertainties regarding the amount of strongly bound water which may remain in the polymer. A final complication is that the T_g of PAM has been reported to be in the range of 153-188°C[4,7-9], well above the temperatures at which polymer decomposition and crosslinking have been indicated to occur.

The objectives of this paper are to determine the influence of drying conditions and water content on the T_g of PAM, HPAM and copolymers, examine the T_g vs. HPAM and copolymer composition relationships in this regard, and explore the possible utility of DSC for the detection of free vs. bound water in these polymers.

43

EXPERIMENTAL

Glass Transition Temperatures (T_g) were determined by DSC using
the DuPont 990 thermal analysis system. Specific details of the
analysis were: atmosphere: nitrogen; sample size: 2-10 mg,
typically 3-5 mg since this was about what could be used with low
bulk density, freeze dried polymers; sample type: open pan (OP),
closed pan (CP) or closed pan which had several holes pierced in
the top (PP). This latter procedure was particularly useful for
analyzing samples which were stored in closed pans prior to DSC
analysis. As standard procedure, oven dried samples were enclosed
prior to removal from the oven and immediately transferred to, and
stored in, a nitrogen purged glove bag. All DSC samples were
prepared in this glove bag. Samples to be run in the OP or PP
mode were covered until insertion into the DSC cell; Heating
rate: typically 10°C/min. Selected samples were run at 5°C/min.
and 2°C/min.

Volatiles Content: (a) Thermal Gravimetric Analysis (TGA) using a
DuPont 990 system and 951 TGA unit: sample size: 10 mg; temperature: 115°C; atmosphere: nitrogen or vacuum; (b) Oven drying:
sample size: 3-10 mg; temperature: 50°C or 115°C; atmosphere:
nitrogen or vacuum.

Water Content: (a) Karl Fischer test: Mitsubishi Moisture Meter
(MCI-Model CA-02); indirect method; sample size: 3-10 mg; temperature: 125°C; typical heating time: 2-10 minutes; (b) DuPont
Moisture Analyzer (26-321-A); sample size: 5-10 mg; temperature:
130°C; typical heating time: 20 minutes.

Ammonia Liberation from Heat Treated PAM: Perkin Elmer Sigma 2
equipped with Thermal Conductivity Detection; column: 6ft.x1/8
in. OD Teflon column packed with 100 to 120 mesh Chromosorb 103;
Carrier gas: helium or air at 30 ml/min.; Temperatures: injector
(230°C), detector (230°C), column (115°C); Detector: thermal
conductivity at 150 milliamps; sample size: one gram; headspace
injection volume: 1 ml of vapor phase above polymer, aging conditions: one hour at each temperature.

Polymer synthesis. The polymers used in this study were from a
variety of sources. Molecular weights are believed to be $> 3 \times 10^6$
in all cases. The HPAMs synthesized in our laboratories were
prepared via conventional processes, i.e., (a) PAM was prepared
via free radical polymerization, (b) HPAM was obtained by the
conventional hydrolysis procedure (NaOH), and (c) the polymers
were converted to the acid form by passage of solutions through
mixed bed ion exchange resins. The polymer was isolated by evaporation under a nitrogen stream followed by drying at 115°C for one

to two hours in a vacuum oven.   HPAM composition was determined by potentiometric titration.

Copolymerization of acrylamide and acrylic acid was also conducted by standard methods. Copolymers 2 and 3 (Table 1) were prepared according to the general method of Klein and Heitzmann[4]. Copolymers 1 and 4 were prepared similarly, except without pH adjustment. Copolymer composition was determined by potentiometric titration.

RESULTS AND DISCUSSION

Influence of Water Content and Drying Conditions on T_g of PAM

Substantial effects due to the water content in an undried PAM sample are shown in Figure 1. In the pierced pan sample mode, Curve A, a large endotherm, presumably due to volatilization of water is detected between about 0 and 150°C. This is followed by T_g at 184° and a large endotherm due to one aspect of the thermal degradation of PAM. In the closed pan configuration, Curve B, one notes the absence of the endotherm due to volatilization of water and a much lower T_g value, presumably indicating plasticization of PAM by water. The effects of drying the sample (1 hr. at 115°C in a vacuum oven) are shown in Curves C and D. Removal of water is indicated by the weak volatilzation endotherm in (Curve C) and the increased T_g in Curve D. These observations indicate the necessity of drying the polymer in order to obtain an accurate T_g value for PAM.

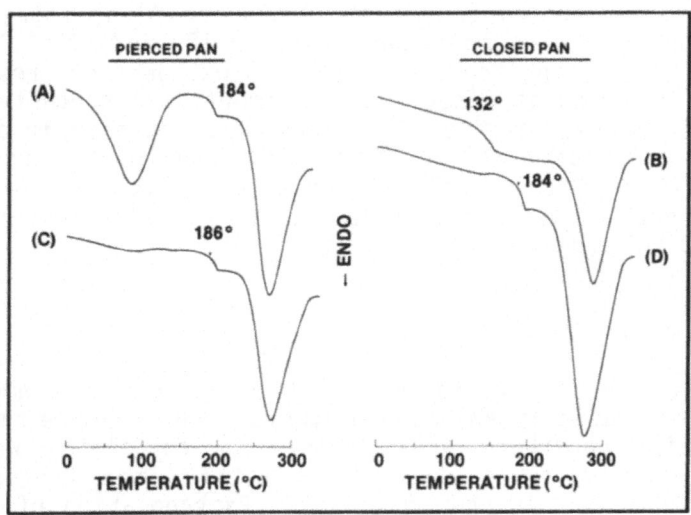

Fig. 1   Effect of water on DSC thermograms of polyacrylamide.

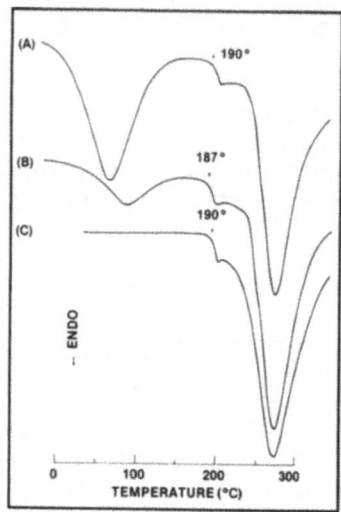

Fig. 2   Influence of heat treatment on DSC thermograms of polyacrylamide: (A) none; (B) 20 min. @ 50°C; (C) 20 min. @
210°C.

The influence of thermal history on PAM stability and T_g was
probed in two experiments.   First, headspace GC of separate samples aged one hour at temperatures between 120 and 190°C indicated
ammonia losses ranging from <0.01 to 0.63 wt. percent, respectively.   Next, a series of annealing experiments was conducted in the
DSC unit (OP mode).   Similar T_g values were obtained in all cases
(Figure 2).   It was concluded from these and related experiments
that the PAM thermal stability problems reported in the literature[1-5] do not significantly influence the T_g values determined by
the present DSC methods.

Evaluation of Water Content by DSC

DSC is a useful means for evaluating both the amount and
nature of the water in AM-based polymers.   For example, the water
volatilization endotherm area (Figure 3) correlates well with
volatiles content (oven drying) or _total_ water content (Karl
Fischer or DuPont Moisture Analyzer).   Extrapolation of the line
in Figure 3 suggested a residual water content of ~1 percent in
the sample which was confirmed by more extensive drying.

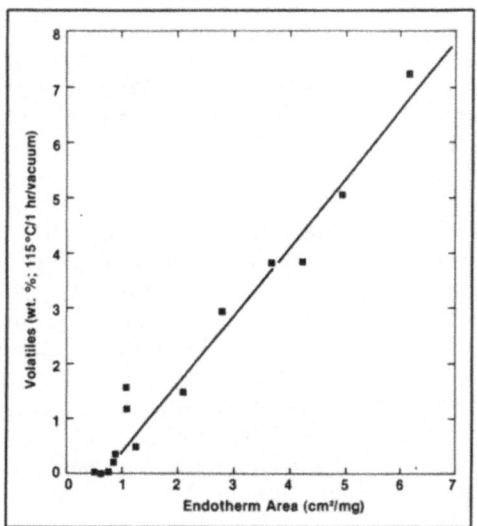

Fig. 3. Correlation of DSC endotherm area with volatiles content
of polyacrylamide.

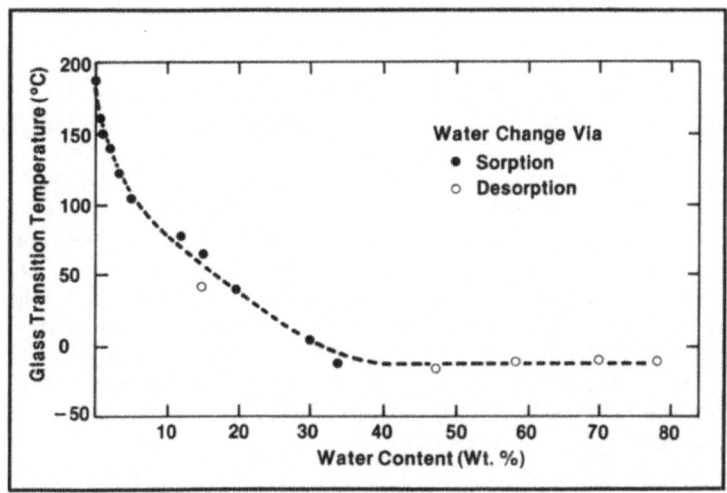

Fig. 4. Influence of water content on T_g of polyacrylamide.

Additional information concerning the <u>nature</u> of the water content was obtained in terms of an approach described by Bair et al[10-11] A preliminary study of the influence of water content on the T_g of PAM is shown in Figure 4. More rigorous determination of this relationship is being conducted to check for possible volatilization effects in the 10 to 25% water content range. This study indicated that: (a) water is an extremely effective plasticizer for PAM, thus careful drying is required to ensure accurate T_g determination in AM-based polymers, (b) T_g decreased up to a water content of about 45 weight percent, at higher concentrations the T_g remained approximately constant at about -10°C; (c) freezing of water was detected at concentrations above about 45 weight percent, (d) in terms of the method of Bair et al[10], the water in PAM would be considered bound up to about 45 weight percent. At higher water concentrations, some of the water is bound and some is free, the latter being indicated by the freezing detected during the cooling step, (e) T_g may be useful for estimating water content in the concentration range where all of the water is bound.

T_g of Partially Hydrolyzed Polyacrylamide (HPAM)

The influence of HPAM composition on T_g was determined for the acid form of the polymer. For some HPAM and AM-<u>co</u>-AA copolymers, it was found that the intensity of the T_g region of the thermogram was very strongly influenced by heating rate, sample size or DSC sample type/mode. These effects may be related to the physical form of the sample (i.e., dense polymer vs. freeze dried polymer vs. thin films). In order to normalize the data for such effects, each of the T_g values shown for HPAM (Table 2) and AM-<u>co</u>-AA copolymers (Table 1) is an average of at least six determinations as indicated.

The T_g of PAM and of PAA was sharp and clearly distinct from the degradation endotherm which is observed close to T_g. For HPAM, however, the T_g became less distinct at about 30 mole% hydrolysis. Evaluations of solution blended, dried PAA and PAM mixtures suggests that this may be due to interaction of carboxyl and amide groups similar to that described for polymer solutions[12]. Strong transitions were again detected at high degrees of hydrolysis. However, the complex nature of some of the thermograms raises some question as to the correct value for T_g. The DSC thermogram characteristics of these HPAMs are summarized in Table 2. As indicated, there are substantial differences in the degradation region of these polymers which suggest a possible basis for compositional analysis.

The T_g vs. composition relationship for this series of polymers is compared in Figure 5 with that of Klein and Heitzmann[4]. A

Table 1. DSC THERMOGRAM CHARACTERISTICS VS. COPOLYMER COMPOSITION[a,b]

		Glass Transition Region		Degradation Region[d]
Sample	AM Mole %	T_g(°C)	Other Features	Endotherm minima (°C)[c]
1	48.8	152	Very weak inflection ca 125°	Shoulder(s):187; Min.(s):235
2	63.0	171	Strong shoulder on degradation endotherm	Min.(s):205 and 217
3	81.5	178	Well resolved T_g region	Min.(s):238 and 268
4	96.7	185	Strong endothermal features	Min.(s):246 and 295
5	PAM	188	Well resolved T_g region	Min.(s):270
6	PAA	124	Well resolved T_g region	Min.(s):235[e]

[a]Prior to DSC analysis, samples were dried in vacuum at 50°C or 115°C for periods up to several weeks or several hours, respectively.
[b]T_g values are averages based on a minimum of six determinations. Various combinations of heating rate (2-10°C/min.), sample mode and sample type. Nitrogen atmosphere in all cases.
[c]s = strong; w = weak; Min. = minimum.
[d]Characteristics of typical thermogram.   [e]Open pan.

Table 2. DSC THERMOGRAM CHARACTERISTICS VS. HPAM COMPOSITION[a,b]

Sample	AM Mole %	Glass Transition Region		Degradation Region[d]
		T_g(°C)	Other Features	Endotherm minima (°C)[c]
1	25.0	131	Strong endothermal features; T_g difficult to evaluate	Min.(s):209; shoulder ca.280
2	36.0	136	Well resolved T_g region Strong endothermal features	Min.(s):209, 236, and 278
3	49.0	152	Well resolved T_g region Strong endothermal features	Min.(s):209 and 232,(w): ca.254
4	69.7	168	Diffuse T_g; weak shoulder on degradation endotherm	Min.(s):217; shoulder(w) ca.271
5	74.0	173	Diffuse T_g region	Min.(s):222; shoulder(w)ca.271
6	88.4	182	Well resolved T_g region	Min.(s):227 and min(s):251
7	PAM	188	Well resolved T_g region	Min(s):270
8	PAA	124	Well resolved T_g region	Min(s): 235

[a] prior to DSC analysis samples were dried in vacuum at 50°C or 115°C for periods up to several weeks or several hours, respectively.
[b] T_g values are averages based on a minimum of six determinations. Various combinations of heating rate (2-10°C/min.), sample mode and sample type. Nitrogen atmosphere in all cases.
[c] s = strong; w = weak; Min. = Minimum.   [d] Characteristics of typical thermogram.

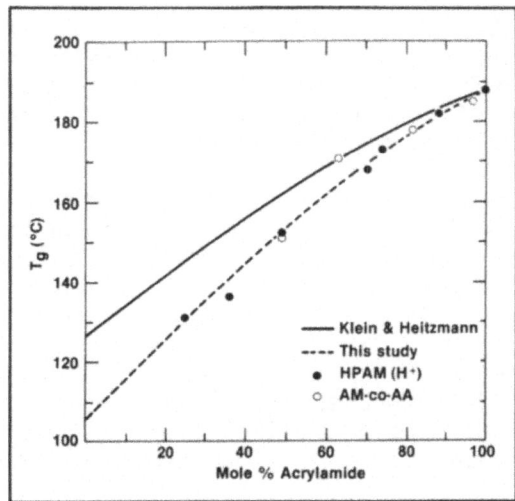

Fig. 5. Glass transition temperature vs. composition for HPAM and AM-co-AA copolymers.

similar trend is evident for the 70 to 100% acrylamide region, in both of these relationships. However, the T_g values observed in this study are generally lower than those of the previous workers. The T_g value (106°C) shown for PAA in the lower curve of Figure 5 was taken from reference 3. DSC measurements indicated a value of 124°C, which agrees with the 126°C value obtained via DSC by Klein and Heitzmann. The reason for this discrepancy (i.e., 106 vs. 124°) has not yet been determined.

Evaluation of Poly (Acrylamide-co-Acrylic Acid) Copolymers

The influence of poly(acrylamide-co-acrylic acid) copolymer composition on Tg was determined for comparison with the HPAM systems. These copolymers present additional challenges with regard to thermal characterization and elucidation of the Tg vs. composition relationship. In some cases two transitions were observed between about 125 and 190°C. Detection and resolution of these transitions was found to be a function of sample type, heating rate and thermal history of the polymer. This suggests that for some HPAMs and AM-co-AA copolymers the observed apparent T_g may be due to more than one transition or structural feature. Careful examination of several variables is therefore required to obtain an adequate and complete description of these copolymers. Evaluations of these poly(acrylamide-co-acrylic acid) copolymers are summarized in Table 1. These data provide additional examples of the influence of copolymer composition on transition behavior and degradation characteristics.

The influence of composition on the T_g of these copolymers is shown in Figure 5 which also includes the data for the series of HPAMs. Both the copolymer and the HPAM data appear to define a similar T_g vs. composition relationship which deviates substantially from that of Klein and Heitzmann[4]. Further, the curve for these systems appears to extrapolate to a T_g value of ~106° for PAA, which agrees well with some literature values for this homopolymer[3,13,14]. This focuses attention on the T_g values for PAA obtained by DSC in this work (124°C) and by Klein and Heitzmann (126°C). The reason for this discrepancy has not yet been established.

Gordon-Taylor-Wood Relationship for HPAM and AM-co-AA Copolymers

The Gordon-Taylor-Wood (GTW) relationship is a common test for randomness in copolymers[8,15,16]. This treatment was applied to the present HPAM and copolymer systems as follows: (1) a ser-

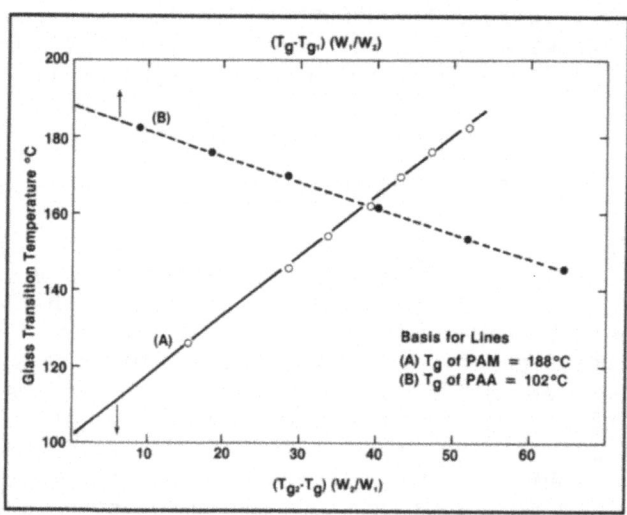

Fig. 6 Gordon-Taylor-Wood relationship for HPAM and AM-co-AA copolymers.

ies of T_g vs. composition "data points" were read from the lower curve of Figure 5; (2) Curve A of Figure 6 was developed using the experimentally determined T_g value of 188°C for PAM. This plot indicated a T_g value of 102°C for PAA; (3) the value of 102° was used as the T_g of PAA in developing Curve B of Figure 6. This plot indicated a T_g value of 188°C for PAM. The observed linearity of these plots sugests that these HPAM and AM-co-AA copolymers behave like random copolymers insofar as the GTW relationship is concerned.

At this point, questions remain regarding the high T_g value obtained for PAA via DSC and the markedly different T_g vs. composition relationships shown in Figure 5. With regard to these problems, Eisenberg et al. noted that T_g values ranging from 88 to 166°C have been reported for PAA[15]. Based on a study of the dehydration kinetics of this polymer, they concluded that the T_g for PAA is about 103°C and that T_g increases with increasing anhydride concentration. Whether such a mechanism could be responsible for the anomalous T_g value observed for PAA or the discrepancy in the T_g vs. HPAM or copolymer composition relationships of Figure 5 is at present unknown. Since the GTW treatment of Figure 6 leads to an extrapolated T_g value of 102°C, there does not appear to be any evidence of a polymer instability problem in the data.

ACKNOWLEDGEMENTS

The authors are pleased to acknowledge the expert experimental assistance of Mr. Gerald D. Harvey, Ms. Phyllis Myer and Mr. Enoch Berluche of the Corporate Research Laboratories, and Dr. Gerald DuPre, Dr. Mark Robillard, Dr. Hugh Huffman and Dr. Michael Hadka of the Analytical Division of Exxon Research and Engineering Company, helpful discussions with Mr. Harvey E. Bair of the Bell Laboratories, Murray Hill, New Jersey concerning water analysis in polymers, and information regarding the measurement of the glass transition temperature of HPAM, kindly provided by Professor Dr. Joachim Klein of the Institut fur Technische Chemie, Technische Universitat Braunschweig.

REFERENCES

(1)   W.-M. Kulicke and J. Klein, The preparation and characterization of high molecular weight polyacrylamide, Angew, Makromol. Chem. 69:169 (1978).

(2)   O. Chiantore, L. Costa and M. Guaita, Glass Temperatures of
      Acrylamide Polymers, Makromol. Chem., Rapid Commun., 3:303
      (1982).

(3)   W. A. Lee and R. A. Rutherford, The glass transition of
      polymers, in: "Polymer Handbook", J. Brandrup and E. H.
      Immergut, eds., Wiley, New York (1975).

(4)   J. Klein and R. Heitzmann, Preparation and characterization
      of poly(acrylamide-co-acrylic acid), Makromol. Chem., 179:
      1895 (1978).

(5)   W. M. Thomas, Acrylamide Polymers, in "Encyl. of Polym. Sci.
      and Tech"., N. Bikales, ed., 1:177 (1964), Wiley-Intersci-
      ence, New York.

(6)   C. L. McCormick, Gow-Sheng Chen and B. H. Hutchinson, Water
      soluble copolymers. V. Compositional determination of random
      copolymers of acrylamide with sulfonated comonomers by
      infrared spectroscopy and C^{13} nuclear magnetic resonance, J.
      Appl. Polym. Sci., 27:3103 (1982).

(7)   W.-M. Kulicke, R. Kniewske and J. Klein, Preparation, char-
      acterization, solution properties and rheological behavior
      of polyacrylamide, in Prog. Polym. Sci., 8:373 (1982).

(8)   K. H. Illers, The glass temperature of copolymers, Kolloid-
      Z, 190:16 (1963).

(9)   J. Brandrup and E. H. Immergut, eds. "Polymer Handbook",
      Wiley, New York (1975).

(10) H. E. Bair, G. E. Johnson, E. W. Anderson and S. Matsuoka,
      Non-equilibrium annealing behavior of poly(vinyl acetate),
      Polym. Eng. Sci. 21:930 (1981).

(11) S. P. Rowland, ed., "Water in Polymers", American Chemical
      Society Symposium Series, No. 127, Am. Chem. Soc., Washing-
      ton, D.C. (1980).

(12) G. Smets and A. M. Hesbain, Hydrolysis of polyacrylamide and
      acrylic acid-acrylamide copolymers, J. Polym. Sci., 40:217
      (1959).

(13) A. Eisenberg, H. Matsura and T. Yokoyama, Glass transition
      in ionic polymers: The acrylates, J. Polym. Sci:A-2, 9:2131
      (1971).

(14)   A. Eisenberg, T. Yokoyama and E. Sambilido, Dehydration
       kinetics and glass transition of poly(acrylic acid), J.
       Polym. Sci., A-1, 7:1717 (1969).

(15)   M. Gordon and J. S. Taylor, Ideal copolymers and the second
       order transitions of synthetic rubbers. I. Non-crystalline
       copolymers, J. Appl. Chem, 2:493 (1952).

(16)   L. A. Wood, Glass transition temperatures of copolymers, J.
       Polym. Sci., 28:319 (1958).

DEFINING B-STAGE AND TOTAL CURE FOR A DICY BASED EPOXY RESIN BY DSC

Bernd K. Appelt and Pamela J. Cook*

General Technology Division
IBM Corporation
Endicott, NY 13760

*Present address: Baker Laboratory
Department of Chemical Engineering
Cornell University
Ithaca, NY 14850

ABSTRACT

The curing reaction of a DICY-based epoxy resin was characterized by DSC. The epoxy resin was composed of a mixture of brominated bisphenol-A diglycidyl ether and of epoxidized cresol novolak with dicyandiamide (DICY) as hardener and tetramethyl butane diamine (TMBDA) as catalyst. The optimum ratio of epoxy/ DICY in terms of the glass transition temperature (Tg) was determined to be about 8:1. It could be shown that the heat of reaction (ΔH) at B-stage is strongly dependent on the amount of unreacted DICY. It was observed that for the optimum formulation, Tg increased linearly with conversion (or ΔH) up to 95% conversion. On the other hand, TMBDA did not affect Tg of the fully cured resin, although it accelerated the cure rate substantially.

Previously, varnish reactivity and B-stage were monitored and adjusted only by empirical tests such as stroke cure time and flow test. It is proposed to monitor calorimetric parameters via DSC and to exercize the appropriate controls for the production process based on these more fundamental and operator independent parameters.

INTRODUCTION

Epoxy resins are presently the most widely used thermosets in the electronics industry. In "first level" packaging, they are used to encapsulate chips and modules providing environmental and electrical protection. In "second level" packaging, they serve as the mechanical support inside printed circuit (PC) cards or boards as well as dielectrics for the incorporated circuitry. The second level package contains the interconnecting circuitry for the first level packages.

The resin system discussed here is used in second level packaging. It is composed of brominated bisphenol-A diglycidyl ether, epoxidized cresol novolak, DICY and TMBDA. This formulation is dissolved in suitable solvents and coated onto glass cloth. While this impregnated cloth moves through a treater tower with multiple heating zones, epoxy groups react partially with amine functionalities of DICY. At the same time, the solvent is driven off and the cloth exits the tower as "prepreg", i.e., the resin is in a state called B-stage where Tg is above ambient. The prepreg can now be cut to size for cards and boards. Subsequently, sheets of prepreg and copper are laminated at elevated temperatures into card and board composites.

The chemical process may be described briefly as follows:[1]

$$R'-NH_2 + \overset{O}{\overset{/\backslash}{CH_2-CH-CH_2}}-R'' \rightarrow R'-NH-CH_2-\overset{OH}{\underset{|}{CH}}-CH_2-R''$$

$$R'-NH-R''' + \overset{O}{\overset{/\backslash}{CH_2-CH-CH_2}}-R'' \rightarrow R'-\underset{\underset{R'''}{|}}{N}-CH_2-\underset{\underset{OH}{|}}{CH}-CH_2-R''$$

amine and epoxide functionalities react forming secondary and tertiary amines under chain extension. Since both reactants are multifunctional, this quickly leads to branched oligomers (B-stage). It is readily apparent that continued reaction of these oligomers leads to a continuous network (a single molecule). This transition is reflected quite sharply in physical properties (such as gel point, viscosity, etc.) and occurs at a very specific reaction extent for a system with given multifunctionality.[2] Subsequently, the remaining free functionalities continue to react leading in time to a tightly cross-linked network (C-stage) exhibiting the final properties.

During lamination at elevated temperatures, the prepreg turns liquid; the viscosity decreases due to heating. In competition, the viscosity rises due to the molecular weight increase from the continuing propagation reaction. Eventually the latter effect takes over and an overall increase in viscosity results. As soon as the gel point is reached, the resin ceases to flow altogether. To generate a quality composite, it is important for the resin from the prepreg sheets to mix and to flow into the copper interface. This can only occur if sufficient time at moderate viscosities is available. Consequently, the prepreg has to be advanced to the same state of B-stage (conversion) if the press conditions are held constant.

Inevitably, the quality of the raw materials (epoxy equivalent weight, reactivity of TMBDA, etc.) vary from batch to batch and adjustments in the formulation and/or treater speed are necessary. Most techniques designed for characterizing chemical reactivity depend on the system remaining soluble. Thus, by their very nature, thermosets become intractable once the gel point has been passed. Calorimetry monitors the inherent energetics associated with any chemical reaction or physical transition and thus is ideally suited for this system.

EXPERIMENTAL

Differential scanning calorimetry: A Perkin-Elmer DSC-2 calorimeter with Thermal Analysis Data Station was used. The calorimeter was calibrated according to manufacturer's specifications. Heats of reaction were calculated from the peak areas using indium as a standard ($\Delta H = 6.80$ cal/g). Tg was taken as the onset of the endothermic deflection. The heating rate was set to 20°/min. For DSC analysis, samples were prepared by two techniques: a) vacuum drying of varnish and b) by flaking off resin from prepreg.

Flow test: Impregnated cloth was cut to a given size and several sheets were laminated under standardized conditions. Thereafter, the laminate was trimmed and the flow was determined as per cent weight loss before and after lamination.

Sample preparation: Varnish was prepared as described elsewhere[3]. The varnish was coated onto glass cloth, air dried, and baked at 140° C for predetermined times. Concentrations of DICY and TMBDA were varied as indicated below. In every case, dilutions were made such that the solid contents of the varnish remained approximately constant.

RESULTS AND DISCUSSION

Amine cured epoxy resins generally employ an epoxide (or mixture of epoxides) which is at least bifunctional and an amine of higher functionality. It is desirable to achieve a high cross-link density in order to obtain a high Tg, low thermal expansion, solvent resistance, etc. Using cross-linking agents of high

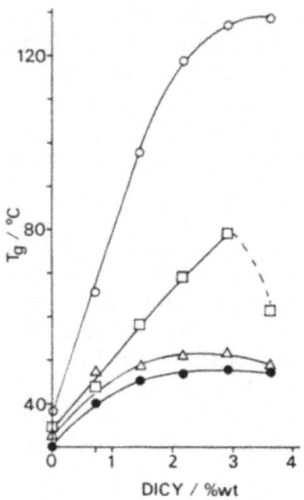

Figure 1: The effect of DICY on Tg: At low levels of conversion
(●3 min and ▲5 min bake at 140° C) Tg is insensitive to DICY.
Once appreciable cross-linking has taken place (☐ 7 min bake at
140° C and ○full cure), Tg rises sharply with DICY.

functionality usually results in non-stoichiometric reactions. A
fraction of functionalities always tends to remain uncross-linked
in any system due to reduced reactivity of the higher functionalities, steric hindrance and side reactions in general.

     These uncross-linked functionalities are undesirable as they
depress Tg, increase moisture absorption, increase the dielectric
constant, etc.

     For the system investigated here, the epoxide equivalent
weight (EEW) was determined to be about 400 g equivalents, the
average functionality was two. DICY has three amine and one
imine proton. Assuming equal reactivity and stoichiometric
reaction, optimum properties were expected for a 4:1 mix. As
observed earlier, the ratio could be increased without affecting
resin performance adversely. Therefore, a calorimetric study for
optimizing the formulation and its cure was of interest.

Examining the effect of DICY first, it was found that Tg of fully cured resin increased sharply with increasing DICY concentration. Above 2% wt., Tg quickly leveled off, suggesting an ultimate Tg of about 125° C (Figure 1). An optimum concentration of about 3% wt. of DICY corresponds to a pseudo-functionality of eight for DICY. This is significantly higher than anticipated. It has been shown[4] that at elevated temperatures the secondary alcohol can react with the nitrile group:

$$R'-\langle\ O\ \rangle-CH_2-\underset{\underset{OH}{|}}{CH}-CH_2-N=C-NR-\underset{\underset{NR_2}{|}}{C}\equiv N \quad + \quad N\equiv C-NR-\underset{\underset{NR_2}{|}}{C}=NR \quad \longrightarrow$$

$$R'-\langle\ O\ \rangle-CH_2-\underset{\underset{\underset{\underset{\underset{R}{|}}{\overset{|}{N}}}{\overset{\|}{C}-NR_2}}{\underset{\underset{R-N}{|}}{\overset{|}{C}=NH}}}{CH}-CH_2-N=C-NR-\underset{\underset{NR_2}{|}}{C}\equiv N \quad \rightleftharpoons \quad R'-\langle\ O\ \rangle-CH_2-\underset{\underset{\underset{\underset{\underset{R}{|}}{\overset{|}{N}}}{\overset{\|}{C}-NR_2}}{\underset{\underset{R-N}{|}}{\overset{|}{C}=O}}}{\underset{\underset{N-H}{|}}{CH}}-CH_2-N=C-NR-\underset{\underset{NR_2}{|}}{C}\equiv N$$

In the first step, an imino group is formed which quickly rearranges into an urea group. Thus a fifth DICY functionality has reacted. Considering the newly generated urea proton, potentially a sixth reaction site has become available. This is still short of the octa-functionality suggested by Tg. In addition, the efficiency of these last two reactions, especially the last one, is probably reduced if only by steric hindrance. Therefore, the effective functionality must be less than six. Consequently, it has to be assumed that epoxide homopolymerizes to some extent which leads to this artifically high functionality (discounting uncross-linked epoxy groups for now).

A typical thermogram of this resin is displayed in Figure 2. The peak onset was always well defined; the peak end was often more difficult to find. Consistent results were obtained by setting peak limits to similar changes in the baseline for corresponding temperature ranges. Frequently, endothermic peaks were riding on Tg of prepreg samples. They were attributed to stress relaxation of the samples.

As may be expected, the heat of reaction is derived mainly from the cross-linking reaction with DICY. This concentration dependence is still very prounounced at the B-stage level (Figure 3).

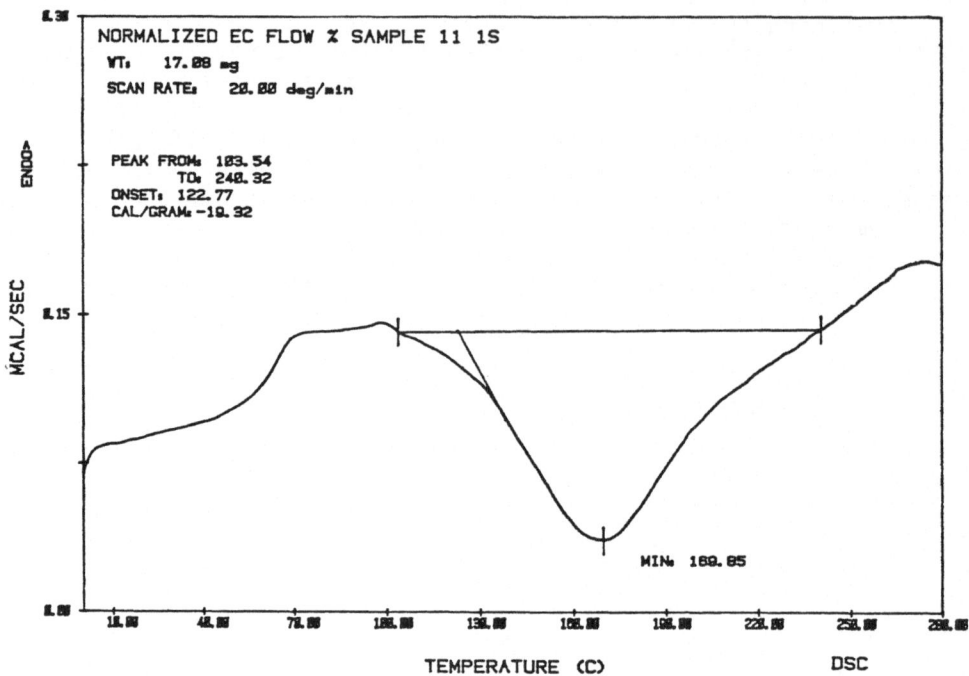

Figure 2: A typical thermogram of this epoxy resin is displayed
to indicate the evaluation of the heat of reaction and Tg.

Here, ΔH reflects the amount of unreacted DICY functionalities.
As can be seen, at lower concentrations ($<1.5\%$) most of the DICY
has reacted and the contribution to ΔH comes from the homopolymerization of epoxide. It should also be noted that at the B-stage
level, Tg is very insensitive to the concentration, c, of DICY
(Figure 1; 3 & 5 min at 140° C) for $c<1.5\%$. The pronounced
dependence of Tg on DICY becomes effective only at more advanced
levels of conversion (Figure 1; 7 min at 140° C and fully cured)
when the cross-link density of the network increases. It is
obvious that ΔH can be employed to assess the consistency of
batches of prepreg. To that end, only ΔH needs to be determined
at B-stage while the conversion can be assessed from Tg. Previously,
this was done via stroke cure time (time for the resin to lose its
stringiness) or flow test. The effect of DICY on flow is minimal
at B-stage (Figure 4; 3 min at 140° C). After some further advancement of the resin, flow decreases quite markedly with DICY. It
appears then that the flow test is extremely sensitive only once
the resin has been advanced close to its gel point. At B-stage
where relatively high flow is still desirable, the flow test seems
to be rather insensitive to conversion and batch variations.

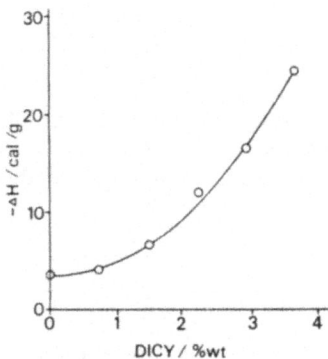

Figure 3: The DICY reaction is exothermic, therefore, the amount
of unreacted DICY functionalities can be determined at any stage
of the reaction (3 min bake at 140° C).

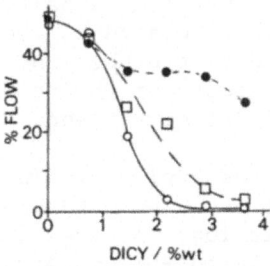

Figure 4: Flow depends strongly on the reaction extent for higher
concentrations of DICY (●3 min, □5 min, and ○7 bake at 140° C).

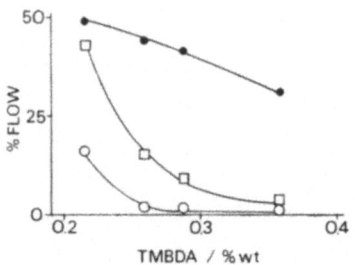

Figure 5:   TMBDA accelerates the cross-linking reaction already at low conversions (● 3 min bake at 140° C). This effect is more obvious for higher conversions where flow decreases drastically (□ 5 min and ○ 7 min bake at 140° C).

It was shown that TMBDA did not have any impact on Tg of fully cured resin. However, it could be shown that the rate of cure was accelerated by TMBDA. This is also reflected in the flow tests (Figure 5). Here, a noticeable gradation in flow can be observed at B-stage already. At higher conversions, the acceleration is even more apparent.

For one formulation (2.9% DICY, .29% TMBDA) the cure reaction was followed to its completion. Conversion was computed from Δ H using vacuum dried varnish as reference. As can be seen (Figure 6), Tg is a linear function of conversion up to 95% or more. At the very end of the reaction a minor advancement effects a significant increase in Tg (about 30° C). Most reactions are very difficult to characterize in their final stages when approaching 100% conversion. Obviously the cross-link density increases significantly and very little energy is liberated in the course of this reaction. It can only be speculated that either the reaction of very few groups tightens up the network or that a different, less exothermic reaction becomes predominant at the end.

It is obvious that for a given formulation the advancement of the cross-linking reaction can be determined accurately over the entire course of the reaction: A-stage (varnish), B-stage (prepreg) or C-stage (laminate). Up to 95% conversion, this can be done more precisely via Δ H. At higher conversions, Tg has to be employed to this end. For laminates, Tg will indicate whether total cure has been achieved. Using both Tg and Δ H in conjunction with each other,

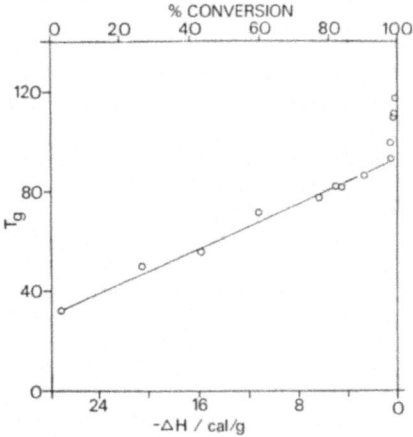

Figure 6:   Tg is linearly dependent on the heat of reaction up to
very high levels of conversion.   Conversion was determined from
Δ H using vacuum dried varnish as a reference.

the batch to batch consistency of a formulation can be assessed as
well.

In the past, controls were exercised based on the results
from stroke cure for varnish reactivity and stroke cure and flow
test for the B-stage of prepreg. As a consequence, either TMBDA
or treater conditions were adjusted to obtain prepreg of similar
quality. While these empirical tests proved to be quite useful,
they are very dependent on operator consistency (stroke cure) and
resin pick-up plus conversion (flow).

As has been demonstrated here, a valuable technique is
available in DSC for monitoring cure reactions independent of
operator dependent variables as often enter into empirical tests.
This technique can be applied towards monitoring of existing
processes and towards optimizing formulations and processes.
Most importantly, DSC can be employed in the development phase of
a new formulation to maximize its properties and again during the
design phase of a new production process to arrive at the most
efficient conditions.

REFERENCES

1   Prime, R. B., Thermosets, in "Thermal Characterization of
     Polymeric Materials", E. A. Turi ed., Academic Press,
     New York (1981).
2   Flory, P. J., "Principles of Polymer Chemistry", Cornell
     University Press, Ithaca (1953).
3   Chellis, L. N., US patent 3,523,037.
4   Saunders, T. F., M. F. Levy and J. F. Serino, J. Polym. Sci.,
     A-1, 5, 1609 (1967).

DIFFERENTIAL SCANNING CALORIMETRY: SIMULTANEOUS TEMPERATURE AND CALORIMETRIC CALIBRATION

W. Eysel and K.-H. Breuer

Mineralogisch-Petrographisches Institut
Universität Heidelberg, Germany

ABSTRACT

With computer equipped DSC instruments it is possible to measure temperatures and enthalpies with one and the same run. For the practical application of this possibility, standards are needed which allow the simultaneous calibration for both types of measurements. Based on numerous T and ΔH measurements a set of 12 standards is proposed in this paper.

INTRODUCTION

Differential scanning calorimetry (DSC) instruments can be used to determine both the temperatures and the enthalpies of phase transitions and chemical reactions. With conventional instruments these two types of measurements have to be carried out individually, i.e., using different runs. The reason is that the temperature difference $\Delta T = T_S - T_R$ between the sample (S) and the reference material (R) has to be recorded as a function of time (t)[*] for optimum ΔH measurement and as a function of T_S for optimum temperature reading (Fig.1).

---

[*]Because of the linear heating rate, t is proportional to T_R.

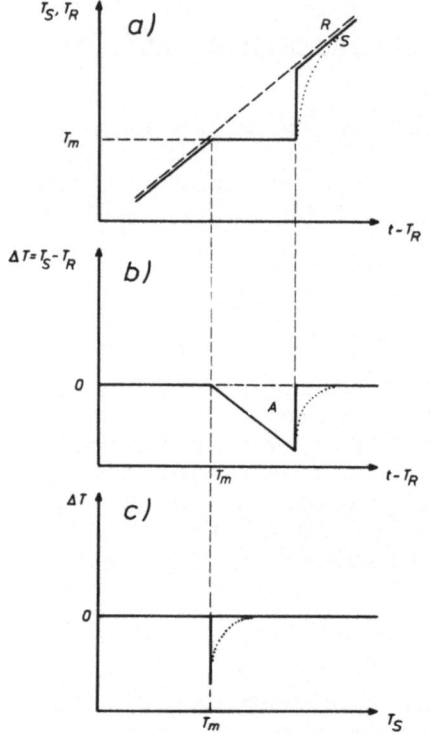

Fig. 1.   Schematic drawing of basic principles
          of DTA and DSC.   a) Individual represen-
          tation of T_S and T_R as a function of t.
          b) ΔT as a function of t.   c) ΔT as a
          function of T_S. Dotted lines = temperature
          equilibrates slowly after the transition.

Illustrating experimental results are shown in Figs. 2
and 3. As a consequence, so far also the calibration had
to be carried out individually fot both types of measurements and different sets of standards were employed.

    With the availability of modern, computer equipped
DSC instruments it is possible to record ΔT as a function of both T_S and t with one run[*] and then get both
results from the disc within two minutes. The time for
a second run, often lasting several hours, is saved this
way.

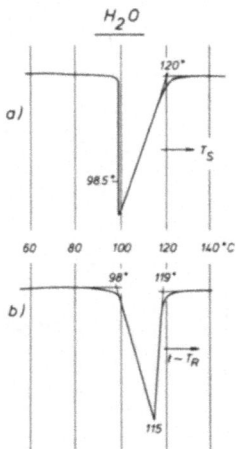

Fig. 2.   Vaporization of water from wet glass
          beads, recorded as a function of T_S

          (a) and t (b).   Du Pont Standard Cell,
          10°C/min. (With the thermocouple in the
          sample this cell demonstrates the effects
          better than the DSC cell).

---

[*]For this type of measurement the sample S must be placed
in the sample holder at which the temperature is measured.

With this possibility, however, the need arises for
standards which are suitable for simultaneous calibration
of enthalpy and temperature measurements. In the present
paper a set of 12 calibration materials is proposed covering most of the temperature range of the employed DSC
cell. They were selected from 25 inorganic compounds
and metals with 32 polymorphic transitions or melting
points. Various sources of error were investigated. To
get a good statistical mean every experiment was repeated
3 to 9 times, always with newly prepared samples.

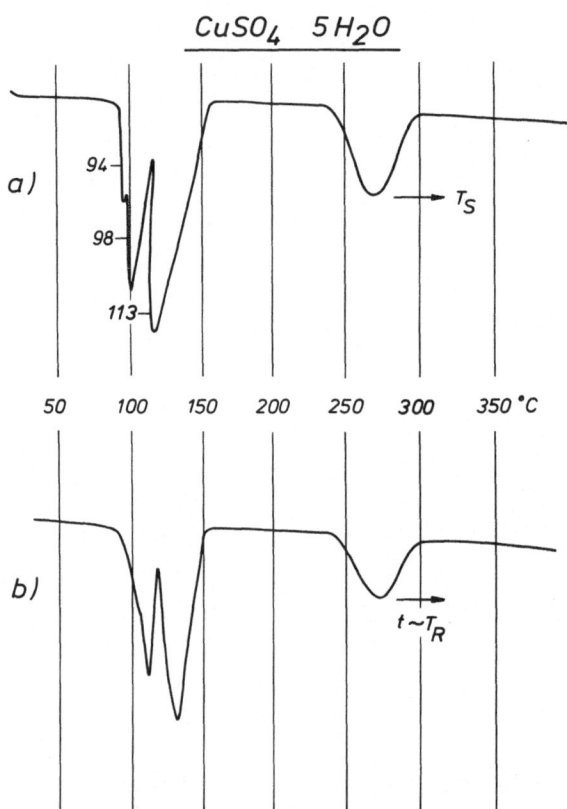

Fig. 3. Dehydration of $CuSO_4 \cdot 5H_2O$ as a function
         of T_S (a) and t (b). Du Pont Standard
         Cell, 10°C/min.

EXPERIMENTAL

A Du Pont 1090 Thermal Analyzer was used with a 910 DSC module. The heating rate ß ranged from 1 to 20°C/min., the sample size from 10 to 11 mg, the measured enthalpies from 1.3 to 25 kJ/mole and the temperature from -50 to 700°C. Samples were weighed with a Sartorius 2434 analytical balance. Only very pure materials were investigated, including the ICTA temperature standards. Purity and origin are listed in previous papers[1,3]. The results are based on about 350 DSC runs. The data in Figs.4,5 and 6 were fitted to straight lines (when possible) by the method of least squares.

CALORIMETRIC CALIBRATION

For enthalpy measurements ΔT is measured as a function of t. The enthalpy ΔH is determined from the peak area A of a signal

$$\Delta H = \frac{K'}{m} A = \frac{K'}{m} \int_{t_1}^{t_2} \Delta T \, dt$$

K' is a calibration constant.

In the previous paper[1] a number of potential calibration materials were investigated and various types of error (sample size, heating rate, background correction, enthalpy size) were considered. As a result the materials in Table 1 were proposed for calorimetric calibration. They consist of metals and inorganic compounds, i.e. materials of very different heat conductivities. The various types of error and detailed experimental conditions are not repeated here.

It was shown that, for the instrument employed, the heat conductivity is without measurable influence. The major proof was that for the majority of materials the same calibration constant was found. Meanwhile, part of the ΔH values in Table 1 were confirmed by measurements with a Perkin-Elmer DSC-2, i.e. a power compensated calorimeter (Breuer, Eysel and Höhne[2]).

Table 1. Standards for ΔH calibration
(after Breuer and Eysel[1])

Compound	Transition	Temp.($^\circ$C)	ΔH (kJ/mole)
In	m.p.	157	3.28 $\pm$ 0.02
RbNO$_3$	p.t.	166	3.87 $\pm$ 0.02
AgNO$_3$	p.t.	168	2.27 $\pm$ 0.01
AgNO$_3$	m.p.	211	12.13 $\pm$ 0.08
RbNO$_3$	p.t.	223	3.19 $\pm$ 0.01
Sn	m.p.	232	7.19 $\pm$ 0.03
Bi	m.p.	271	11.09 $\pm$ 0.12
Pb	m.p.	327	4.79 $\pm$ 0.07
Zn	m.p.	419	7.10 $\pm$ 0.04
Ag$_2$SO$_4$	p.t.	426	15.90 $\pm$ 0.16
CsCl	p.t.	476	2.90 $\pm$ 0.03
Li$_2$SO$_4$	p.t.	578	24.46 $\pm$ 0.07
K$_2$CrO$_4$	p.t.	668	6.79 $\pm$ 0.10

TEMPERATURE CALIBRATION

In contrast to calorimetric standards, there exists
an official set of temperature standards for DTA and DSC
measurements (Table 2). They were developed by the International Confederation for Thermal Analysis (ICTA). The
materials can be obtained from this organization. These
standards were created several years ago, when accurate
T and ΔH measurements were still not possible to be made
simultaneously. Unfortunately some of them are not suitable for calorimetric calibration[1], indicated by "-" in
Table 2 and have to be replaced by others for this
purpose.

For temperature measurements ΔT has to be recorded
as a function of the sample temperature T_S. Provided
a good and reproducible instrumental arrangement exists,
the measurement will depend predominantly on sample

Table 2.   ICTA Temperature Standards (^{O}C)

Compound	Extrapolated Onset	Peak	Equilibrium[3]
KNO_3	128	135	127.7
In	154	159	157
Sn	230	237	231.9
$KCLO_4$	299	309	299.5
Ag_2SO_4	424	433	?
SiO_2	571	574	573 "-"
K_2SO_4	582	588	583 "-"
K_2CrO_4	665	673	665 "-"

properties. Only spontaneous transitions are suitable
for accurate temperature measurements. Unfortunately,
many transitions are more or less sluggish resulting in
a broadening of the signal and/or a shift of the peak to
higher temperatures. An example is the dehydration peak
of $CuSO_4 \cdot H_2O$ at 270°C in Fig.3. The resulting errors, in
addition, are strongly dependent on the heating rate.
The sluggishness can have two origins: i) Retardation of
nucleation of the new phase; the single grains start to
transform at different times. ii) The transition or reaction, once a nucleus has formed, proceeds very slowly
through a grain. The major task for finding good calibration materials is to find compounds with spontaneous
transitions.

In this paper the well known melting points of
metals (Table 3) are the basis for a temperature calibration curve, which then is used to establish additional
inorganic compounds as standards. Only heating runs are
considered, since some metals show a considerable under-
cooling when crystallized from the melt.

Table 3.   Melting points of metals
(Equilibrium temperatures T_E
after Allard[4])

Metal	T_E ($^\circ$C)	Metal	T_E ($^\circ$C)
Hg	-38.9	Cd	321.0
In	156.6	Pb	327.4
Sn	231.9	Zn	419.5
Bi	271.4		

Since for most transitions the extrapolated peak onset temperature* was reproducible significantly better than the peak temperature**, it was decided to use only the onset temperature.

The best method to test the behaviour of a transition is to measure it with variable heating rates ß ($^\circ$C /min.). Since it was found that the extrapolated peak onset temperatures (and maximums) vary with the sample size, approximately the same sample masses should always be used.

From the results the following conclusions are drawn:

a) The measured temperatures of the extrapolated onsets and peaks increase with increasing ß.

b) For all metals the measured values can be approximated by a straight line, which allows a temperature extrapolation towards ß = 0.

---

*   In agreement with the ICTA definition the extrapolated onset is defined as the point of intersection of the tangent drawn at the point of greatest slope on the leading edge of the peak with the extrapolated base line.

** This point will be discussed in more detail in a subsequent paper.

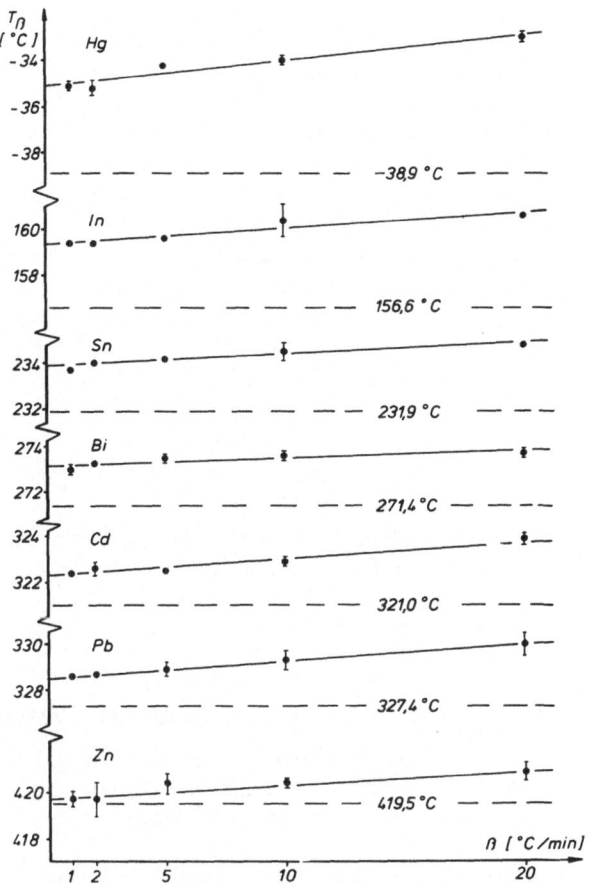

Fig. 4. Melting of metals as function of
the heating rate ß (peak onsets).

c) For the rather fast melting of the metals the slopes
   are very small and of comparable size.

d) In no case is $T_{ß=0}$ identical with the equilibrium
   temperature T_E in Table 3. The difference $T_{ß=0} - T_E$
   is a function of temperature (due to the thermocouple
   and cell employed) and heating rate.

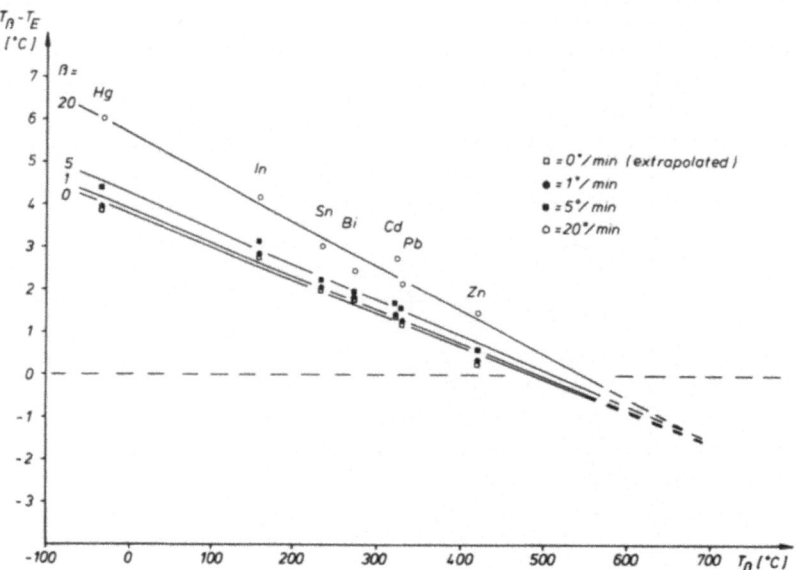

Fig. 5.   Calibration curves based on melting
          points of metals (peak onsets).

In Fig.5 these differences $T_\beta - T_E$ are plotted against
the temperature T_β for β = 0,1,5 and 20°C/min.   The result
is a set of calibration curves for the applied DSC cell.
Obviously for every heating rate another calibration
curve with a specific slope is valid.

The same measurements were carried out for inorganic materials selected from Tables 1 and 2.   With the
exception of CsCl and K_2CrO_4 all transitions allow a
linear extrapolation but on an average the slopes are
steeper than those of the metals.   The stronger dependence
on the heating rate indicates lower transition rates.
Correspondingly for temperature calibration KNO_3 and
$KClO_4$ should be applied at low heating rates only.   The
transitions of CsCl and K_2CrO_4 are completely rejected.

Using the extrapolated temperatures $T_{\beta=0}$ and the
corresponding calibration curve in Fig.5, approximate
equilibrium temperatures T_E for the inorganic compounds
were derived.   Besides the enthalpies these are given
in Table 4.

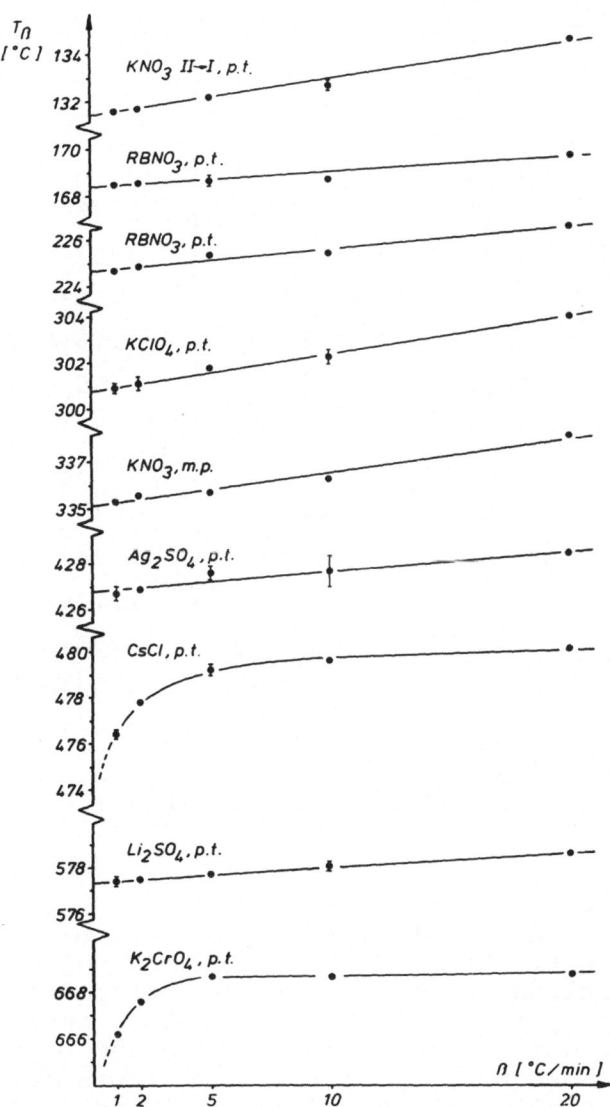

Fig. 6. Transitions of inorganic compounds as a
function of ß (peak onsets).

Table 4. Selected standards for simultaneous T and
          ΔH calibration. m.p. = melting point, p.t. =
          polymorphic transition.

Material	Transition	$T_E(^oC)^a$	ΔH (kJ/mole)
Hg	m.p.	-38.9	2.35$\pm$0.02 (0.8%)
KNO$_3$(II$\rightarrow$ I)	p.t.	128.7	5.05$\pm$0.04 (0.8%)
In	m.p.	156.6	3.28$\pm$0.02 (0.6%)
RbNO$_3$	p.t.	166.0	3.87$\pm$0.01 (0.4%)
RbNO$_3$	p.t.	222.7	3.19$\pm$0.01 (0.4%)
Sn	m.p.	231.9	7.19$\pm$0.03 (0.4%)
Bi	m.p.	271.4	11.09$\pm$0.12 (1.1%)
KClO$_4$	p.t.	299.4	14.53$\pm$0.15 (1.0%)
Cd	m.p.	321.0	6.21$\pm$0.11 (1.7%)
Pb	m.p.	327.4	4.79$\pm$0.07 (1.4%)
KNO$_3$	m.p.	334.0	9.94$\pm$0.04 (0.4%)
Zn	m.p.	419.5	7.10$\pm$0.04 (0.6%)
Ag$_2$SO$_4$	p.t.	426.4	15.90$\pm$0.16 (1.0%)
Li$_2$SO$_4$	p.t.	578.1	25.02$\pm$0.16 (0.6%)

[a]Temperature after Table 2 or determined in this paper.
Estimated maximum error for the latter: $\pm$ 1^oC.

     These proposed temperature standards were chosen on
the basis of their rather fast transition behaviour.
The T calibration curves therefore are valid only for
fast transitions. Corrections by extrapolation towards
ß = O for sluggish transitions are less accurate and
depend on the transition speeds. In the opinion of the
authors it does not pay to derive calibration curves for
sluggish transitions. In these cases accurate temperature determinations are less accurate if not impossible
at all by using dynamic methods.

     To measure optimum temperatures the described cali-
bration procedure has to be carried out individually for
every instrument.

SIMULTANEOUS T AND ΔH CALIBRATION

The task for establishing a list of suitable standards for combined temperature and calorimetric calibration was to select those materials from Tables 1, 2 and
3 which best fulfill the requirements for simultaneous
work. In Table 4 the 12 selected standards (14 transitions) for both types of measurements are listed. The
temperature data are either taken from Table 3 or derived
in this paper. The ΔH data are mostly taken from the
previous paper[1], some of which were improved (KNO_3,
Li_2SO_4). The enthalpies of Hg and $KClO_4$ were newly
determined.

Some of the inorganic materials were investigated
by Gray[5]. The ΔH values are in good agreement with our
results.

For calorimetric measurements the melting of Zn could
only be used with $\beta \geq 10°$/min. At lower rates the sample
reacted with the Al sample holders and also vaporizes
partly (0.3%), resulting in incorrect (smaller) ΔH values.

As pointed out before[1], the K_2CrO_4 and K_2SO_4 transitions contain part of a λ-point transition resulting in
a rather bad background correction. Moreover, for K_2CrO_4,
the transition temperature is extremely dependent on the
heating rate, especially between 0 and 5°C/min. (comparable with CsCl). Therefore, peak onsets of the NBS-
ICTA standards K_2CrO_4 as well as CsCl are not recommended
here. K_2SO_4 was replaced by Li_2SO_4. Thus, the temperature interval between 600 and 700° remains uncovered for
the present.

For the first simultaneous calibration of an instrument it is recommended to use all or many of the proposed
standards. Depending on the desired accuracy fewer
standards may be sufficient for subsequent controls.

ACKNOWLEDGEMENTS

The authors thank Mr. H. Maltry for technical
assistance. The investigations were supported by the
Land Baden-Württemberg and the Deutsche Forschungsge-
meinschaft.

REFERENCES

1. K.-H. Breuer and W. Eysel, The calorimetric
   calibration of differential scanning calorimetry cells, Thermochimica Acta 57, 317-
   329 (1982).
2. K.-H. Breuer, W. Eysel and G.W.H. Höhne, Enthalpy standards for heat flux calorimeters,
   Paper presented at the 5. Ulmer Kalorime-
   trietage, Ulm, Germany (1983).
3. NBS Circ. 500 (1952).
4. S. Allard,ed., Int. Tables of selected constants
   16, Pergamon Press, Oxford, N.Y., Paris,
   Braunschweig (1969).
5. A.P. Gray, The calorimetry of the NBS-ICTA
   temperature standards, in: Thermal Analysis,
   Proceedings Fourth ICTA, J. Buzas, ed.,
   Akademiai Kiado, Budapest, Vol.3, 991-1003,
   (1975).

DYNAMIC MECHANICAL ANALYSIS APPLICATION

TO DYNAMIC THERMOMECHANOMETRY

Takayuki Murayama

Monsanto Triangle Park Development Center
Post Office Box 12274
Research Triangle Park, North Carolina 27709

INTRODUCTION

The dynamic mechanical behavior of viscoelastic materials has been discussed as part of mechanical properties such as stress-strain behavior, strength of material, creep, and stress relaxation. However, in recent years the technology of investigating the dynamic mechanical properties of materials has advanced greatly due to new electronic circuits, improved transducers, signal receiving systems, and computerized controlling systems. The interpretations and applications of dynamic data have stimulated interest from both the practical and scientific standpoint. Consequently, widespread research work on these topics has established diverse results and principles.[1,2]

The commonly-used dynamic mechanical instruments measure the deformation of a material in response to vibrational forces. The dynamic modulus, the loss modulus, and a mechanical damping or internal friction are determined from these measurements. The modulus indicates stiffness of material, and it may be a shear, a tensile, or a flexile modulus, depending upon the experimental equipment. The mechanical damping (internal friction) gives the amount of energy dissipated as heat during the deformation. The internal friction of material is important, not only as a property index, but also for environmental and industrial application. Since noise is radiated by the vibration of an object -- $(0.001 \sim 0.004)$ -- the application of damping materials to the vibrating surface will convert the energy into heat, which is dissipated within the damping materials rather than being radiated as airborne noise.

81

Amorphous viscoelastic polymers are good damping materials, having high internal friction (0.1∿0.3). High damping or internal friction is essential in decreasing the effect of undesirable vibration, in reducing the amplitude of resonance vibration to safe limits, and in all kinds of structures from airplanes to buildings.[3] The investigation of the dynamic modulus and internal friction over a wide range of temperatures and frequencies has proven to be very useful in studying the structure of high polymers and the variations of properties in relation to end-use performance. These dynamic parameters have been used to determine the glass transition region, relaxation spectra, molecular orientation and structural or morphological changes resulting from processing. Dynamic mechanical spectroscopy tools and techniques such as the torsion pendulum, torsional braid analysis, the duPont 981 Dynamic Mechanical Analyzer, the PL-Dynamic Mechanical Thermal Analyzer, the Dynastat and Mechanical Spectrometer are used for these studies. One of the most commonly-used instruments of this type has been the direct-reading standard Rheovibron®.[4] In order to improve the range, accuracy, and convenience, several groups, including the manufacturer, have automated both the operation of various Rheovibron® units and the data processing as well.[5-8] This paper presents results of trials with the Autovibron™ DDV-II-C[9] and discusses the expanded capability of this instrument as the dynamic thermomechanometry.

ANALYTICAL DEVELOPMENT

The dynamic instrument uses the method of sinusoidal excitation and response. In this case, the applied force and the resulting deformation both vary sinusoidally with time, the rate usually being specified by the frequency f in cycles/sec or $\omega = 2\pi f$ in radians/sec. For linear viscoelastic behavior, the strain will alternate sinusoidally but will be out of phase with the stress. This phase lag results from the time necessary for molecular rearrangements and is associated with relaxation phenomena.[10] The stress σ and strain ε can be expressed as follows:

$$\sigma = \sigma_o \sin(\omega t + \delta) \tag{1}$$

$$\varepsilon = \varepsilon_o \sin \omega t \tag{2}$$

where ω is the angular frequency, and δ is the phase angle. Then

$$\sigma = \sigma_o \sin \omega t \cos \delta + \sigma_o \cos \omega t \sin \delta \tag{3}$$

The stress can be considered to consist of two components, one is in phase with the strain ($\sigma_o \cos \delta$) and the other 90° out of phase ($\sigma_o \sin \delta$). When these are divided by the strain, we can separate the modulus into an in-phase (real) and out-of-phase (imaginary) components. These relationships are

$$\sigma = \varepsilon_o E' \sin \omega t + \varepsilon_o E'' \cos \omega t \qquad (4)$$

$$E' = \frac{\sigma_o}{E_o} \cos \delta \text{ and } E'' = \frac{\sigma_o}{E_o} \sin \delta \qquad (5)$$

where E' is the real part of the modulus and E'' the imaginary part. The complex representation for the modulus can be expressed as follows:

$$\varepsilon = \varepsilon_o \exp i\omega t \qquad (6)$$

$$\delta = \delta_o \exp i(\omega t + \delta) \qquad (7)$$

then

$$\frac{\sigma}{\varepsilon} = E* = \frac{\sigma_o}{\varepsilon_o} e^{i\delta} = \frac{\sigma_o}{\varepsilon_o} (\cos \delta + i \sin \delta) = E' + iE'' \qquad (8)$$

and similarly for the other deformation types. For example, the ratio of the peak stress to peak strain for shear is[11]

$$\left| G* \right| = G'^2 + G''^2, \qquad (9)$$

where G* is the shear complex modulus, G' is the real part of the modulus and G'' the imaginary part, and the phase angle δ is given by

$$\tan \delta = G''/G' \qquad (10)$$

$$G' = \left| G* \right| \cos \delta \text{ and } G'' = \left| G* \right| \sin \delta \qquad (11)$$

The real parts of the moduli E' and G', are the storage moduli, because they are related to the storage of energy as potential energy and its release in the periodic deformation. The imaginary parts of the moduli, E'' and G'', are the loss moduli and are associated with the dissipation of energy as heat when the materials are deformed. These dynamic moduli can also be expressed in terms of a complex compliance

$$J* = 1/G* = J' = iJ'' \qquad (12)$$

where J' is the storage compliance and J'' is the loss compliance. The loss tangent tan δ is internal friction or damping, and is the ratio of energy dissipated per cycle to the maximum potential energy stored during a cycle.

These dynamic parameters (E', E'' and tan δ) can be determined by using the Rheovibron®. The mathematical expressions in relation to the instrumental parameter and automated systems are developed.[12,13]

This system is expanded to the thermomechanical analysis capability which requires the measurement of length in a penetration, expansion, or extension mode. The basis of this analysis has been discussed in the literature.[14] The relationships of the thermal expansivity α and the length 1 of a sample with respect to temperature and tensile force f are expressed as follows

$$\alpha \equiv (1/V)(\partial V/ \partial T) \qquad (13)$$

$$dl = (\partial l/\partial T)_f dT + (\partial l/\partial f)_T df \qquad (14)$$

$$df = (\partial f/\partial T)_l dT + (\partial f/\partial l)_T dl \qquad (15)$$

where V is the volume of sample, and T is temperature. The changes of the sample length are detected as a function of temperature, the dynamic modulus E', loss modulus E'', internal friction tan δ and length changes in expansion and contraction dl are obtained by this system.

EXPERIMENTAL

The Autovibron^{TM} system[9] is designed to measure the temperature dependence of the complex modulus (E*), dynamic storage modulus (E'), dynamic loss modulus (E'') and dynamic loss tangent (tan δ) of viscoelastic materials at specific selected frequencies (0.01 to 1 Hz, 3.5, 11, 35, 110 Hz) of strain input. During measurement, a sinusoidal tensile strain is imposed on one end of the sample, and a sinusoidal tensile stress is measured at the other end. The phase angle δ between strain and stress in the sample is measured. The instrument uses two transducers for detection of the complex dynamic modulus (ratio of maximum stress amplitude to maximum strain amplitude) and the phase angle δ between stress and strain. From these two quantities, the real part (E') and the imaginary part (E'') of the complex dynamic modulus (E*) can be calculated.

The amplitude of the sinusoidal deformation is made small (1.0 x 10^{-3} cm, approximately 0.03 to 0.05 percent strain) to ensure linear viscoelastic response. The frequency is maintained constant for each data point.

Data are usually taken over a range of temperature and sometimes over a limited range of frequencies. A typical range of temperature is -150°C to as far as above the glass transition temperature.

The changes of sample length as a function of temperature are obtained over very small temperature intervals (< 1.5°C). The detailed Autovibron™ system is described elsewhere.[6,9,15] The automation system in this experiment is designed around a computer program that is implemented with a Hewlett-Packard 9825A calculator, interfaced for control and data acquisition to a Hewlett-Packard 6940B Multiprogrammer. Commands in the program instruct the instrument through the multiprogrammer to initiate controls over the experiment.

Other program commands instruct the multiprogrammer to acquire data and return it to the calculator for manipulation and/or storage. A separate program is provided for data reduction and a line printer and/or plotter can be provided as optional equipment.

The standard phase angle measuring system has been replaced with a lock-in amplifier (Princeton Applied Research Model 5204) to simplify automation of the measurement, to improve resolution of small angles, and to extend the range of tan δ measurements.

A Rheovibron® interface, on command from the calculator, uses a signal from either the load or displacement transducer to the lock-in amplifier where it is measured. The measured result is passed through the Amp, Phase, Temp, and Relay cards to the Voltage Monitor where it is digitized and stored in the calculator memory. The two signals are alternately measured throughout the test run. The interface also passes the load transducer signal through the tension amplifier and relay cord to the voltage monitor for digitizing and to the calculator where it is compared for tension limits. If the tension requires correction, a signal from the calculator through the relay card commands the motor to increase or decrease sample tension the proper number of steps. Temperature at the sample is detected by a platinum resistance thermometer. The signal is read by the program through the temperature translator, the Amp, Phase, Temp card, the Relay card, and voltage monitor to the calculator. The measured temperature is compared to values in the calculator program which control power to the environmental chamber heaters.

The Tension Amplifier, Amp, Phase, Temp card, Relay card, Motor Power Supply and Amplifier, and Voltage Monitor are all contained in the Multiprogrammer.

The following data values are measured and recorded on a magnetic tape cassette for eventual data reduction:

> In-phase dynamic displacement
> Quadrature dynamic displacement
> Sample length, cm
> Average static tension, dynes
> Temperature °C
> In-phase dynamic load
> Quadrature dynamic load.

MEASUREMENT FEATURES

1. Temperature

Temperature is detected by a 100-ohm platinum resistance thermometer (RDF #80RB), and is measured and transmitted through the multiprogrammer to the calculator. Accuracy is +1°C over a full-scale range (-180°C to +650°C).

The temperature chamber is designed with sufficient thermal mass to make control unnecessary from -150°C to approximately -60°C. Natural heat absorption after removal of liquid N_2 cooling causes the system to increase in temperature at a rate very close to 1°/min, which is about ideal for data acquisition. Above -30°C some power must be supplied to the heaters to maintain the rate of temperature rise. This is handled by a heater control system consisting of two solid-state relays which proportion power to the heaters in amounts depending on needs. The system maintains a rate of temperature increase of approximately 1°C/min.

2. Modulus and Phase Measurement

A lock-in amplifier (Princeton Applied Research Model 5204) is used to measure in-phase and quadrature amplitudes of both the load and displacement signals and phase relation of each signal with respect to a reference from the Rheovibron®. The calculator program sends a command to the multiprogrammer to switch on the load signal to the lock-in amplifier which, after a program delay for settling, reads the in-phase and quadrature components with respect to the reference, and stores them in the calculator memory.

After these values are stored, the multiprogrammer receives a command to switch the dynamic displacement signal into the lock-in amplifier where it repeats the same measurements on these signals and stores the results in the calculator.

The calculator program then computes the voltage magnitudes of both the load and displacement signals, P and X, and the phase angle, δ, between the load and displacement.

The load and displacement signals are scaled by the calibration factors stored in the program and the complex modulus, E, is computed and printed by the calculator. At the same time the tangent of the loss angle, tan δ, is printed with the temperature. This modulus is uncorrected for instrument compliance but is corrected for changes in sample length which will be discussed in the section on tensioning the sample.

3. Sample Tension Control

The type of sample most frequently used (thin films, fibers, monofilaments, etc.) on the Rheovibron® will not support a compressive load without buckling. This condition would, of course, distort the negative portion of the sine wave making measurements impossible. To alleviate this situation, the sample must be maintained under sufficient tension to eliminate buckling but insufficient to introduce non-linearity or excessive creep into the experiment. The amount of tension is critical to the measurement.

The tension control system uses a stepping motor to add or subtract small increments of tension to the sample. The motor receives its commands from the calculator. Each step of the motor represents a change in length of the sample of 1.25×10^{-4} cm. The program computes the modulus of the sample and uses this value to establish maximum and minimum limits of tension. It then calculates the magnitude of a single step to determine its load value and computes the number of steps required to put the sample tension within limits. The program then instructs the motor, through the multiprogrammer, to initiate the calculated number of steps, either + or -. The tension is measured and corrected twice during the taking of each datum point. As the sample softens, the program changes tension limits. An up-and-down count is stored in a register which, when multiplied by the value of each step, is added to, or subtracted from, the initial sample length.

These controlling features for the Rheovibron®, a combination of multiprogrammer, programmable calculator, and lock-in analyzer, are used to provide a programmed heating rate, continuous sample tensioning, measurement of the phase angle between stress and strain, and acquisition and reduction of the data.

A number of polymer films and fibers have been examined as a function of temperature. Dynamic measurements were made at 11 Hz, 1°C/min heating rate in nitrogen atmosphere. Data point of dynamic tensile modulus and tan δ were obtained at 1°C increments. This is essentially a continuous monitoring of the changes in storage modulus, loss modulus, and tan δ and length of sample.

Results and Discussion

Figure 1 shows the loss tangent (tan δ) and the dynamic modulus (E') at 0% RH as a function of temperature for unoriented and oriented (4.74X) nylon 66 film. The dynamic moduli change and the α damping peak are clearly associated with the glass transition in which polymer chain segments acquire considerable mobility. The temperature of the maximum of the tan δ peak and a maximum tan δ value of the α peak have been useful for determination of the glass transition. However, the shape of the α peak of the semi-crystalline polymer has been associated with the region of the glass transition. Thus, in order to characterize this region, the new terms, ΔT_1 and ΔT_2 are introduced. The ΔT_1 indicates the half power widths of the peak in the loss tangent, and ΔT_2 is the width of tan δ peak at one-half tan δ_{max}. The values for these samples are summarized in Table I.

The dynamic thermomechanometry parameter, length changes, dℓ, as a function of temperature are shown in Figure 4. The dℓ at the temperature of the maximum of the tan δ peak for these samples are indicated in Table I. The occurrences of the glass transition, denoted by the maximum in tan δ at 85°C for unoriented nylon 66 and at 110°C for the highly-oriented nylon 66 film, are typical of the values obtained by other workers for similarly-prepared nylon 66.[16,17] The effect of annealing, drawing, and molecular weight on the α peak of nylon 66 is reported in our previous papers.[1,17,18] We concluded that the α transition is little affected by the annealing treatment, but an increase in orientation causes the glass transition to move to a higher temperature. The values of ΔT_1 and ΔT_2 of oriented film are much larger than that of unoriented nylon 66 film. The oriented nylon 66 indicates that the glass transition is distributed in a wide temperature range.

Figure 2 shows the dynamic modulus (E') and the loss tangent (tan δ) at 0% RH as a function of temperature for unoriented and oriented (5.5X) poly(ethylene terephthalate), (PET) fibers. The glass transition parameters (α peak temperature, tan δ_{max}, ΔT_1, and ΔT_2) and length changes (dℓ) at the glass transition temperature are summarized in Table I. The glass transition occurs at about 82°C for unoriented amorphous PET, while this transition is found near 135°C for oriented PET. The shift in glass transition is due to the combined effect of crystallinity and orientation. These data are in good agreement with the previous results.[1]

The ΔT_1, ΔT_2 and dℓ show large differences between oriented and unoriented PET. These values indicate, not only the range of glass transition, but also the range of order and crystallinity in the structure. The width of the α peak increases as crystallinity and order increase.

Figure 3 shows the other example of the dynamic mechanical measurements by the Autovibron™. The dynamic modulus and the loss tangent of the acrylic fiber are obtained at very small temperature intervals (< 1.5°C). The glass transition temperature, maximum tan δ value of the α peak, and dynamic moduli – temperature curve show similar results in the published paper.[6] These parameters and length changes (dℓ) at the glass transition are summarized in Table I.

The ΔT_1 and ΔT_2 values show a good indication of the distribution of the α peak with intensity and the peak temperature. In general, the polymers having higher tan δ peak have higher length changes as a function of temperature. The added capability of measurement of the dℓ values in this system is useful for studying of shrinkage and dimensional stability of polymers.[19]

The Autovibron™ system gives an improved measurement accuracy and sensitivity by a programmed heating rate, continuous sample tensioning, and measurement of the phase angle between stress and strain. These features are provided by using a combination of a multiprogrammer, programmable calculator, and lock-in analyzer.

Measurements on the manual Rheovibron® require the constant attention of a skilled operator throughout the run. In the major transition, the operator is not able to observe data at sufficiently small intervals, and so, some accuracy on transition location must be sacrificed when the temperature intervals are taken 5–10°C apart. However, the Autovibron™ can be observed over very small temperature intervals (< 1.5°C) that essentially provide a continuous monitoring of the changes in storage modulus, loss modulus, tan δ and length.

As a result, the Autovibron™ has expanded capability, not only for measurements of the dynamic mechanical properties of materials, but also as a thermal analyzer. The glass transition of polymers can be determined by the loss tangent peak parameters, such as the maximum in loss tangent, peak temperature, and widths of tan δ peak.

One of the strengths of the dynamic mechanical analyzer (DMA) is that tan δ is a very sensitive method for detecting secondary transitions and relaxation in polymers. In addition, the Autovibron™ (DMA) system also has the ability to monitor the length of sample at each temperature during the dynamic measurement. This capability can be used for thermomechnometry and will be useful for studies on shrinkage and relaxation of polymers.

Table 1.

Samples	Temperature at Which Maximum Occurs in tan δ Peak °C	Maximum tan δ	ΔT_1 Width of Loss Tangent Peak at 1/2 tan δ_{max}	ΔT_2 Width of Loss Tangent Peak at 1/2 tan δ_{max}	$d\ell$ Length Changes at Tg %
Nylon 66					
Unoriented	90	0.095	22	32	4.5
Oriented (4.74X)	110	0.084	58	84	3.0
Polyester					
Unoriented	82	0.86	5	6	45
Oriented (5.5X)	135	0.135	42	68	12.2
Acrylic					
Fiber	123	0.300	43	62	7.2

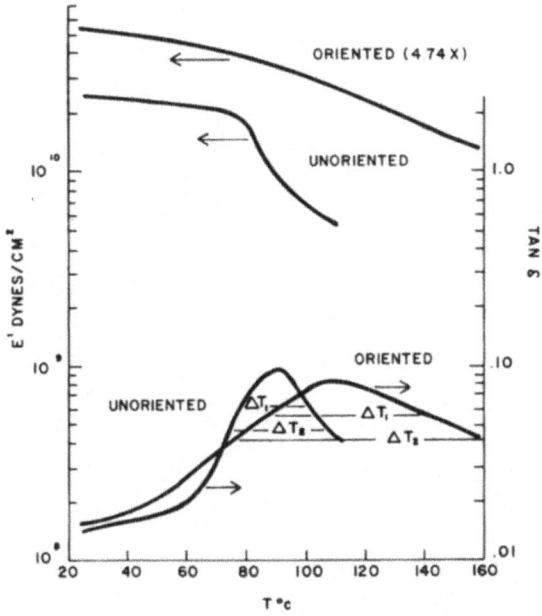

Figure 1.

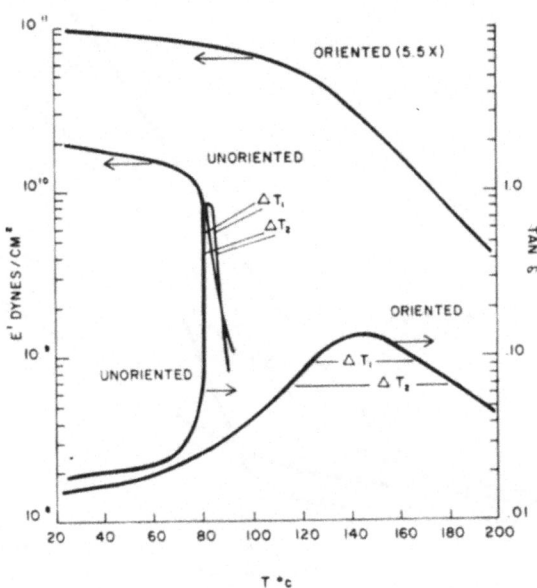

Figure 2.

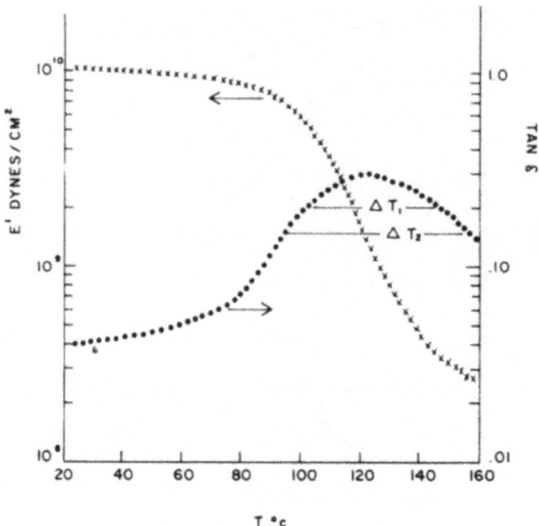

Figure 3.

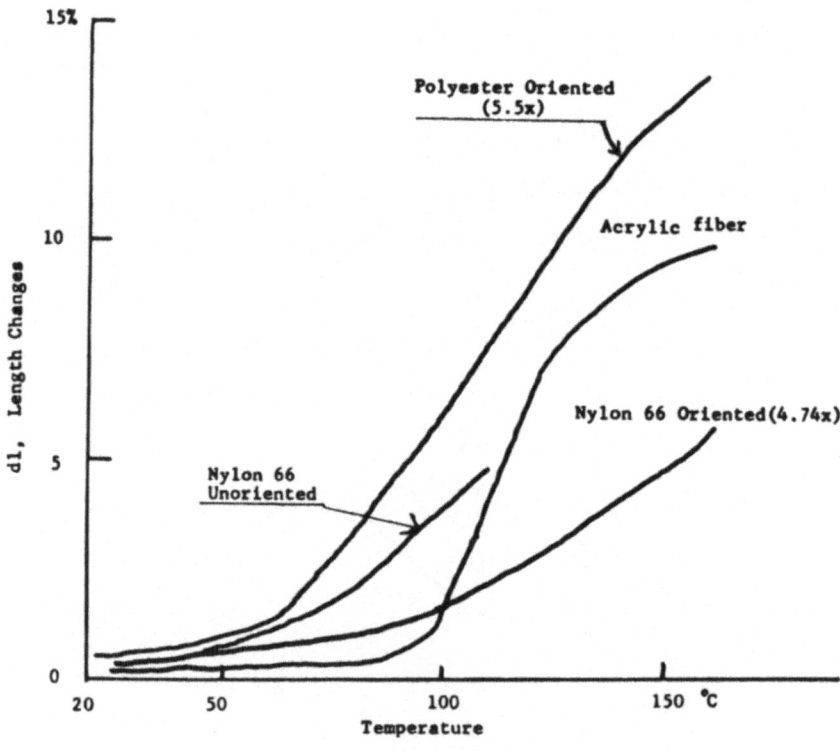

Figure 4.

REFERENCES

1. T. Murayama, "Dynamic Mechanical Analysis of Polymeric Material,"
    Elsevier Scientific Publishing Co., Amsterdam-Oxford,
    New York (1978).
2. B. E. Read and G. D. Dean, "The Determination of Dynamic
    Properties of Polymers and Composites," Wiley, New York
    (1978).
3. M. P. Blake and W. S. Mitchell, "Vibration and Acoustic
    Measuring Handbook," Spartan, New York (1972).
4. M. Takayanagi, Viscoelastic Properties of Crystalline Polymers,
    Mem. of the Fac. of Eng., Kyushu Univ., No. 1, Vol. 23
    (1963).
5. Toyo Baldwin Co., Tokyo, Technical Literature (1979).
6. A. S. Kenyon, et al., J. Macromol. Sci. Phys., B13, 553 (1977).
7. A. F. Yee and M. T. Takemori, J. Appl. Polym. Sci., 21, 2597
    (1977).
8. S. M. Webler, J. A. Manson and R. Lang, Am. Chem. Soc., Div.
    Polym. Chem. Polym. Prepr., 22, 257 (1981).
9. Imass, Inc., Product Bulletin, Accord, Massachusetts (1980).
10. N. G. McCrum, B. E. Read, and G. Williams, "Anelastic and
    Dielectric Effects in Polymeric Solids," Wiley, London (1967).
11. J. D. Ferry, "Viscoelastic Properties of Polymers," Wiley,
    New York (1969).
12. D. J. Massa, J. Appl. Phys., 44, 2595 (1973).
13. A. R. Wedgewood and J. C. Seferis, Polymer, 22, 966 (1981).
14. B. Wunderlich, "Thermal Characterization of 'Polymeric Materials,"
    Chapter 2, p. 122, Academic Press (1981).
15. T. Murayama, Polymer Engineering and Science, (August 1982),
    Vol. 22, No. 12, 788.
16. A. E. Woodward, J. M. Crissman, and J. A. Sauer, J. Polym. Sci.,
    44, 23 (1960).
17. J. H. Dumbleton and T. Murayama, Kolloid-Z, 238, 410 (1970).
18. T. Murayama and B. Silverman, J. Polym. Sci., Polym. Phys. Ed.,
    11, 1873 (1973).
19. M. Jaffe, "Thermal Characterization of Polymeric Material,"
    Ed. by Edith A. Turi, Academic Press, (1981) Chapter 7,
    709-792.

APPLICATIONS OF DIFFERENTIAL RHEOMETRY

P. H. Foss and M. T. Shaw

Institute of Materials Science
and Department of Chemical Engineering
University of Connecticut
Storrs, Connecticut 06268

INTRODUCTION

In many material characterization situations the variables of
interest are the relative properties of a set of materials as a
function of aging, composition or some other independent parameter. In these situations, the inherent variance of measurements
on commercial rheometers may be of the same order of magnitude as
the expected property difference. Clearly, reaching conclusions
about the effect of the independent variable on the properties of
the material would be impossible under these circumstances, even
though the effect may be quite real. In a differential rheological technique one hopes to eliminate many of the interrun errors
by comparing one material against the other during the same run.
In this way one may emphasize the importance of relative properties and suppress the effect of absolute properties. In this
manner one may obtain a better estimate of the differential properties than could be obtained by the subtraction of two very
large, independently measured, numbers.

A few differential methods of rheometry have appeared in the
literature over the past 50 years. In 1941 J. Pryce-Jones (1)
reported on a twin Couette system in which the two outer cylinders
rotated in opposite directions while the two inner cylinders were
connected together and to a torsion wire. Sakamoto et al. (2)
described another dual Couette system with two identical cylinder
sets one above the other. The upper-outer and the lower-inner
cylinders were connected together while the upper-inner was fixed
to the support. The lower-outer cylinder was driven at a constant

speed. The difference in viscosity between the solution in the
upper and lower gaps is related to the speed of rotation of the
connected cylinder pair. A differential Couette rheometer consisting of three concentric cylinders was designed by Gudim et
al. (3). In their system the innermost cylinder was rotated at a
constant speed while the outermost was fixed. The speed of rotation of the middle cylinder was related to the viscosity difference between the two solutions.

In the differential rheometers discussed above, the material
property of interest was the steady or zero-shear-rate viscosity.
While viscosity is an important property of materials, it can only
be measured on fluids. The dynamic mechanical properties, on the
other hand, can be measured equally well on solids or fluids and
can be very sensitive to changes in the material structure. Generally, the low frequency properties are the most sensitive to
small changes in structure (4). Thus, the objective of this work
was to investigate the theoretical response of an opposed squeeze
flow geometry (5,6,7) and compare it to the experimental results
for representative viscoelastic materials. While the experimental confirmation of the analyses of these problems was confined to
a limited number of well-characterized materials, the general purpose of the combined theoretical and experimental approach was to
demonstrate the applicability of the rheometer to the study of
the viscoelastic properties of any material within the instrument's force, size and speed capabilities.

INSTRUMENTATION

The opposed squeeze flow differential rheometer, constructed
by the Instrument and Electronic Shops of the Institute of
Materials Science, consists of two gear-driven pushrods and a
load cell blade, Figure (1). The basic premise of the instrument
is that if the two rods are forced to move in a programmed manner
then the response of the load cell will be a function of the
differences in the material properties of the two samples and, to
a lesser extent, the absolute properties. The first design of
the instrument was with a fairly stiff (6000 N/m) cantilevered
load cell blade, the motion of which was measured with strain
gauges cemented to the beam near the point of attachment to the
frame. For the sinusoidal mode an analog controller was used to
track the rod position to an externally inputed sine wave. The
feedback loop was closed by an LVDT (Shavitz Engineering, 050HM)
attached to the end of one of the pushrods. The region around the
pushrods and load cell blade can be enclosed in an oven to regulate the sample temperature. Data collection and input signal
generation were performed with both a digital microcomputer
(APPLE II+) and/or a digital frequency generator/autocorrelator
(Schlumberger Frequency Response Analyzer).

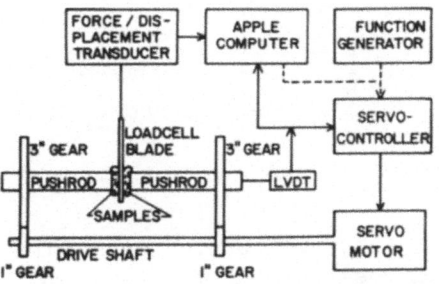

Fig. 1. Differential Rheometer-Schematic

THEORY

The theoretical solution for a viscoelastic fluid in dynamic
squeezing flow is based on the equations of Kramer (8). Kramer's
equations for a viscoelastic fluid in small amplitude sinusoidal
squeezing flow are:

$$F' = 3/8\ \pi R^4\ a/h_o^3\ G' \tag{1a}$$

$$F'' = 3/8\ \pi R^4\ a/h_o^3\ G'' \tag{1b}$$

Here F' and F'' are the in-phase and out-of-phase components of
the dynamic force for a sample with half-thickness h_o and amplitude a. The solution of Equations 1a and 1b for a dual opposed
squeeze flow geometry is complicated by the motion of the load
cell blade. The blade motion changes both the amplitude and the
relative phase angle of the two sample gaps. Figure (2) shows
the effects of the blade motion on the sample gaps; note that the
effects have been greatly exaggerated for clarity. If the resultant gap amplitudes are used in Equation (1) and the two force

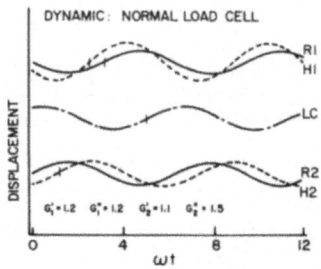

Fig. 2. Effect of load cell motion on sample gaps where R_i is
         pushrod i position, H_i is sample gap i and LC is the
         load cell position.

vectors are subtracted, taking the gap phase angles into account,
the result is:

$$G_1' = \frac{(EA - FB)}{A^2 + B^2} \tag{2a}$$

$$G_1'' = \frac{(AF + BE)}{A^2 + B^2} \tag{2b}$$

Where:

$$A = -\tilde{d} \cos \theta - 1 \tag{3a}$$

$$B = \tilde{d} \sin \theta \tag{3b}$$

$$E = \frac{-C_2}{C_1} \tilde{G}_2'(A+2) - \frac{C_2}{C_1} \tilde{G}_2''B - \frac{\tilde{d}}{C_1} \sin \theta \tag{3c}$$

$$F = \frac{C_2}{C_1} \tilde{G}_2' B - \frac{C_2}{C_1} \tilde{G}_2''(A+2) - \frac{\tilde{d}}{C_1} \sin \theta \tag{3d}$$

$$C_1 = \frac{3}{8} \pi R^4 \frac{1}{2k(P_1 a)^3} \tag{3e}$$

$$C_2 = \frac{3}{8} \pi R^4 \frac{1}{2k(P_2 a)^3} \tag{3f}$$

EXPERIMENTAL

A series of viscoelastic fluids based on SE-30 (General Electric, polydimethyl siloxane gum, M_w = 450,000) were used in the evaluation of the Differential Rheometer (DR) for viscoelastic fluids. The samples were prepared with a serial dilution technique starting with a 8.874% mixture of low molecular weight PDMS (Scientific Polymer Products, Inc., M_w = 103,400) in SE-30. Portions of this mixture were serially diluted with pure SE-30 to obtain four different viscoelastic fluids. The samples were characterized on a Rheometrics Mechanical Spectrometer (RMS) in the oscillating parallel plate mode. The measurements were taken at 22 C using 25-mm fixtures and a strain amplitude of less than 5 percent.

The samples were evaluated with the differential rheometer in the following manner: Samples of SE-30 and the material of interest were weighed to 0.3 ± 0.0005 g. For each experimental run the samples were loaded into the sample gaps and the control system activated at the initial sample thickness. The loading of the samples invariably introduced an initial deflection in the load cell blade. The samples were therefore allowed to relax until the blade had returned to the zero position (2-3 min). The input waveform from the function generator was then started at the frequency and amplitude of interest. The APPLE-based data collection system was started at the end of 1 cycle and continued for 4-10 cycles at 50-250 points/cycle. After data collection the samples were removed, the pushrods cleaned and fresh samples loaded. The second pair of samples were inserted with the sample sides reversed and data was collected as before. After all runs were completed the pair of data files for each sample set were subtracted point by point to eliminate much of the coherent noise caused by nonuniformity of the two rod motions. The resultant voltage file was transmitted to the University of Connecticut, Department of University Computer Systems IBM 3081. The voltages

were converted into rod and load-cell position and then a sine
wave was fit (least squares) to determine amplitude and phase
angle of both signals. The results were inserted into Equations
2a and 2b to calculate G' and G''.

RESULTS

    The results of the experimental characterization of the vis-
coelastic fluid samples with the Rheometrics Mechanical Spectrometer (RMS) are presented in Table 1. The reported standard deviations for each sample at a given frequency is for three independent
sample loadings. Since the frequency was swept for a given sample
loading, the observations at different frequencies cannot be considered strictly independent. The theoretical equations for the
opposed squeeze flow of two viscoelastic fluids were tested by
varying sample modulus, thickness and strain amplitude. In addition, the ability of the DR to resolve small differences was compared with that found with the RMS.

Table 1.   Viscoelastic Fluid Samples –
Characterization by RMS

Sample	Diluent Content,%	f mHz	G' Pa	S* Pa	G'' Pa	S* Pa
SE-30	0.000	50	1090	150.	4300	30.
SE-30	0.000	100	2800	120.	7200	200.
SE-30	0.000	200	6050	250.	11300	400.
SE-30	0.000	500	14100	500.	18100	600.
3	8.874	50	1080	80.	4000	45.
3	8.874	100	2650	60.	6850	35.
3	8.874	200	5800	30.	10800	60.
3	8.874	500	13500	90.	17400	80.
1	2.627	50	780	155.	3400	100.
1	2.627	100	2220	70.	5800	230.
1	2.627	200	4800	150.	9050	430.
1	2.627	500	11300	500.	14800	360.

* standard deviation

In Table 2 are listed the DR results for sample 3 run against SE-30, the standard. These should be compared with the Rheometrics results of Table 1. The results in Table 2 are based on six experimental runs with the frequency swept from 50 to 500 mHz. Samples were allowed to rest for a few moments between frequencies. The 500 mHz runs were taken at a 2.05% strain amplitude rather than 5.09%. At high strains significant distortion of the outside (free) surface was seen.

In Table 3 the effect of nominal strain amplitude (a/h) was evaluated. In the runs reported in the first half of Table 3, sample 1 (run against SE-30) was subjected to amplitudes from 0.52 to 5.09 percent while the second half lists results for sample 3 and higher amplitudes (2.66-7.7%).

Table 2. Differential Rheometer 5% Strain Results - Sample 3

f mHz	G' Pa	S Pa	G'' Pa	S* Pa
50	1070	8.	4000	65.
100	2760	55.	6600	150.
200	5820	17.	10360	80.
500	13600	290.	16200	240.

* standard deviation

Table 3. Stain Amplitude Evaluations - DR

100-mHz   2.12-mm Sample Thickness

Sample	a/h_o %	d/a	θ rad	G' Pa	G'' Pa
1	0.52	0.0678	-0.21	1840	6460
	1.24	0.0537	0.35	2440	6280
	2.55	0.0515	0.37	2470	6310
	5.09	0.0662	0.44	2450	6040
3	2.66	0.0416	0.57	2690	6430
	5.09	0.0405	0.60	2700	6440
	5.09	0.0284	0.59	2750	6660
	5.09	0.0238	0.71	2820	6740
	7.70	0.0276	0.61	2760	6670

The final check of the Kramer relations was to decrease the sample thickness while keeping the strain amplitude constant. The sample volume was decreased proportionally. These results are presented in Figure 3. In Figure 3 the modulus values are reduced by the SE-30 modulus at the same frequency as measured on the RMS.

DISCUSSION

One of the prime factors in the evaluation of the differential rheometer is the breadth of the experimental distribution of moduli (differences) for a sample. On examining the standard deviations for sample 3 obtained by differential rheometry (Table 2) and by standard rheological instrumentation (RMS, Table 1), one might conclude that the resolving power of the two instruments was essentially the same. However, the variances for sample 3 in Table 1 are atypically small; those for sample 1 and pure SE-30 are much more normal for the RMS. An F test on the variances for sample 3 by DR and either sample 1 or SE-30 by RMS, rather than the abnormally small sample 3 variances, indicates that the distributions by DR were generally significantly narrower than the RMS results (95% confidence level).

In the majority of cases the sample 3 results by the two rheological techniques Tables 1 and 2 are not statistically different at any reasonable confidence level (<80%). If normally expected distributions for the RMS sample 3 data (broader than was fortuitously observed) had been used, the differences between the means by the two methods would be even less significant. It is encouraging to conclude that the differential rheometer and the RMS substantially agree and that the DR prototype gives as low or lower variances despite its relatively crude machining.

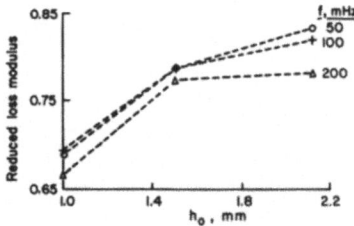

Fig. 3.   Sample aspect ratio effect on calculated low modulus
          (reduced by SE-30) moduli from RMS characterization.

The remainder of the results concern the testing of the
Kramer (8) relations (eq. 1a and 1b). Table 3 displayed the results of holding the sample thickness constant (h_0) while increasing the pushrod amplitude. As we saw, the calculated modulus
values are independent of strain amplitude for this material over
the range covered. The 0.52% result is considered to be in error
for the following reasons: With the experimental setup used in
this study, 0.52% of a 2.12-mm sample (half-thickness) corresponds
to a rotation of the 7.5-cm drive gear of less than 13 degrees.
At this small level of motion, backlash in the drive control system became extremely important. At the highest strains (7.7%) the
sample size used (0.3±0.0005g) just filled the gap when the pushrods were at their most retracted position.

The only remaining variable of interest in the Kramer relations is the sample aspect ratio (h_0/R). To test this the sample
thickness was varied at constant strain amplitude. As we saw in
Figure 3 there appeared to be a dependence of the calculated loss
modulus on sample thickness. Some portion of the effect of sample
thickness is not being corrected for by the Kramer (8) relations.
A similar effect was observed for the storage modulus. (In
Figure 3 the loss modulus values are divided by the SE-30 value
as measured on the RMS.) While it appears that the frequency is
a parameter in this effect, it is probably the sample modulus
(which depends on frequency) that is influencing the placement of
the three curves, and not frequency itself. The effect suggested
by Figure 3 may be caused by a breakdown of the assumption that
the force is sinusoidal as the sample thickness is oscillated
around h_0. If the amplitude of the oscillation is too high, one
may see the effect of the cubic dependence of force on gap common
to squeezing flows. In this case the sinusoidal assumption breaks
down. In the Kramer relations the sample thickness is treated as
a constant; i.e., the strain is assumed small. If the cubic dependence of force on sample thickness becomes important, a higher
experimental force amplitude than predicted will result for both
samples. This in turn will cause the calculated modulus of the
"unknown" to be too low (more different from the "known" material).

It is also well known that the lubrication assumptions
commonly used in squeeze flow begin to break down as the flow
field changes qualitatively with increasing sample thickness. As
the sample thickness increases the Kramer force prediction would
rapidly (cubicly) decrease. However, the observed force differences would again be expected to be larger than predicted due to
the proportionally greater importance of extensional stresses and
the calculated modulus of the "unknown" sample would again become
more different from the "known" material. No data was taken at
higher sample thickness to support this speculation. Obviously
for qualitative, relative measurements it would be best to keep

the sample thicknesses constant and well within the region of
validity of the lubrication approximation.

This work seemed to show that the Kramer (8) force-strain
relationship was applicable for a viscoelastic fluid in low amplitude oscillating squeeze flow. The effects of frequency, modulus
and strain amplitude were accounted for quite nicely. The sample
thickness (aspect ratio) effect was not totally accounted for.
More study is needed in this area, perhaps in a single-sample
geometry for simplicity. While all measurements were conducted
at a fixed temperature, it is a relatively trivial matter to program temperature and examine differential changes in relaxation
transitions. In this case, the infinitely stiff loadcell (or
equivalent servo system) would provide the most straightforward
analysis of the absolute differences in properties between sample
and reference.

REFERENCES

1.   J. Pryce-Jones, Rev. Sci. Inst., 18, 39 (1941).
2.   R. Sakamoto, K. Iso and T. Takeda, Nihon Zairyo Gakkai, 9,
      313 (1960).
3.   L. Gudim, S. P. Papkov, V. G. Kulichikhin, A. Ya. Malkin,
      V. D. Kalmykova, Vysokomol Soedin., 14B, 244 (1972).
4.   S. Middleman, Flow of High Polymers, Interscience Publishers,
      New York, 1968, Ch. 5.
5.   W. F. Gebhart, University of Connecticut, Dept. of Chem. Eng.
      CHEG 299 Senior Independent Study Report, Dec. 1980.
6.   W. F. Gebhart, University of Connecticut, Dept. of Chem. Eng.
      CHEG 299 Senior Independent Study Report, May 1981.
7.   P. H. Foss, University of Connecticut, Dept. of Chem. Eng.,
      Masters Thesis, 1983.
8.   J. M. Kramer, Appl. Sci. Res., 30, 1 (1974).

MATERIALS CHARACTERIZATION USING THERMOGRAVIMETRY COUPLED TO TRIPLE QUADRUPOLE TANDEM MASS SPECTROMETRY (MS/MS)

Bori Shushan, Bill Davidson   R. Bruce Prime

Sciex®   IBM Corporation
55 Glen Cameron Road   San Jose, CA
Thornhill, Ontario L3T 1P2

ABSTRACT

Thermogravimetric analysis (TGA) coupled with sequential mass spectrometry (MS/MS) offers unambiguous structural identification of compounds evolving during thermo-degradative processes without interrupting the TGA-mass spectrometer analytical procedure. MS/MS identification of co-evolving compounds, combined with the weight-loss data obtained from TGA, furnishes a means of delineating complex thermolytic pathways. The compatibility of TGA (an atmospheric pressure technique) with Atmospheric Pressure Chemical Ionization (APCI), as offered on the TAGA® 6000 MS/MS system, permits an interface which is easily coupled and decoupled making the system "user-friendly" and well suited to routine quality control and trouble-shooting applications.

INTRODUCTION

Thermogravimetric Analysis (TGA) has become an indispensible tool in the routine characterization of polymetric materials. However, the complexity of thermal degradative processes and the great variety of additives which can be present in polymer formulations requires the combination of TGA with other analytical techniques before a reasonably clear picture of the chemistry involved can be drawn. Of all the available combinations, that of TGA-mass spectrometry (MS) appears to hold the greatest promise for the rapid, on-line analysis of gases evolving from the TGA apparatus. Although TGA/MS combinations have been used successfully for over a decade [1,2] the technique still suffers from the inability to unambiguously identify chemicals co-evolving

during the thermo-degradation of polymers. With the recent introduction of commercially available triple-quadrupole tandem mass
spectrometers (MS/MS)[3] real time structural identification of
single components within complex evolved gas mixtures can be carried out rapidly under computer control. This paper will describe
the coupling of TGA with MS/MS for the identification of individual compounds produced during the thermolytic decomposition of a
representative polymer and how this information can be used to
elucidate degradative pathways.

EXPERIMENTAL

    A schematic of the TGA/MS/MS interface is shown in Figure 1.
The TAGA® 6000 triple quadrupole mass spectrometer is fitted with
an atmospheric pressure chemical ionization (APCI) ion source.
This ion-source uses air as the CI reagent gas permitting the routine use of oxidative purge gases within the coupled TGA apparatus. Using this interface, several commercially available TGA instruments can readily be coupled to the TAGA® 6000 MS/MS including
the Perkin-Elmer TGS-2 (top and middle, Figure 1) and DuPont 951
(bottom, Figure 1). In both cases a ball-and-socket joint connects the TGA to the all-glass transfer line permitting easy
coupling and decoupling.

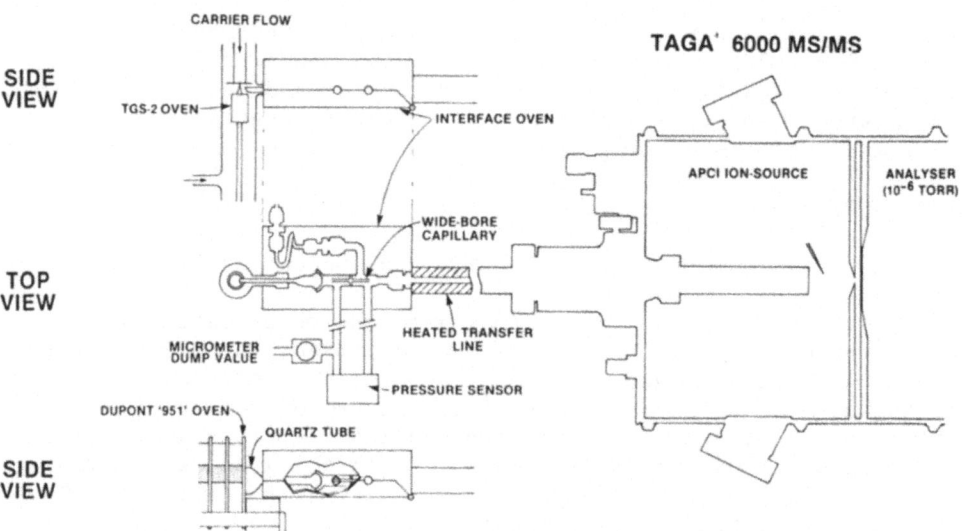

Figure 1.   Schematic of the TGA/MS/MS interface

After passing through a short (ca.20mm) section of wide-bore capillary (0.5mm I.D.), a portion (0.1%-100%) of the TGA effluent (100-200 ml/min.) is carried into the APCI ion source by a high velocity (ca 2 l.min^{-1}) stream of carrier/reagent gas. The amount of TGA effluent allowed into the ion source is accurately ($\pm$30 ul/min) controlled by a micrometer 'dump' valve connected in series with a device which measures the pressure on either side of the capillary restriction (this pressure is directly proportional to the flow across the capillary). The relatively high gas velocities as well as the fact that the whole interface assembly is heated (250-350°C) avoids condensation of less volatile materials on the interface walls.

Chemicals entering the ion source from the transfer line are converted to their quasi-molecular or molecular ions via well characterized APCI reactions[4]. The ions are then focussed into the high vacuum analyzer portion for MS and/or MS/MS analysis (see Figure 6).

RESULTS AND DISCUSSION

To illustrate the basic operation of the TGA/MS/MS system a 12.4 mg sample of cellulose acetate (C.A.) was heated in a DuPont 951 TGA from 100 to 600°C at 20°C/min using air as purge gas (200 ml/min) where 1.3% of the TGA effluent was transferred to the mass spectrometer ion source. The TGA response shown in Figure 2 consists of the weight loss curve ('TG') and its negative first differential ('DTG').

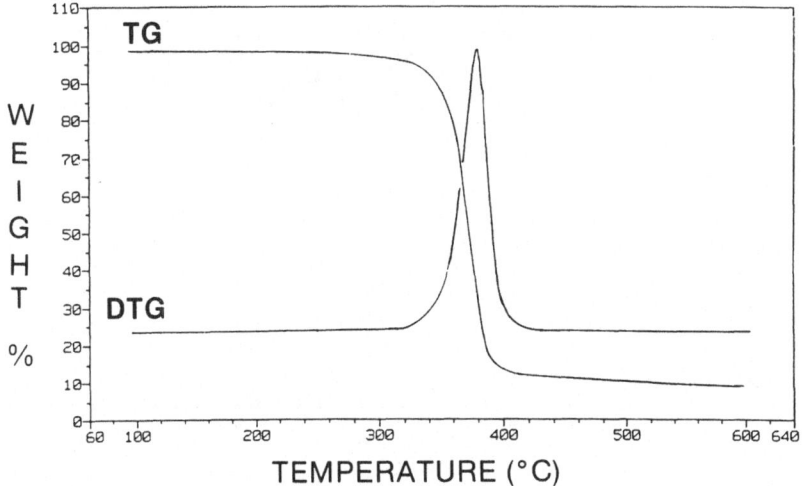

Figure 2. DuPont 951 TGA response to cellulose acetate

In the single MS mode the first of three axially aligned
quadrupoles is automatically and repetitively scanned and the mass
spectra recorded using SCIEX® MS/MS data system. The total Ion
Thermogram for this TGA run is shown in Figure 3 (top) and the
average of mass spectra recorded between 360 and 410°C is shown
below it. The complexity of this spectrum underscores one of the
limitations of conventional TGA/MS. Since each peak corresponds
to the protonated molecular ion of a neutral evolved during therm-
olysis, only molecular weight information is available. Structural assignment on the basis of molcular weight, however, can be
a tenuous proposition. In order to unambiguously identify a component without forgoing direct TGA-mass spectral integrity, MS/MS
techniques must be employed.

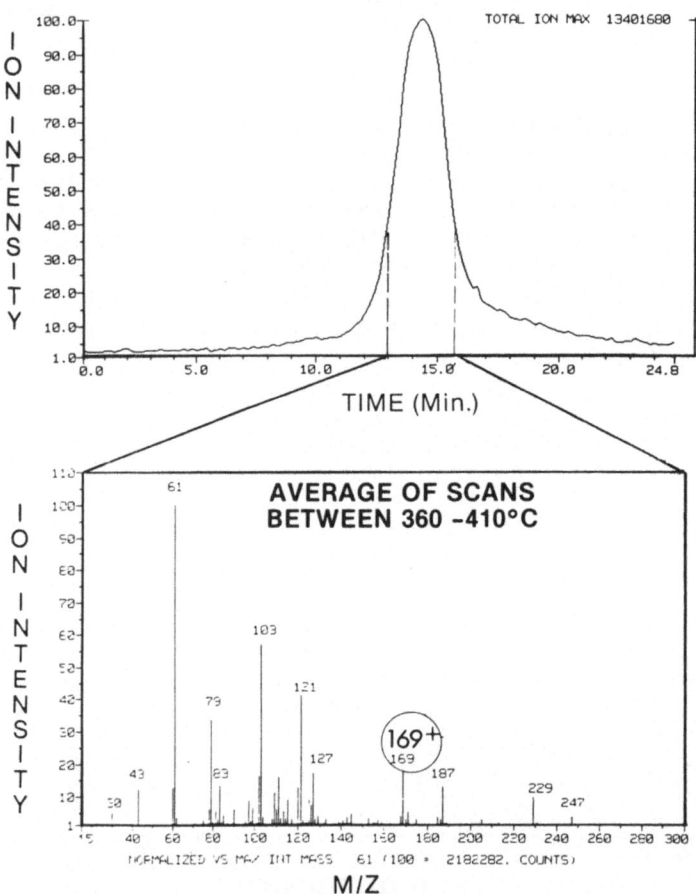

Figure 3. Total ion thermogram of single MS scans obtained during
          TGA run (top). Average of single MS APCI mass spectra,
          (bottom).

Figure 4 (top) shows the Reconstructed Ion Thermogram for the 169+ ion (notice how closely this RIT resembles the DTG trace of Figure 1). Below the thermogram is the MS/MS daughter ion spectrum of 169+ ion obtained during the evolution of this compound from the TGA. In order to generate this spectrum the first quadrupole (Q1) was set to filter 169+ ions only. The ions were induced to fragment in the second quadrupole by collisions with a neutral gas (Argon) at a pressure of ca.10^{-4} torr. While fragmentation proceeded, the third quadrupole (Q3) was scanned between 10 and 171 amu to yield this spectrum of fragment (daughter) ions originating from the 169+ precursor. The whole process takes a few seconds under computer control and the resulting fragmentation pattern is highly indicative of the precursor ion's structure (see inset Figure 4). Many such daughter ion spectra can be obtained sequentially and repetitively during a

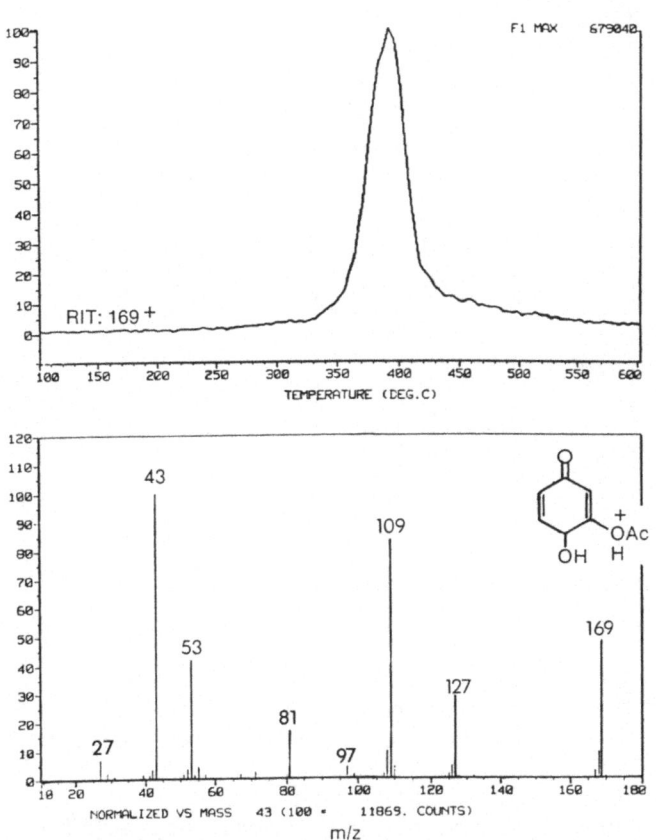

Figure 4.   Reconstructed ion thermogram for 169+ ion (top). MS/MS daughter ion spectrum for 169+ ion (bottom).

single TGA run providing complete structural information of the evolved compounds. A summary of results for the TGA/MS/MS analysis of cellulose acetate is provided in Table 1. (It is instructive to note that an earlier TGA/MS study[5] of cellulose acetate had wrongly identified the 103^+ ion as protonated valeric acid on the basis of molecular weight data. The present MS/MS analysis has demonstrated this ion to be due to acetic anhydride instead.) Based on these results a scheme for the thermolytic breakdown of cellulose acetate is proposed in Figure 5.

Table 1.   Some products of the thermal degradation of cellulose acetate analysed by TGA/MS/MS

ION	STRUCTURE	NAME
61^+	CH_3CO_2H	(ACETIC ACID)
103^+	$CH_3COOCOCH_3$	(ACETIC ANHYDRIDE)
109^+	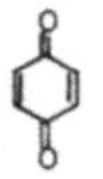	(BENZOQUINONE)
127^+	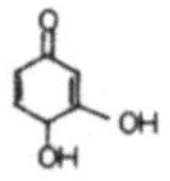	
169^+	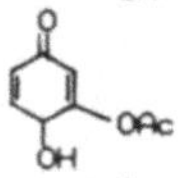	
187^+	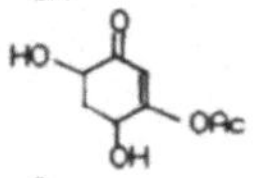	
229^+	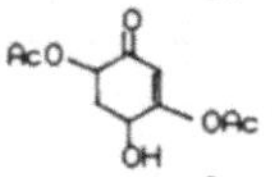	
247^+	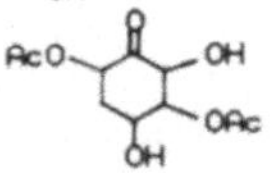	

CELLULOSE ACETATE

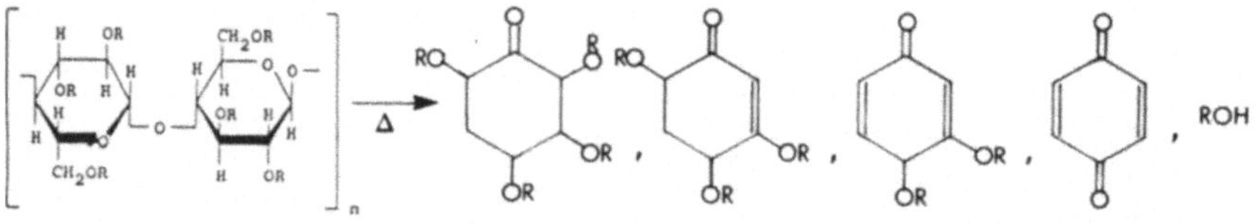

R = -H, -COCH₃

Figure 5.   Proposed scheme for the thermal degradation of cellulose acetate based on TGA/MS/MS analysis

CONCLUSIONS

The technique of Thermogravimetry coupled to APCI/MS/MS can provide a vast amount of chemical information in only one or two experiments. The rapid identification of co-evolving compounds by MS/MS, combined with the weight-loss data from TGA, furnishes a means of delineating complex thermodegradative processes. Such an analytical procedure is not only useful in the analysis of polymers such as cellulose acetate, but would be of great value in the identification of copolymer substrates, and the various polymer additives such as antioxidants, stabilizers, and plasticizers.

The unique APCI ion source design permits an easily coupled and uncoupled interface making the SCIEX® TGA/MS/MS system very "user-friendly" and particularly well suited to routine quality assurance and trouble shooting applications.

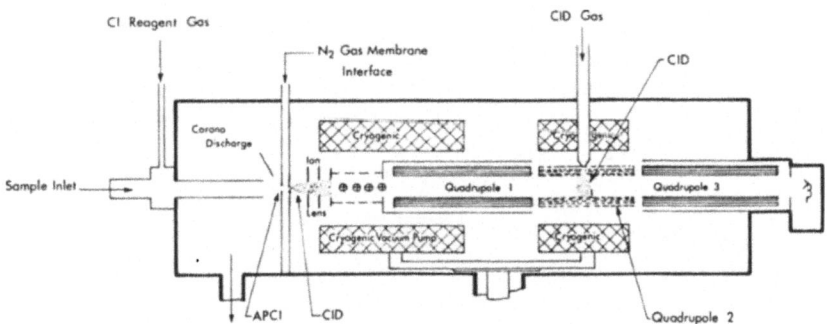

Figure 6.   Schematic of the TAGA 6000 MS/MS

REFERENCES

1. F. Zitomer, Analytical Chemistry, 40, 1091-1095 (1968)

2. J. Mitchell, Jr. and J. Chiu, Analytical Chemistry Annual Review, 45, 273-332R (1973)

3. J.A. Buckley, D. Douglas, D. Simmons and J.B. French, Pittsburgh Conference, 1981.

4. A.M. Lovett, N.M. Reid, J.A. Buckley, J.B. French and D.M. Cameron, Biomedical Mass Spectrometry 6(3), 91-97 (1979).

5. R.G. Beimer, Pittsburgh Conference, March 1974.

CALORIMETRIC RESPONSE OF FOSSIL FUEL SYSTEMS: PROSPECTS AND PROBLEMS

Krishnan Rajeshwar

Department of Electrical Engineering
Colorado State University
Fort Collins, CO 80523

INTRODUCTION

It is now clear that world production of conventional fossil fuels (mainly oil and gas) will be declining by the end of the century. Energy demands for the future, therefore, will have to be met, at least partially by alternative fuel supplies. In this respect, the United States and many other parts of the world contain substantial resources of coals, oil shales and oil sands which can be profitably tapped.

Since the majority of state-of-the-art conversion technologies for processing these materials utilize the application of heat in one form or the other, thermal analysis (TA) techniques lend themselves readily as characterization tools. As shown below, a judicious combination of calorimetric probes in general and differential scanning calorimetry (DSC) in particular, with other TA techniques such as thermogravimetry (TG), offers a convenient means of rapidly evaluating the resource potential of a given depositional area. Measurement problems imposed by the heterogeneity of coals, oil shales and oil sands and by the complexity of the thermal behavior of these materials, however, will have to be recognized.

In this paper, we present two examples of the use of DSC in the study of coals, oil shales and oil sands. Results culled from previous work in this laboratory are used as illustrative examples. The complementary nature of information obtainable from TG analyses of these materials is also demonstrated. Some typical problems in the study of these materials are discussed as

113

are relevant data reported by other investigators on similar samples.

EXPERIMENTAL

Instrumentation for DSC and TG was fairly standard and details may be found in previous publications from this laboratory.[1,2]

The coal samples were HVB (ASTM rank) and originated from the Allegheny, Pottsville and Conemaugh formations in Ohio. Proximate analyses of these samples have been presented elsewhere.[2] Oil shale samples from the Green River formation originated from the Anvil Points mine near Rifle, Colorado.[1] Some samples from the Logan Wash, Rock Springs and Vernal deposits were also used.[3] Oil sand samples were from the Athabasca formation in Canada and from the N.W. Asphalt Ridge, P.R. Spring and Circle Cliffs deposits in Utah. Further details on these samples may be found elsewhere.[4]

RESULTS AND DISCUSSION

Variables Affecting Calorimetric Response of Fossil Fuel Systems

Before discussing the applicability of TA techniques to the study of coals, oil shales and oil sands, it is pertinent to examine the influence of experimental variables on thermal behavior.

As reviewed elsewhere,[5] much of the early conflict in the differential thermal analysis (DTA) data on coals, can be resolved if allowance is made for the influence of the varying sample geometry and ambient atmosphere employed in these studies. While the mode of sample containment is certainly critical for all TA studies in general, the sensitivity of thermal behavior to this variable is extreme for coals, oil shales and oil sands. As an illustrative example, consider the results presented in Fig. 1 for the Ohio bituminous coal sample.[2] The shift from endothermicity to a net exothermic behavior is readily apparent in the presence of even traces of O_2 in the ambient atmosphere. Furthermore, the baseline drift is less drastic for the sample in a sealed pan although in this case, one is faced with the question of self-generated atmospheres. Obviously, a controlled atmosphere leak (via a pinhole) would be beneficial, while at the same time a fairly rapid flow of purge gas and the use of a thin layer of sample minimize undesirable effects arising from diffusion limitations. Such limitations can distort the picture considerably and in the extreme case may even alter the kinetics. An example is

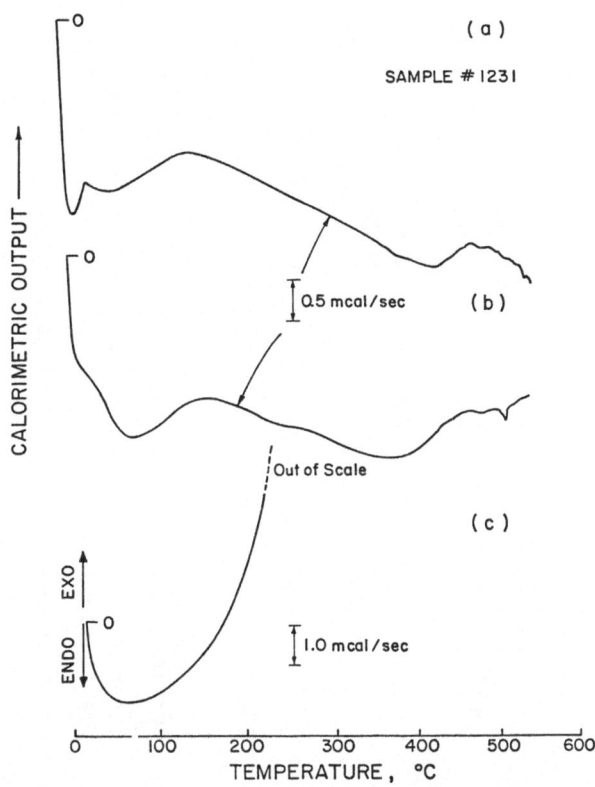

Figure 1:   Influence of ambient atmosphere and sample containment on DSC thermograms for a representative Ohio bituminous coal sample.   Data in (a) and (b) were obtained with open and sealed pans respectively in N_2 atmosphere.   The scan in (c) was obtained with open pans exposed to 1 percent O_2/N_2 atmosphere.   Heating rate: $10°C\ min^{-1}$ (Copyright: Elsevier   Publishing   Co., 1982).

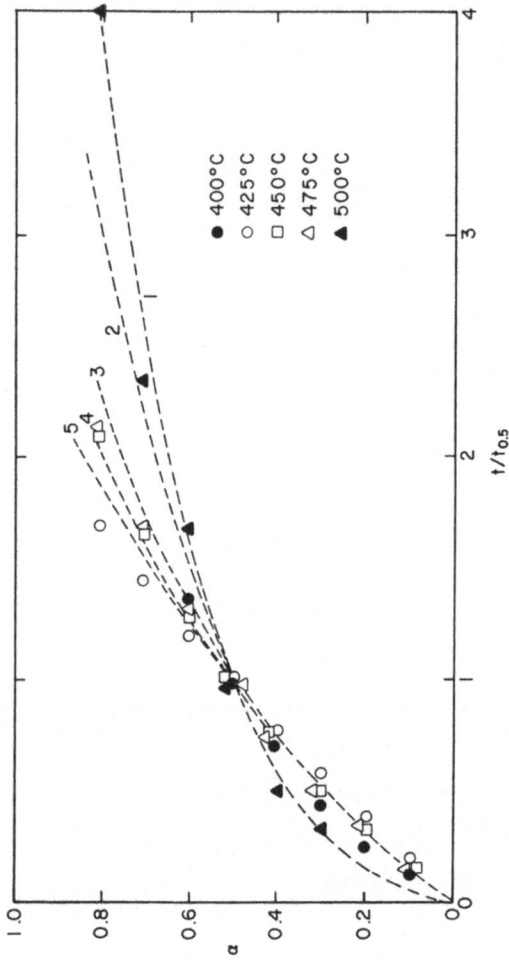

Figure 2:   Reduced-time plots for a Green River oil shale sample
             ($108 \, \ell \, \text{tonne}^{-1}$). α is the fractional conversion and
             $t_{0.5}$ is half-life for decomposition. Dashed lines
             marked 1-5 represent theoretical curves for diffusion-
             controlled reactions in a sphere and cylinder, first-
             order reaction and phase-boundary controlled reaction
             in a contracting sphere and cylinder respectively.
             More details may be found in Ref. 6 (Copyright:
             Elsevier Publishing Co., 1982).

presented in Fig. 2 for a Green River oil shale sample.[6] Data for model calculations were taken from an early study on oil shale cores.[7] The transition from intrinsic kinetics behavior at low temperatures to pyrolysis limited by diffusive transport of product gases through the reacting oil shale matrix at high temperatures is readily apparent. The situation is admittedly exaggerated in this case, by the use of solid cores, although similar behavior ensues to a varying degree on powdered samples depending on particle size and sample containment. Thus the use of large sample sizes and coarse particles would be obviously deleterious in terms of masking the intrinsic thermal behavior of the sample. In this respect, early DTA and TG results on these materials utilizing large sample mass and deep cavities for containment, must be viewed with caution as discussed by other authors.[8]

In the case of oil shales, for example, control of experimental variables, taking into account the above effects, has been shown to yield fairly simple behavior adequately described by first-order kinetics.[9]

Yet another effect on DSC is the continual drift in the baseline imposed by the gradual weight loss from coals, oil shales and oil sands. This problem has been elegantly tackled on coal samples by the simultaneous use of TG to correct for the influence of varying sample mass.[10]

Applicability of DSC to Resource Evaluation

The energy content of a given coal, oil shale and oil sand sample is clearly a primary parameter in determining its resource potential. Apart from the obvious importance of its organic content (which manifests as an endothermic or exothermic effect in DSC depending on the choice of heating conditions), an often neglected parameter is mineral content which also manifests in characteristic thermal effects. For example, sodium and aluminum bearing minerals in oil shales are value-added commodities; their estimation by TA methods, therefore, assumes added significance. The use of DTA and TG for the estimation of dawsonite, nahcolite and nordstrandite content of oil shale has been described by previous authors.[11] The main advantage of TA methods for estimation of organic and mineral content over traditional methods (e.g., Fischer assay, ASTM) lies in their amenability to routine use and automation.

Figure 3 shows model DSC data on a suite of Green River oil shale samples illustrating the sensitivity of the pyrolysis enthalpy to oil yield.[1] Data such as those shown in this figure

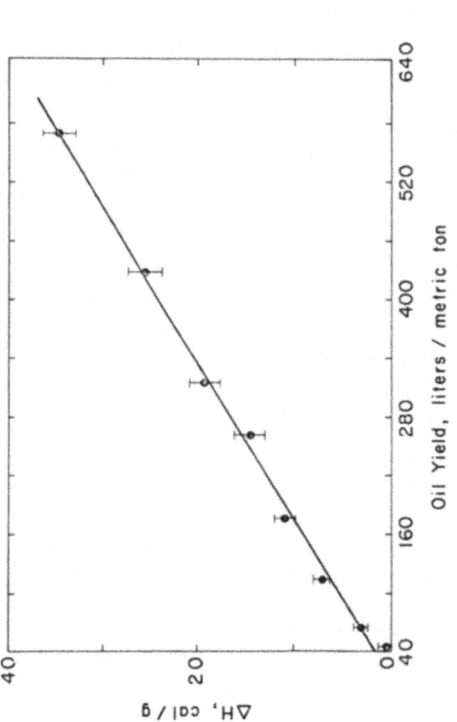

Figure 3:   Correlation of the enthalpy of decomposition, ΔH, as
            determined by DSC, with oil yield for Green River oil
            shales. The sample with an oil yield of 572 L/metric
            ton is a kerogen concentrate (Copyright: American
            Chemical Society, 1981).

underline the potentiality of the DSC technique as an assay tool
for rapid screening of oil shales. The samples selected for this
study were free of interfering minerals (e.g., nahcolite, dawsonite) so that the measured enthalpy for the pyrolysis endotherm
could be simple assigned to the indigenous organic matter in the
shale. Cases when these minerals are present have been treated by
other authors in DTA and TG studies on these materials.[11]

Note that the data in Fig. 3 were obtained under an inert
atmosphere such that oxidation of the shale organic matter was
precluded. Combustion calorimetry on oil shale also permits
estimation of its organic content, although in this case, care has
to be exercised to ensure complete combution of the samples.[12,13]

Estimation of the shale organic content also follows in a
straightforward manner from TG analyses. The TG weight loss at
500°C has been correlated with the oil yield in this manner
(Fig. 4).[14] More important, it has been shown that a single
master plot describes all shales from a given depositional area
(e.g., the tri-state region in the Green River formation) for
strata wherein interfering mineral effects are again absent.[3]
Similar results have been presented for Australian and Green River
oil shale samples.[15,16]

Attempts at correlating volatile matter content of coal with
DSC endotherm areas (similar to the approach described above for
oil shale) have not been successful for Ohio bituminous coals
(Fig. 5).[2] While the scatter is less for as-received samples, the
complicating effects introduced by sample heterogeneity are quite
well illustrated by these data.

By way of contrast, TG can readily provide a rapid proximate
analysis of coal (Fig. 6).[2] Very good correlation has been
obtained between the ASTM method and the results from TG analyses.
A slight variant of the method illustrated in Fig. 6 utilizes a
first nonisothermal scan followed by isothermal TG at 950°C for
7 min. to yield the fixed carbon and ash content.[17,18]

Similar to the case of oil shale, DSC analyses may be carried
out under conditions wherein the organic matter in coal undergoes
combustion to yield its calorific value. Such estimates are
reported to accord well with ASTM results.[17]

Pyrolysis enthalpies for oil sand samples are less readily
measured by DSC because of their weak thermal effects. One way to

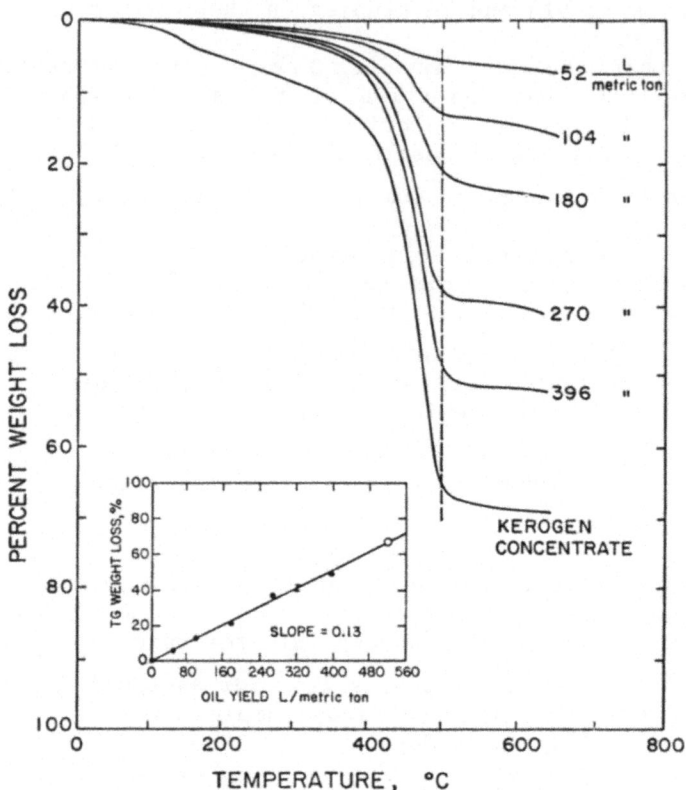

Figure 4:   Nonisothermal TG plots for Green River oil shale
             (heating rate: 10°C min^{-1}, N$_2$ atmosphere). Inset:
             Correlation of TG weight loss at 500°C with oil yield.
             The open circle denotes a sample of kerogen
             concentrate (Copyright, Elsevier Publishing Co.,
             1983).

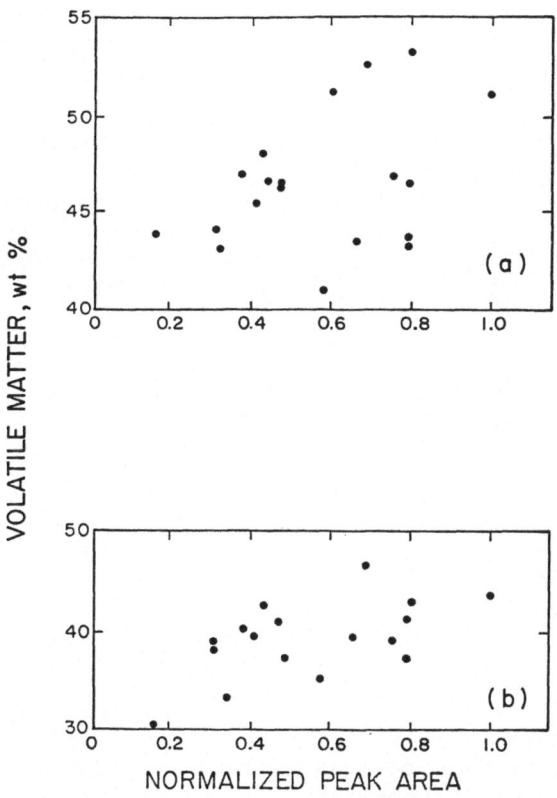

Figure 5:   Correlation of normalized DSC endotherm areas with
            volatile matter for Ohio bituminous coal samples.
            Data in (a) and (b) are for volatile matter expressed
            on daf basis and "as received" respectively.
            Samples were in open pan and a heating rate of
            $10°C \text{ min}^{-1}$ was employed (Copyright: Elsevier Publish-
            ing Co., 1982).

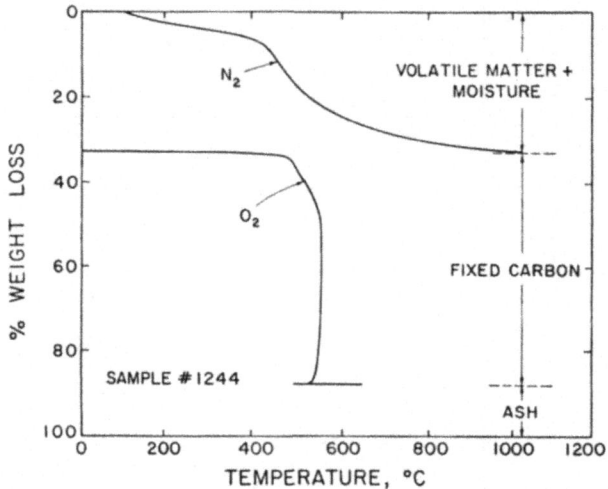

Figure 6:   Nonisothermal TG (heating rate: $10°C \ min^{-1}$) on Ohio
bituminous coal illustrating application to proximate
analysis (Copyright: Elsevier Publishing Co., 1982).

enhance the sensitivity is to work with bitumen extracts (prepared from the starting samples by solvent extraction). Pyrolysis enthalpies of a few hundred J/g have been measured by DSC in this laboratory on selected samples (Table 1).[19] Our results for Athabasca samples show good agreement with values reported recently by other authors.[20]

Table 1. Parameters from thermal analyses[a] on oil
          sand bitumen extracts.

Sample	Parameter			
	TG wt loss at 300°C %	TG wt loss at 500°C %	TG wt loss at 800°C %	ΔH[b,c] J/g
N.W. Asphalt Ridge	16	85	91.5	672
P.R. Spring	4	71.5	79	634
Circle Cliffs	13	71	78	406
Athabasca	27 (21)[d]	88 (87)	91.5 (91)	474 (478)

[a] Heating rate: 20°C min^{-1}, N_2 atmosphere.

[b] Based on starting weight (daf) of sample.

[c] Nominal temperature range of measurement: 25-550°C.

[d] Values refer to duplicate sample.

Figure 7 compares the influence of pyrolysis temperature on volatile matter yield for an Athabasca oil sand bitumen and a Utah (N.W. Asphalt Ridge formation) bitumen extract.[5] These data were obtained from isothermal TG. The volatile matter yield was obtained by extrapolation of the instantaneous weight loss values to infinity under conditions wherein the pyrolysis rate approached a linear, steady-state value. The inset in Fig. 7 shows an example of such a plot. Data such as those shown in Fig. 7 are also useful in assessing the relative energy richness of a suite of oil sand samples at a given temperature.

Applicability of DSC to Specific Heat Determination

The method originally suggested by O'Neill for measurement of specific heats (C_p) by DSC,[21] has proven to be of particular value for coals, oil shales and oil sands.

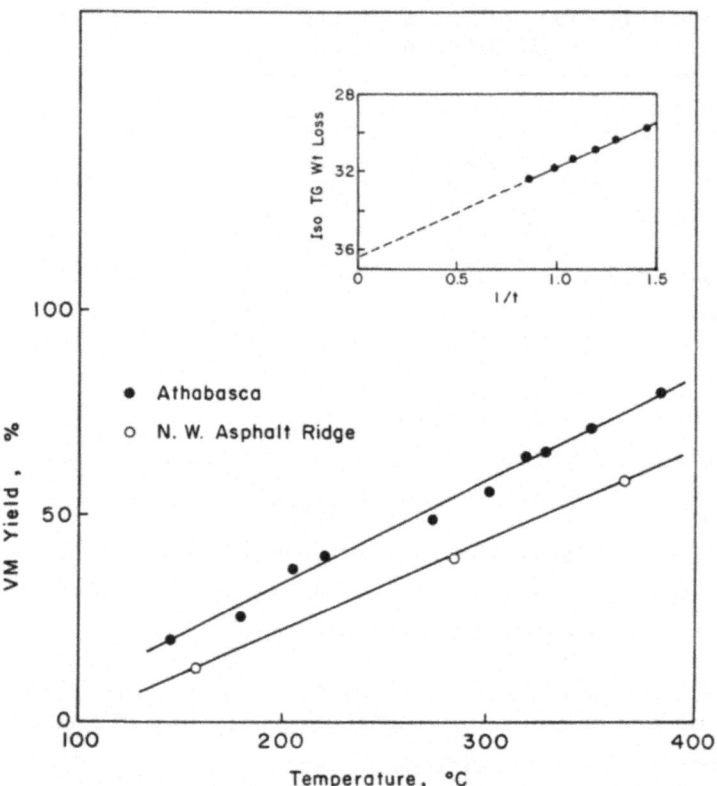

Figure 7:   Dependence of volatile matter (VM) yield on tempera-
            ture for (a) Athabasca and (b) N.W. Asphalt Ridge
            specimen.   Inset: Typical   plot   of   isothermal   TG
            weight loss vs 1/t (t = time of heating) for an
            Athabasca   bitumen   sample   poised   at   207 ± 1°C
            (Copyright: Elsevier Publishing Co., 1983).

Figure 8 shows typical data obtained in our laboratory on Green River oil shale samples for three levels of organic content.[22] While the method works well for low-grade samples, samples with high levels of organic content pose problems because of weight-loss and attendant base-line drift (vide supra). This problem is again obvious when attempts are made to correlate C_p values with shale organic content (Fig. 9). A general increase of C_p with increasing organic content, however, is apparent. By a similar method, stratigraphic variations in the oil shale enthalpy have been examined by other authors for samples from the Federal lease tract C-a.[23] The oil shale in this tract is not uniform from top to bottom of the formation but comprises alternating rich and relatively lean zones.

Values of C_p in the range $0.67\text{-}1.57$ J g^{-1} K^{-1} have been measured for Athabasca and Utah oil sand samples by DSC.[4] Removal of the mineral matrix from the oil sand sample results in an enhancement of C_p as shown in Fig. 10 for N.W. Asphalt Ridge specimen as a typical example. Incidentally, our C_p values for the Athabasca samples as measured by the DSC technique accord well with those presented by other authors using conductive electrical heating at 60 Hz.[24]

The O'Neill method has been employed by other authors for measurement of C_p of bituminous and subbituminous coals.[25] The C_p values, as measured, are reported to be sensitive to the coal particle size, moisture content and the nature of pretreatment (whether wet- or dry-screened). Only minor differences, however, were noted between the two sets of samples.

CONCLUSIONS

Two major applications of DSC techniques for the study of coals, oil shales and oil sands have been identified, namely resource evaluation (i.e., assay and proximate analyses) and specific heat determinations. The information content from DSC is much enhanced when TG is also used. The method of sample containment and choice of experimental variables are critical in the study of these materials.

ACKNOWLEDGMENTS

The author gratefully acknowledges help from Don Jones and Robert Rosenvold in the experiments described herein. This

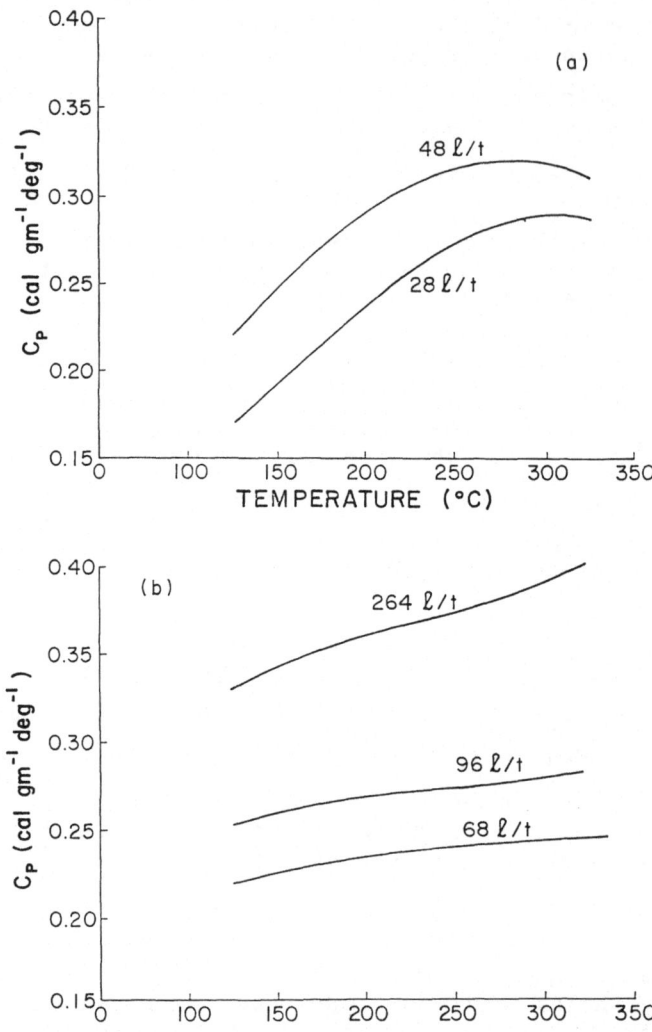

Figure 8:   Variation of specific heat   (C_p)   with temperature for
            Colorado oil shales. Data in (a), (b) and (c) are for
            lean shales, shales of medium grade and rich shales
            respectively (Copyright: American Chemical Society,
            1980).

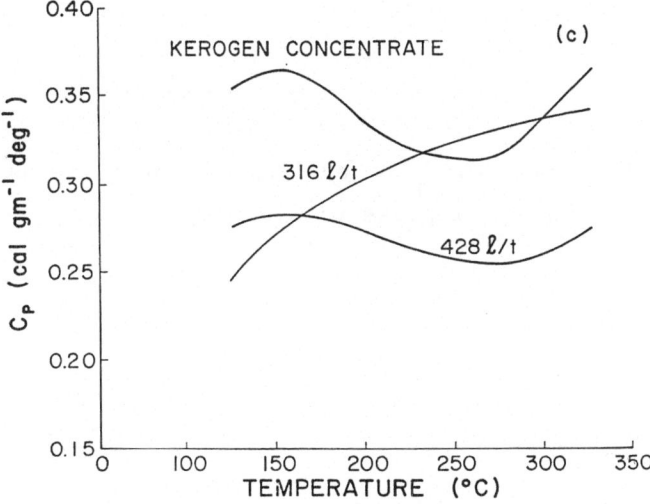

Figure 8. (continued).

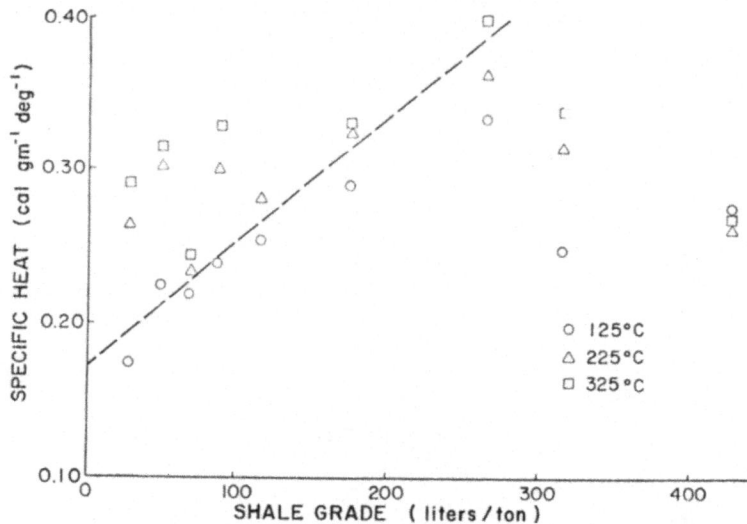

Figure 9:   Dependence of C_p on shale grade for Colorado oil shale
             at three representative temperatures (Copyright:
             American Chemical Society, 1980).

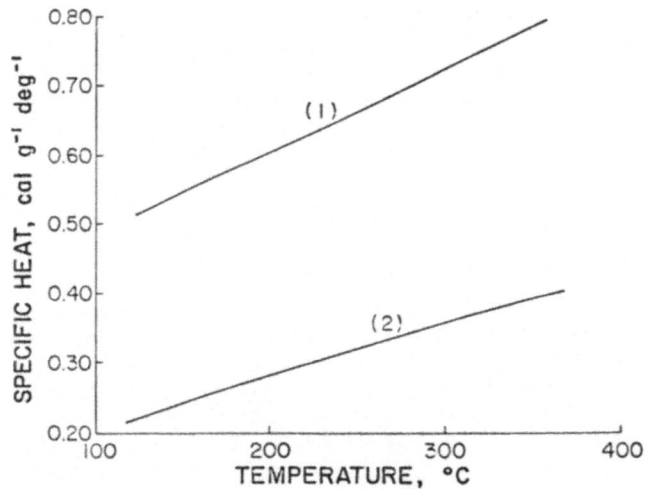

Figure 10:   Comparison of C_p vs temperature behavior for (1) oil
             sand bitumen from N.W. Asphalt Ridge and (2) parent
             oil sand from which the bitumen was extracted.

research was partially supported by a grant from the U.S. Department of Energy and the Laramie Energy Technology Center. Copyright permission from Elsevier Publishing Co. and the American Chemical Society to reproduce the figures cited in the text, is also gratefully acknowledged.

REFERENCES

1. K. Rajeshwar, D. B. Jones, and J. B. DuBow, "Characterization of Oil Shales by Differential Scanning Calorimetry," Anal. Chem. 53:121 (1981).

2. R. J. Rosenvold, J. B. DuBow, and K. Rajeshwar, "Thermal Analyses of Ohio Bituminous Coals," Thermochim. Acta 53:321 (1982).

3. R. J. Rosenvold, K. Rajeshwar, and J. B. DuBow, "On the Correlation between Thermogravimetric Response and Potential Oil Yields for Green River Oil Shales," Thermochim. Acta 57:1 (1982).

4. K. Rajeshwar, D. B. Jones, and J. B. DuBow, "Thermophysical Characterization of Oil Sands: 1. Specific Heats," Fuel 61:237 (1982).

5. K. Rajeshwar, "Thermal Analysis of Coals, Oil Shales and Oil Sands," Thermochim. Acta 00:000 (1983).

6. K. Rajeshwar and J. B. DuBow, "On the Validity of a First-Order Kinetics Scheme for the Thermal Decomposition of Oil Shale Kerogen," Thermochim. Acta 54:71 (1982).

7. A. B. Hubbard and W. E. Robinson, "A Thermal Decomposition Study of Colorado Oil Shale," U.S. Bureau of Mines Report of Investigations 4744 (1950).

8. J. W. Smith and D. R. Johnson, "Thermal Analysis of Natural Fuels," in: Proc. 2nd Toronto Symposium on Thermal Analysis, H. G. McAdie, ed., Chemical Institute of Canada, Toronto (1967).

9. K. Rajeshwar, "The Kinetics of the Thermal Decomposition of Green River Oil Shale Kerogen by Non-isothermal Thermogravimetry," Thermochim. Acta 45:253 (1981).

10. O. P. Mahajan, A. Tomita, and P. L. Walker, Jr., "Differential Scanning Calorimetry Studies on Coal: 1. Pyrolysis in an Inert Atmosphere," Fuel 55:63 (1976).

11. D. R. Johnson, N. B. Young, and J. W. Smith, "Thermal Analysis on Oil Shale: Determination of Potential Oil Yields and Dawsonite, Nahcolite and Norstrandite Content," LERC Report of Investigations (RI-77/6) (1977).

12. P. C. Crawford, D. L. Ornellas, R. C. Lum, and P. G. Johnson, "Combustion Calorimetry of Oil Shales," Thermochim. Acta 34:239 (1979).

13. D. E. Rogers and D. M. Bibby, "A Study of a New Zealand Oil Shale by Differential Thermal Analysis," Thermochim. Acta 30:303 (1979).

14. K. Rajeshwar, R. J. Rosenvold, and J. B. DuBow, "Thermogravimetric Assay of Oil Shale," Thermochim. Acta 00:000 (1983).

15. J. D. Saxby, "Thermogravimetric Analysis of Oil Shales," Thermochim. Acta 47:121 (1981).

16. C. M. Earnest, "Thermogravimetry of Selected American and Australian Oil Shales in Inert Dynamic Atmospheres," Thermochim. Acta 58:271 (1982).

17. J. P. Elder, personal communication (1982).

18. J. W. Cumming and J. McLaughlin, "The Thermogravimetric Behavior of Coal," Thermochim. Acta 57:253 (1982).

19. R. J. Rosenvold, J. B. DuBow and K. Rajeshwar, "Thermophysical Characterization of Oil Sands: 4. Thermal Analysis," Thermochim. Acta 58:325 (1982).

20. C. R. Phillips, R. Luymes and T. M. Halahel, "Enthalpies of Pyrolysis and Oxidation of Athabasca Oil Sands," Fuel 61:639 (1982).

21. M. J. O'Neill, "Measurement of Specific Heat Functions by Differential Scanning Calorimetry," Anal. Chem. 38:1331 (1966).

22. D. B. Jones, K. Rajeshwar, and J. B. DuBow, "Specific Heats of Colorado Oil Shales. A Differential Scanning Calorimetry Study," I&EC Prod. Res. Develop. 19:125 (1980).

23. D. R. Johnson, J. W. Smith and N. B. Young, "Stratigraphic Variation of Oil Shale Enthalpy of Retorting through the Green River Formation on the Colorado C-a Tract," LETC Report of Investigations (RI-79) (1979).

24. M. R. Cervenan, F. E. Vermeulen, and F. S. Chute, "Thermal
     Conductivity and Specific Heat of Oil Sand Samples," Can. J.
     Earth Sci. 18:926 (1981).

25. E. I. Vargha-Butler, M. R. Soulard, H. A. Hamza, and A. W.
     Neumann, "Determination of Specific Heats of Coal Powders by
     Differential Scanning Calorimetry, Fuel 61:437 (1982).

MULTIPLE SAMPLE DIFFERENTIAL SCANNING CALORIMETRY

Robert C. Johnson and Valdis Ivansons

E. I. Du Pont De Nemours & Company
Central Research & Development Department
Experimental Station
Wilmington, DE 19898

We shall describe the design and performance of multiple sample instruments for differential thermal analysis which do runs on several samples simultaneously[1,2,3]. The first of these is a five-sample DTA system that served as a basis for the DSC system. Following this is a description of the principles of operation of the isothermal boundaries that were formed as integral parts of the heating block and thermoelectric disk of a heat flow type multiple sample DSC cell.

Although excellent differential thermal analysis instruments are available from several manufacturers, most have limitations that can lead to major problems in a busy analytical laboratory. Some of the problems are:
1. Limited rate at which samples can be run and analyzed;
2. Uncertainty of intercomparison of runs on some materials which are sensitive to thermal history;
3. Uncertainty in calorimetric quantities because they are computed using a "cell constant" which is derived from a separate run and assumed to be the same for many runs;
4. Uncertainty in evaluation of specific heat because three separate runs under identical conditions are required.

The approach that has been taken for the solution of these problems has been to develop an instrument that will run several samples simultaneously inside the same furnace with a single reference while storing the information in a data system. Following the run, the necessary computations and the generation of hard copies of the thermal analysis curves are done by the data system.

Experimental

A five-sample DTA system that was built first is illustrative of most of the system that was designed for the DSC and will be described in detail.

The data system provides teletype control of all aspects of the experiment except the furnace program; it stores the data, makes identification labels, and draws the DTA curves. It is a "stand-alone" system comprised of:
1. A five-sample cell and six low-level amplifiers for temperature differences ΔT and reference temperature T_R signals.
2. A microprocessor-based data system controlled by a teleype keyboard using conversational commands.
3. An independent furnace controller-programmer of a Du Pont 990 thermal analyzer.
4. X-Y recorder of a Du Pont 990 thermal analyzer.

The DTA cell shown in Fig. 1 was made from the parts of a Du Pont standard DTA cell. Holes were drilled in the silver heating block to accommodate six 2mm o.d. glass capillary tubes which hold the reference and five samples. The Chromel-Alumel thermocouples in each of the five samples are connected in a circuit to the Chromel-Alumel thermocouple in the reference to provide five differential thermocouple voltages of the samples with respect to the reference. These five differential voltages, as well as the reference thermo-couple voltage, are each amplified by one of the six low-level amplifiers in the data system to provide outputs representing ΔT_1, ΔT_2, ΔT_3, ΔT_4, ΔT_5 and reference temperature T_R, which are multi-plexed by the data system.

The data system, Fig. 2, is comprised of an Intel SBC-80/10 single board computer, two Intel SBC-016 16k (8-bit byte) RAM memory boards, an Analog Devices RTI-1200 (12-bit) analog/digital converter interface, a teletype keyboard, and a recorder of a Du Pont 990 thermal analyzer. The responses of the system to the commands entered by the operator are controlled by the program stored in the 5k of programmable read-only memory (PROM) that reside on the SBC-80/10 and the RTI-1200 circuit boards. The data system responds to operator commands under three modes of operation: SETUP, RUN, or PLOT. SETUP information is used to make the labels for the DTA curve. RUN information selects data storage rate, sample to be monitored, and starts the data logging. PLOT information selects the sample number, the $Y(\Delta T)$ and $X(T_s)$ scales, and origin X_0, Y_0. The data system then prints the label with the teletype and draws the curve with the recorder without any intervention by the operator.

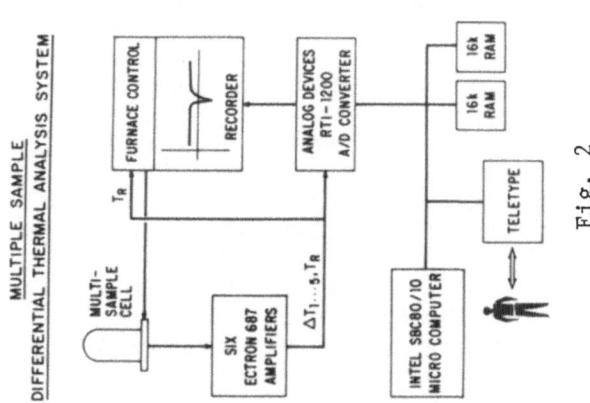

Fig. 2

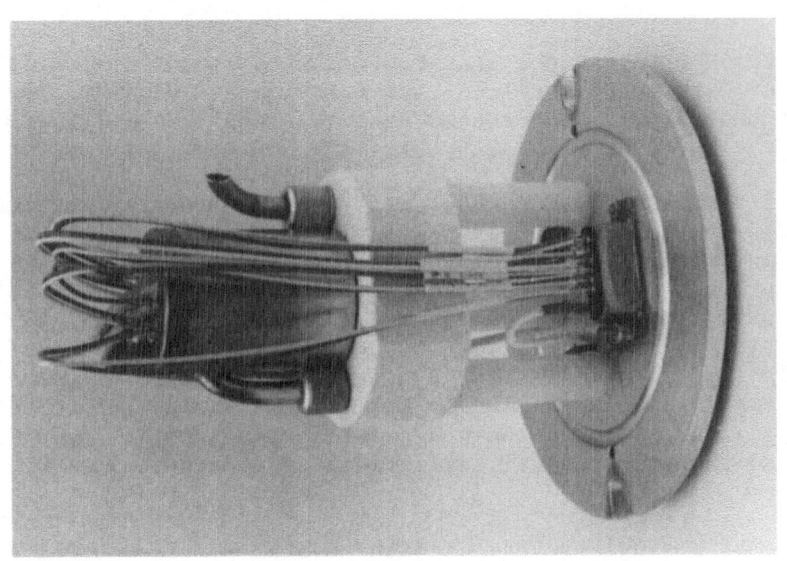

Fig. 1. Five-sample DTA cell.

In this form, the system:

- Stores reference temperature T_R and all five ΔT data in semi-conductor RAM;
- Traces $\Delta T\text{-vs-}T_R$ curve on the recorder for any one sample during run. The operator selects which sample is to be monitored;
- Traces each DTA curve ($\Delta T\text{-vs-}T_S$) after the run while retaining data in RAM. The operator selects the desired DTA curve;
- Traces DTA curves with any of six ΔT or T_S scales selected by the operator;
- Places origin of ΔT and T_S at any position selected by the operator so that portions of the DTA curves can be magnified;
- Types a complete label for the DTA curve from data entered by the operator;
- Detects DTA information in one sample channel with no crosstalk from any others.

An interesting example of the capability of this system is shown in Fig. 3. Five samples of different compositions of the anthracene-phenazine binary solid system were run simultaneously. The endothermic transitions that corresponded to the melting of the eutectic and the melting at the liquidus were detected. This single run gives five points on the liquidus and three points on the solidus for the anthracene-phenazine binary system.

The development of a multiple sample DSC required the design and construction of a DSC cell and the operation of it using the data system that had already been run with the DTA cell. We shall describe a DSC cell that successfully runs three samples simultaneously.

The cell is based on a Du Pont differential scanning calorimeter cell. A view of the interior of the top of this DSC cell, showing the sample and reference platforms, is given in Fig. 4. The Du Pont DSC is a heat flow type DSC cell in which a constantan sheet ("thermo-electric disk") that supports the sample and reference serves as the major heat flow path for transferring heat to the sample and reference pans and also as the common element of the differential thermocouple. This thermoelectric disk is mounted inside a silver heating block which has a silver lid. The sample and reference are in sealed metal pans so that the thermal environment is reproducible from run to run.

The use of a single reference with several samples on the same thermoelectric sheet introduces special requirements for the DSC cell design. These are:

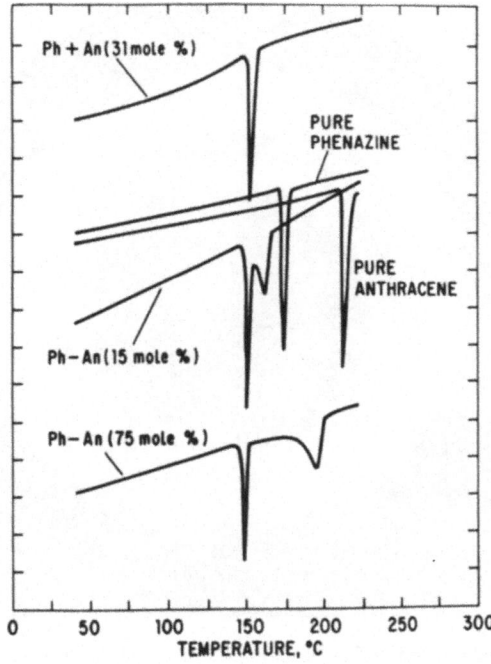

Fig. 3   Multiple DTA of anthracene-phenazine binary system.

(1) Thermal equivalence of all positions for the sample and
    reference.

(2) The absence of crosstalk between sample outputs; crosstalk
    is the appearance of a feature on the DSC curve for one
    sample caused by a transition occurring at another position.

The first of these requirements means that the reference cannot
bear any special relationship with any of the samples; interchange
of reference and sample positions must result in the same thermal
structure. The second means that thermal communication between the
positions of the sample and reference must be reduced to a very low
level.

A three-sample, one-reference DSC cell was built using the
thermoelectric disk formed by a die-press that made four sample and
reference platforms similar to the two that are provided in the
commercial cell. The new form of cell was made to be free of crosstalk problems by introducing, symmetrically, isothermal boundaries

Fig. 4.   Du Pont single sample DSC cell, top view, showing
          the sample and reference platforms in the thermo-
          electric disk.

between the sample and reference positions while retaining the continuous sheet of thermoelectric material. This is shown in Fig. 5.
The isothermal boundary in this case was made by soldering a thick
copper cross to the wall of the silver block and to the tops of the
ribs separating the sample and reference platforms. In this way, an
isothermal boundary separated each of the platforms into identical
compartments which were in poor thermal communication with each other.

   Several advantages follow from this concept of isothermal boundaries established on a single continuous thermoelectric sheet:
   • reduced crosstalk;
   • possibility of providing separated atmospheres in each
     compartment and reference compartment;
   • possibility of making linear or annular arrays of sample
     and reference compartments.

   Figure 6 provides an example of the DSC traces for the melting
endotherm of polyethylene samples. The figure is a reproduction of
the traces from two runs, one with one sample of polyethylene and
the other with three samples. The top trace in both cases is

Fig. 5. Three-sample DSC cell with isothermal boundaries.

the monitor of the melting of the sample at one position which is
recorded during the progress of the run. The curves on the left
show the absence of crosstalk in the traces from the empty sample
pans that might have been produced by the melting transition in the
third sample pan. The traces on the right are examples of the simultaneous recording of the melting of three polyethylene samples.

Future Development and Applications

     Advantages that this multi-sample system offers over existing
DTA or DSC equipment are:
     1. Increased sample throughput.
     2. Better basis for intercomparison of data because all
         samples are run under identical conditions.
     3. Inclusion of reference sample materials for quality control.
     4. Inclusion of temperature or transition-heat standards
         among samples for calibration purposes.
     5. More dependable calorimetry if one sample is a sapphire
         standard because the complete temperature dependence of
         the cell constant can be evaluated for every run.
     6. Provision from a single run of the data for a complete
         quantitative specific heat C_p versus T_S curve.

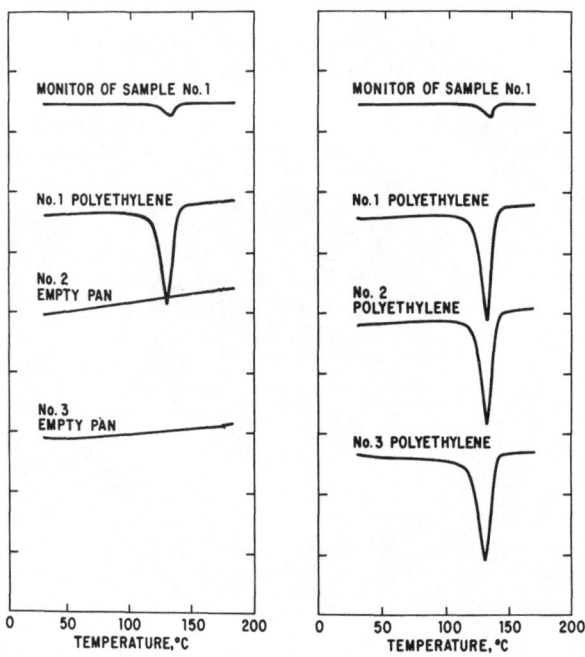

Fig. 6.   Multiple DSC curves, showing the absence of cross-
          talk (left) and the simultaneous recording of the
          melting of three polyethylene samples (right).

A two-sample DSC cell that incorporates these principles has
recently become available[4]. The isothermal boundary ribs are situ-
ated entirely below the thermoelectric disk as part of the heating
block structure and function in the same way as in the three-sample
cell.

Summary
      A new multiple sample DSC system incorporating a microprocessor-
based data system has been developed and tested. It does DSC runs
on three samples in a single furnace simultaneously in the time
formerly required to do one. Because all samples experience the same
thermal history, more dependable intercomparisons of samples may be
made by including reference or quality control materials and calibration standards. The data system provides teletype control of all

aspects of the experiment except the furnace program; it stores the data, makes identification labels, and draws the DSC curves. The system comprises a multiple sample, single reference DSC cell, a data system, a furnace-controller and a recorder. The design and performance of the system have been described.

REFERENCES

1. J. L. Kulp and P. F. Kerr, Am. Mineralogist, 34, 839 (1949).
2. C. M. A. de Bruyn and H. W. van der Marel, Geol. en Mijnbouw [N.S.] 16, 407 (1954).
3. T. Horlin, T. Niklewski and M. Nygren, Chem. Commun. Univ. Stockholm, Nr. 9 (1975).
4. Du Pont Company, Clinical and Instrument Systems Division, Concord Plaza, Mc Kean Building, Wilmington, Delaware 19898.

QUANTITATIVE EVALUATION OF THE GIBBS-DIMARZIO THEORY

OF THE GLASS TRANSITION FOR POLYSTYRENE

A. R. Greenberg

R. P. Kusy

Dept. of Mechanical Engineering
University of Colorado
Boulder, CO 80309

Dental Research Center
Univ. of North Carolina
Chapel Hill, NC 27514

ABSTRACT

The applicability of the Gibbs-DiMarzio (G-DM) theory of the glass transition (T_g) is quantitatively evaluated for polystyrene (PS). The analysis was conducted under the assumption that both the inter- intramolecular energy ratio (r) and the effective chain segment density (n) remain constant while the fractional free volume at T_g(V_o) varies as a function of the reciprocal degree of polymerization ($10^3/\overline{P}$). Based upon reduced parametric plots of T_g/T_{g_∞} versus $10^3/\overline{P}$, the results showed that the G-DM equations were satisfactory for PS; when $0.015 \leq V_o \leq 0.045$, optimum agreement occurred at n = 1.80, r = 1.05.

INTRODUCTION

Over the past 30 years a number of theories have been developed to account for the variation in the glass transition temperature (T_g) of polymeric materials as a function of molecular weight (MW). Three of the most notable of these approaches have included the straight line technique of Fox and Flory[1] and the statistical mechanical theories of Gibbs[2] and Gibbs-DiMarzio[3] (G-DM). Because of the universal applicability claimed for the latter approach, the G-DM theory has received considerable attention.[4-7]

Recently Kusy and Greenberg[6] introduced a reduced variables technique which simplified the use of the rather unwieldy statistical mechanical equations. Subsequent expansion of this work in terms of the inter- intramolecular energy ratio (r) and the fractional free volume at T_g(V_o) explicitly characterized the functional dependence

143

of each of these parameters on MW.[8] In addition, an index (n) which related the effective number average of chain atom segments ($\bar{x}$) per degree of polymerization ($\bar{P}$) was incorporated into the reduced variable equations in order to better account for structural differences among polymers. By assigning reasonable values to the parameters n, r and V_O, the theoretical predictions of the statistical mechanical equations were compared with T_g data obtained from the literature for four well documented polymers. Qualitative assessment of these results suggested that the G-DM theory adequately described the PMMA, PS and PVC cases but was unsuccessful for PαMS.

Because of the complexities inherent in a theory which incorporates multifunctional dependencies, further examination has indicated that a qualitative evaluation is not sufficient to establish the validity of a particular approach. Recently a statistical analysis technique was utilized in a comparison of the Gibbs theory with experimental results obtained for PMMA, and the advantages of such a quantitative determination were clearly demonstrated.[9] The present effort applies these quantitative analytical techniques to the PS data set in order to make an unbiased assessment of the G-DM theory.

THEORETICAL DEVELOPMENT

The system of equations which comprises the G-DM theory has been presented in detail elsewhere.[3,6,8] However, the approach can be summarized by consideration of the following two equations:

$$\frac{2\beta \exp\beta}{1+2\exp\beta} - \ln[1+2\exp\beta] \tag{1}$$

$$= \frac{\bar{x}}{\bar{x}-3}\left\{\frac{1}{1-V_o}\left(\ln V_o + (1+V_o)\ln\left(\frac{(\bar{x}+1)(1-V_o)}{2\bar{x}V_o}+1\right)\right) + \frac{\ln[3(\bar{x}+1)]}{\bar{x}}\right\}$$

$$\beta = -\frac{1}{r}\left[\frac{\ln V_o - 2\ln\left(\frac{2\bar{x}V_o}{2\bar{x}V_o+(\bar{x}+1)(1-V_o)}\right)}{\left(\frac{(\bar{x}+1)(1-V_o)}{2\bar{x}V_o+(\bar{x}+1)(1-V_o)}\right)^2}\right] \tag{2}$$

Here β is a dimensionless parameter equal to $-\varepsilon/kT_g$ where ε is the flex energy and k is Boltzmann's constant. In accordance with the assumption that V_o decreases monotonically with decreasing MW, equations (1) and (2) can be solved iteratively by maintaining n$\bar{P}$ and r constant while simultaneously varying V_o until a unique β derived

from equation (2) satisfies equation (1). This procedure is repeated
for different combinations of $n\overline{P}$ and r.

Using the reduced variables method, a plot of $T_g/T_{g\infty}$ vs. $\overline{P}^{-1}$
generates a family of curves each corresponding to a specific value
of n (cf. Figure 1). Each value of n is in turn plotted for the
cases where r = 0.8, 1.0 and 1.2. These results suggest that large
variations in r have relatively little effect on the family of curves;
however, this latitude is not possible unless V_o is allowed to float
over rather wide limits as a function of $\overline{P}^{-1}$ (cf. Figure 2). Close
examination of Figures 1 and 2 indicates that if V_o is to be restricted to reasonable limits within the range $1 \leq 10^3/\overline{P} \leq 100$ then only
certain combinations of n and r can be considered (cf. Figure 3).
For example, if n = 1.4 and V_o is to be kept within 0.015–0.045, then
r must remain within 0.94–1.07. If these restrictions are somewhat
relaxed such that $0.010 \leq V_o \leq 0.050$, then under the same conditions
(i.e., n = 1.4) r may range from 0.89–1.23. Only n-r combinations
corresponding to $0.015 \leq V_o \leq 0.045$ will be utilized in the comparison which follows.

RESULTS

The details of the statistical analysis technique have been
discussed in reference (9). The method utilizes a transformation of
variables and a linear regression of the differences between the predicted and experimental values (δ) as a function of $\overline{P}^{-1}$.[10] If the
theory perfectly described the data, then all of the δ values would
be zero and both the slope and intercept of the resulting regression
line would also equal zero (null hypothesis).[11] Since in general
this is not the case, the F statistic is utilized to test the null
hypothesis (H_o) at the p = 0.05 level. If H_o cannot be rejected,
the theory is regarded as having fit the data set for a particular
combination of n and r.

The first step in the procedure is to obtain F vs. n curves
for applicable values of ι. A typical plot for the case of PS is
shown in Figure 4. Here the four parabolas correspond to r = 0.95
and r = 1.10. The dotted line at F = 3.1 indicates the critical
level, p = 0.05, which depends upon the number of points in the data
set. The best fit between the predicted and experimental values is
assumed to occur at the minimum value of F, i.e., at the vertex of
the parabola. Based on this criterion, the results indicate that for
PS the best fit occurs at two points: n = 1.80, r = 1.05 and
n = 1.82, r = 1.10.

A necessary condition for the use of this statistical methodology
is that the data points not be preferentially distributed about the
regression line.   In order to verify that this requirement is met,
a scatter diagram must be plotted for each of the polymer data sets
for which H_o was accepted. Typical results for PS are shown in

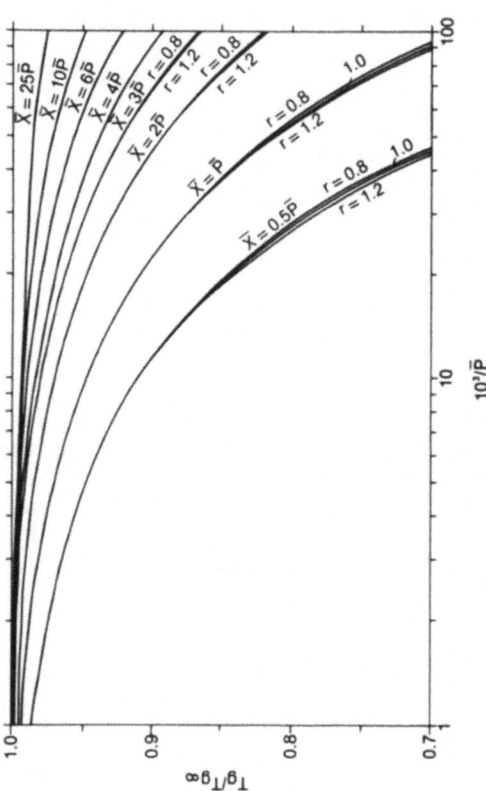

Fig. 1. Reduced variables plot indicating the dependence of the glass transition (T_g) upon the logarithmic reciprocal degree of polymerization $(P)^{-1}$ as a function of constant values of the ratio of hole energy to flex energy ($r = 0.8$, 1.0, and 1.2) and number average of chain atom segments per $\bar{P}$ ($n = 0.5, 1, 2, 3, 4, 6, 10$ and 25). The relationships assume any constant value of ε and require the variation in V_o indicated in Figure 2 (cf. reference (8)).

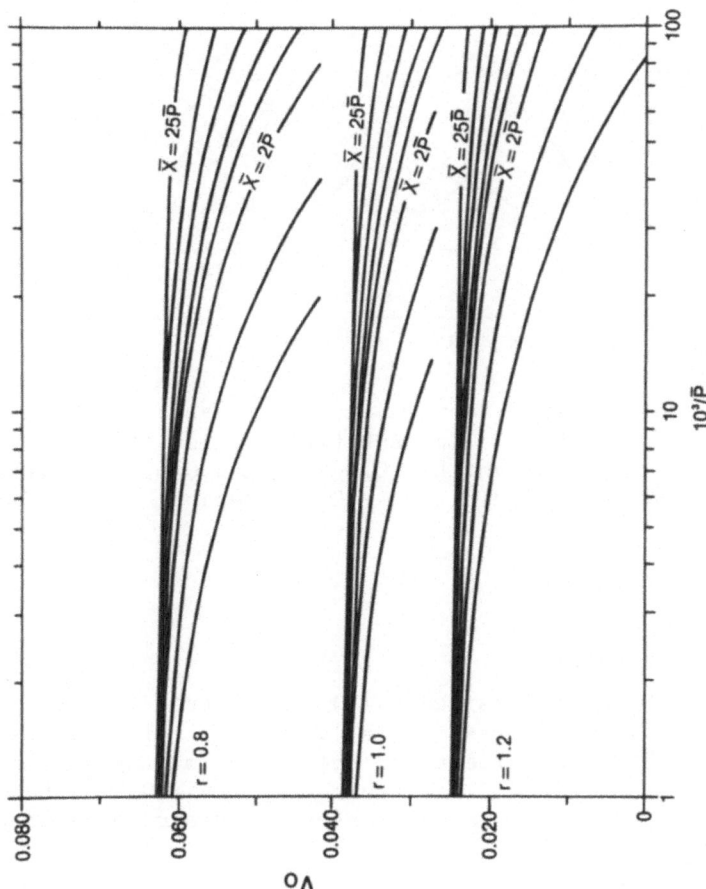

Fig. 2. Relationship between the parameter V_O and $(\overline{P})^{-1}$ as determined from equations 1 and 2. Values for r and x correspond to those indicated in Figure 1 (cf. reference (8)).

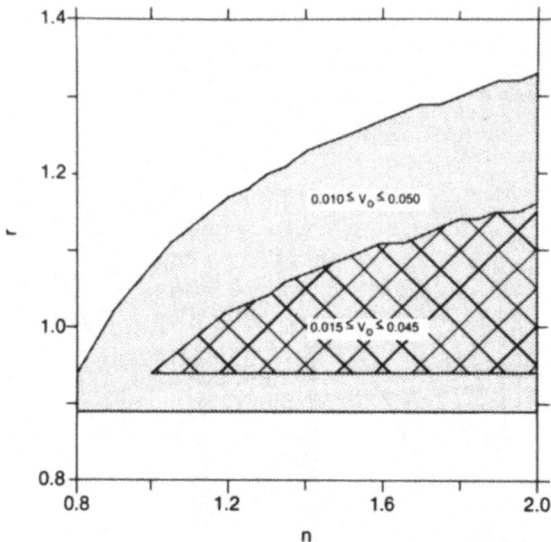

Fig. 3. Interrelationship among n, r, and V_0 as determined from
         equations 1 and 2. Cross-hatched region indicates n-r
         combinations for which $0.015 \leq V_0 \leq 0.045$. Stippled area
         designates the $0.010 \leq V_0 \leq 0.050$ region.

Figure 5.[12-20] Here the residuals (δ) are plotted as a function of
$10^3/\overline{P}$ for the "best fit" case of n = 1.80 and r = 1.05 as determined
from Figure 4. The two salient features of this diagram are that the
regression line does approximate the equation $\delta = 0$ and that the
individual data points are indeed reasonably distributed around the
regression line.

DISCUSSION

     Based upon the results of the statistical analysis the Ġ-DM
theory can be regarded as satisfactory for PS. The necessity for
such a quantitative evaluation becomes apparent when the information
contained in the F-n curves (cf. Figure 4) and the scatter diagram
(cf. Figure 5) is presented in a reduced variables format. This
plot is shown in Figure 6 where the individual data points are

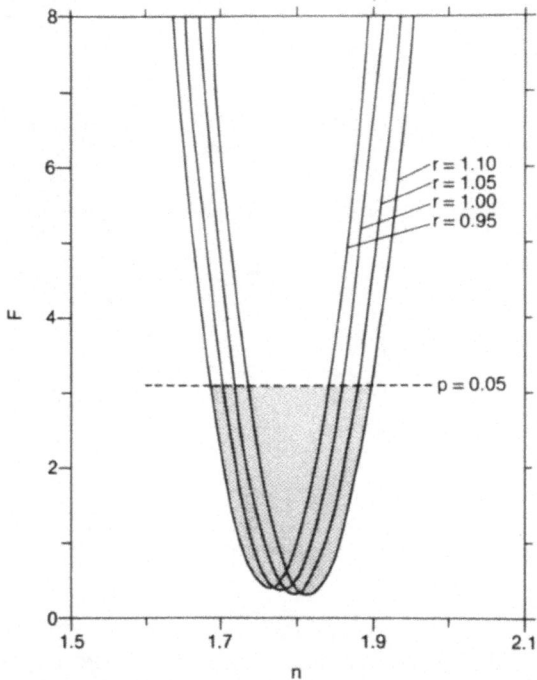

Fig. 4.   F-test results as a function of n and r for PS using the
G–DM theory.

superposed by a theoretical curve derived for the "best fit" situation.

   Visual examination of Figure 6 would indeed suggest that there
is a good agreement between the theory and the data. However opin-
ions concerning the applicability of any T_g vs. MW theory should be
based upon an unbiased quantitative technique. Moreover some caution
must be exercised in the use of such methodology inasmuch as the
conclusions drawn are only as good as the information on which they
are based. As previously described, the current data sets include
all published results for both pure polymers and blends, irrespective

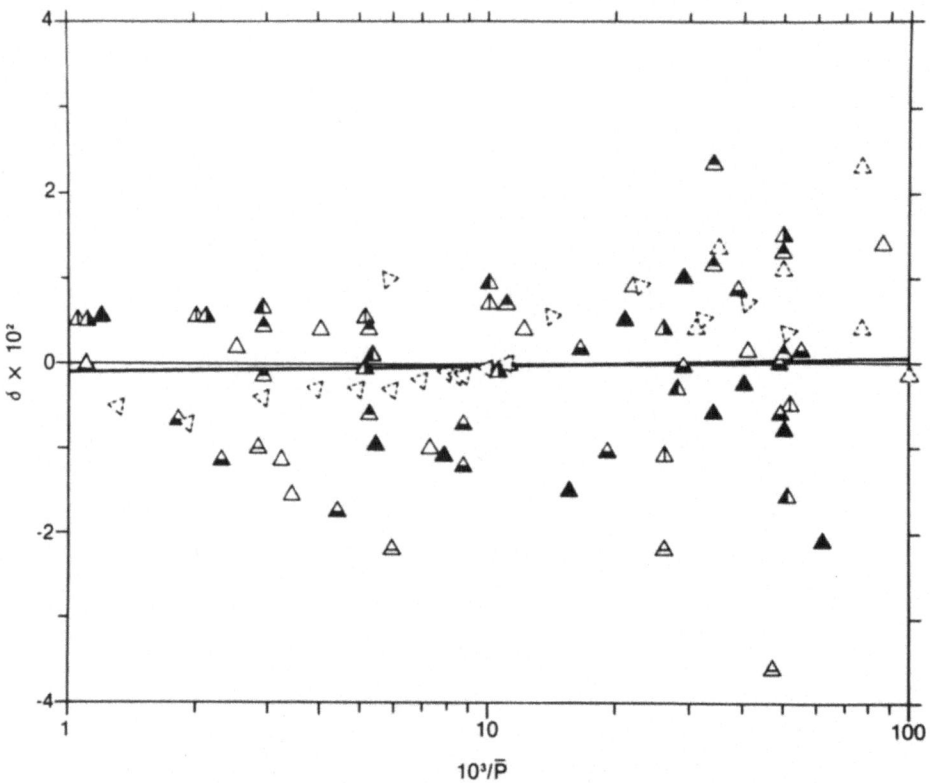

Fig. 5.   Scatter diagram for PS using the G-DM theory with n = 1.80
          and r = 1.05: △ , Enns et al.;[12] ▲ , Fox and Flory;[13]
          ◁ and ▷,Glandt et al. on blends;[14] ▲ , Krause and Iskan-
          dar;[15] ▲ , Richardson and Savill;[16] ▲ , Rudin and
          Burgin;[17] ▲ and ▲ , Stadnicki et al.;[18] △ ,
          Ueberreiter and Kanig;[19] and △ , Ueberreiter and Kanig
          on blends.[20]

of tacticity, test methdology, physical form or thermal history.
Figure 6 indicates that each of the data sets contains a substantial
amount of scatter. The relative contributions of each of the above
mentioned factors to the overall variation cannot presently be
determined. Indeed the relative importance of each of these factors
may vary for different polymers. Nevertheless these contributions
should be established since such information could prove critical
with regard to the acceptance or rejection of a particular theoretical approach.

    The present results have been evaluated under the assumption
that r remains constant while V_0 varies as a function of MW. This
situation is in agreement with an iso-viscosity viewpoint.[21]

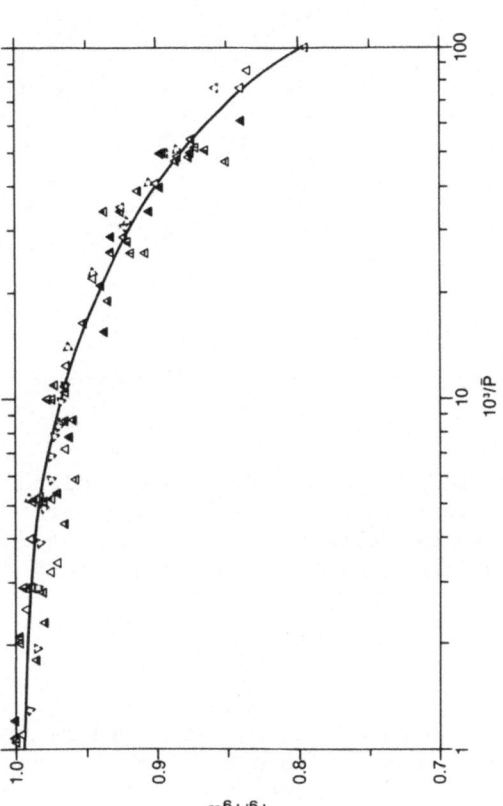

Fig. 6.   Reduced variables plot for PS. Curve represents solution for equations 1 and 2 with n = 1.80 and r = 1.05 (cf. Figures 4 and 5).

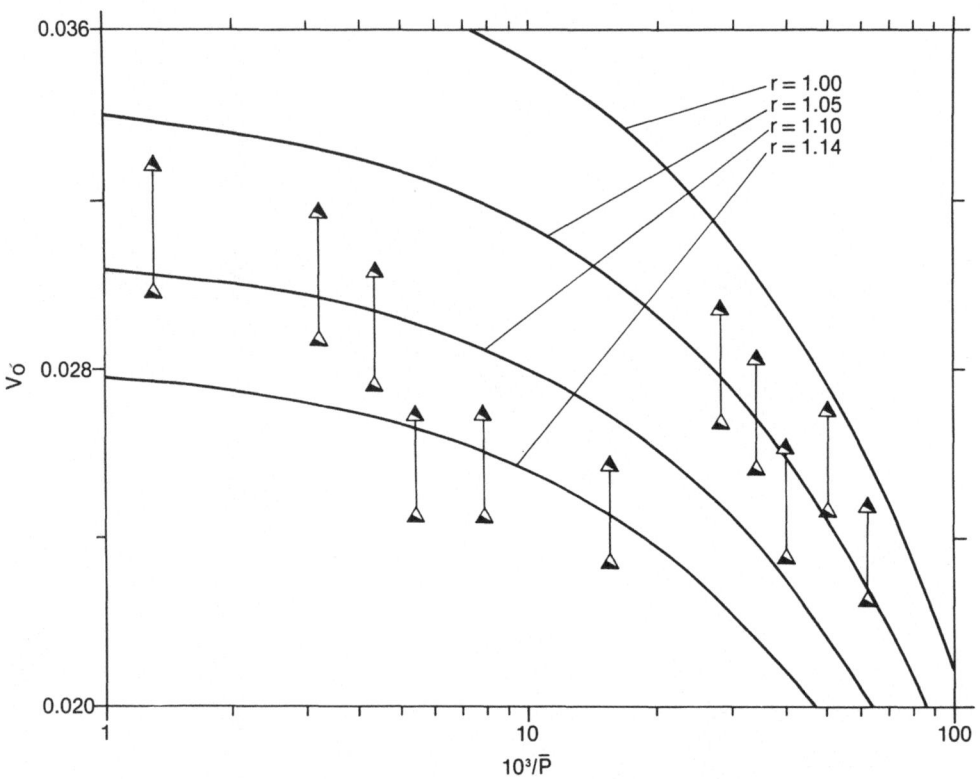

Fig. 7.   Dependence of V_O on the logarithmic reciprocal degree of
polymerization for PS. Data from Williams[23] (▲) and as
reinterpreted by Miller[21] (▲). Curves demonstrate the
influence of the parameter r with n = 1.80 (cf. Figures 2,
4-6).

However from the standpoint of the iso-free volume theories one could
argue for a constant V_O and variable r.[22] While the same analytical
techniques could be utilized to obtain solutions within the constraints of the latter assumption, there is a strong basis for taking
the former approach.[21,23,24] Unfortunately relatively little information is available concerning the dependence of V_O on MW. For PS,
measured values of V_O over a a wide range of MW have been reported
by Williams.[23] Starting with this same data set, Miller[21] subsequently presented somewhat different Vo values such that the two sets
differ by a constant amount. This situation is shown in Figure 7 in
conjunction with the "best fit" curves obtained from Figure 4
(n = 1.80). Here the data points center on the r = 1.10, 1.14 and
1.05 curves for $(10^3/\bar{P})$ = 1-5, 5-20 and 20-100, respectively.
Whether the measured dependence of V_O on MW would have supported the
statistical mechanical approach or would have substantiated a

different functional relationship cannot be determined because of
the inconsistency in the distribution of the data. Nonetheless the
information suggests the validity of the iso-viscosity assumptions
made at the outset.

To place the present results in their proper context, the nature
of certain terms in the G-DM equations needs to be more fully elab-
orated. Originally the G-DM equations were expressed in terms of
"$\bar{x}$", the number of monomer segments.[3] The term "monomer" was defined
as a unit which occupied one lattice site such that it normally
included one carbon backbone atom plus the attendant side groups.
Hence the chain length could be given in terms of $\bar{x}$, the number of
carbon atoms in the main chain. Within the current terminology the
chain length can be expressed via $\bar{x} = n\bar{P}$, where n = 2 for vinyl
monomers having two carbon backbone atoms per repeat unit. The
qualitative results of a previous study have demonstrated that if "n"
were restricted to a value of 2.0, the G-DM theory could possibly
apply to only a relatively small number of materials.[8] For this
reason the index "n" was utilized as a variable dependent upon the
chemical nature of the polymer.[25] Relatively large values of the
index, $n \gtrsim 10$, were associated with flexible molecules such as PDMS,
while low values, $n \approx 1$, were representative of much stiffer materials such as PαMS.[8] The quantitative techniques utilized in the
present study have demonstrated that the G-DM relationships satisfy
the PS case when n = 1.80. If a fit were to be obtained for PαMS and
PMMA, much lower values of n would be required.

ACKNOWLEDGMENT

This investigation was supported in part by NIH Research Grant
No. DE02668 (RPK).

REFERENCES

1. T. G. Fox and P. J. Flory, J. Appl. Phys., 21:581 (1950).
2. J. H. Gibbs, J. Chem. Phys., 25:185 (1956).
3. J. H. Gibbs and E. A. DiMarzio, J. Chem. Phys., 28:373
    (1958).
4. J. Moacanin and R. Simha, J. Chem. Phys., 45:964 (1966).
5. A. Eisenberg and S. Saito, J. Chem. Phys., 45:1673 (1966).
6. R. P. Kusy and A. R. Greenberg, Polymer, 23:36 (1982).
7. E. A. DiMarzio, Annals N. Y. Acad. Sci., 371:1 (1981).
8. A. R. Greenberg and R. P. Kusy, Polymer, in press.
9. R. P. Kusy and A. R. Greenberg, Polymer, accepted for
    publication.

10. R. R. Sokal and F. J. Rohlf, in "Biometry", W. H. Freeman, Ch. 7, 13 and 14 (1969).

11. J. Neter and W. Wasserman, "Applied Linear Statistical Models," R. D. Irwin, Inc., Homewood, Illinois, Ch. 5 (1974).

12. J. B. Enns, R. R. Boyer and J. K. Gillham, Polym. Preprints, 18(2):475 (1977).

13. T. G. Fox and P. J. Flory, J. Polym. Sci., 14:315 (1954).

14. C. A. Glandt, H. K. Toh, J. K. Gillman and R. F. Boyer, Polym. Preprints, 16(2):126 (1975).

15. S. Krause and M. Iskander, Proc. 10th N. Amer. Therm. Anal. Conf., Boston, 51 (1980).

16. M. J. Richardson and N. G. Savill, Polymer, 18:3 (1977).

17. A. Rudin and D. Burgin, Polymer, 16:291 (1975).

18. S. J. Stadnicki, J. K. Gillham, and R. F. Boyer, Polym. Preprints, 16(1):559 (1975).

19. K. Ueberreiter and G. Kanig, Z. Naturforsch., 6A:551 (1951).

20. K. Ueberreiter and G. Kanig, J. Colloid Sci., 7:569 (1952).

21. A. A. Miller, J. Polym. Sci., A2:1095 (1964).

22. M. L. Williams, R. F. Landel and J. D. Ferry, J. Amer. Chem. Soc., 77:3701 (1955).

23. M. L. Williams, J. Appl. Phys., 29:1395 (1958).

24. Y. S. Lipatov, V. F. Rosovizky and V. F. Babich, Eur. Polym. J., 13:651 (1977).

25. J. M. G. Cowie and S. A. E. Henshall, Eur. Polym. J., 12:215 (1976).

STUDY ON ADHESIVES PERFORMANCE BY

DYNAMIC MECHANICAL TECHNIQUES

Lecon Woo

Material Development
Travenol Laboratories, Inc.
Round Lake, IL   60073

INTRODUCTION

Pressure sensitive adhesives (PSA's) are evaluated by several key performance tests: tack, shear adhesion (holding power), peel, and for melt processed systems, their molten viscosity. Tack measures the debonding mechanical forces after a momentary contact where shear adhesion is determined by the resistance to shear creep. Tack can be measured by the stopping distance for a rolling steel ball with a known amount of initial kinetic energy, or alternately by a probe tack tester measuring the peak debonding normal force on a fixed area probe surface. Shear adhesion is frequently measured as the time of failure for a given adhesive surface area loaded under shear.

Dahlquist (1) first proposed there is a minimum compliance below which no significant tack can develop. In their comprehensive study of tack and tackifiers, Krauss and Rollman (2) further expanded the Dahlquist criterion to include a broader definition of mechanical compliances. The phenomenon of peel was thoroughly studied by Kaeble (3) and Gent et al. (4) Dynamic mechanical analysis covering a wide temperature and frequency ranges lend itself naturally to the understanding of adhesives performance. In this study we will concentrate mainly on shear adhesion and tack. By using a combination of the state of the art commercial instrumentations as well as significant refinements on adhesive tests, we will attempt to generate a

thorough understanding on certain aspect of the adhesive
performance and relate it to fundamental properties. Two
adhesives were chosen for this study. One; a commercial
pressure sensitive tape, the other a formulation based on a
commercial polymer, are designed to yield general understandings
applicable to wide classes of generically similar adhesive
systems.

ADHESIVE SYSTEMS

The commercial adhesive tape studied will be identified as
Sample A, 12mm in width and manufactured by Lepage Packaging Co.
Pittsburgh, Penna. under the name Lepage preminum tape. The
adhesive layer is measured to be 0.153 $\pm$ 0.005mm thick.
Infrared analysis identified it as a vinyl acetate based
polymer. Some of the adhesives were striped from the tape by
dissolving in toluene and recovered in a vacuum oven at 80°C.
The DMA spectrum (Figure 1) indicated it is substantially free
of tackifier or other additives by virture of the relative
narrowness of the main relaxation peak (Tg) at around -10°C.

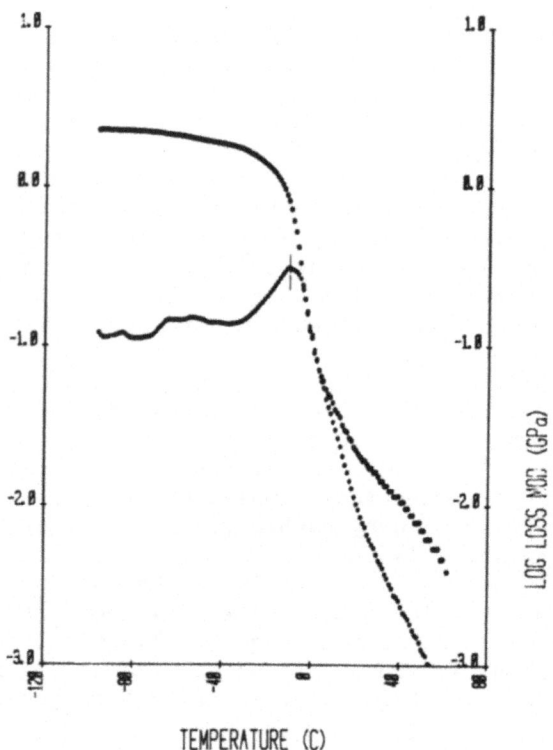

Figure 1.   Sample A DMA Spectrum, E' and E" vs. Temperature,
Ca 20 HZ.

In the second system, a commercial styrene isoprene block copolymer was formulated according to:

	Formulation	Parts
Polymer	Kraton-1107(a)	100
Tackifier	Wingtak Plus (b)	100
Plasticizer	Shellflex 371	20
Antioxidant	Irganox 1076(c)	3
	BHT	1

(a) Shell Chemical Co.

(b) Goodyear Tire and Rubber Co.

(c) Ciba-Geigy Co.

40% solutions were first made in toluene, casted on Mylar substrates with a knife coater, and dried to constant weight at 80°C in a vacuum oven. The addition of hindered phenol type antioxidants was found to extend significantly the shelf life of unprotected laboratory prepared tapes. Half inch (1.2mm) wide strips were cut from the Mylar substrate for testing. The thickness of the adhesive was found to be about 0.10 ± 0.02mm.

EXPERIMENTAL

A. Dynamic mechanical measurements: A resonantly driven instrument, the Dupont 981 DMA was used for wide temperature scans. It is used in conjunction with a Hewlett Packard 9825S desk top computer with 6940 interface. A Hewlett Packard 9872 multicolor plotter was the output device. Design principle and operation of the DMA were described elsewhere and will not be repeated (5,6,7). Samples of approximately 1.5mm thick by 12mm in length were used. Since the adhesive samples were tacky and difficult to handle at room temperature, all samples were first molded in a hot compression molder and cooled to dry ice temperature (-78°C), cut and mounted on precooled DMA fixtures. A Rheometrics Mechanical Spectrometer Model RMS-605 was used for frequency scanning, forced driven mechanical testing. Temperature was varied in steps between 25°C and 80°C and frequency varied between 10^{-1} and 10^{+2} radian sec-1. Strain levels of about 2-5% were used. In this experiment elastic and loss shear moduli G' and G" are directly measured and $\eta*$, the dynamic viscosity, is calculated by:

$$\gamma^* (\omega) = \frac{[G'^2 + G''^2]^{1/2}}{\omega} \quad \ldots\ldots(1)$$

Where (ω) is the testing frequency.

Due to sample expansion between temperatures, it was found necessary to readjust the gap setting to maintain the normal force on the sample at near zero.

B. Automated shear adhesion apparatus: In the standard industry test for shear adhesion (Pressure Sensitive Consul PSTC-2) a fixed load is applied to a defined area of adhesive in a vertical plane. Performance is measured as the total time to failure.

However, in order to gain insight to the shear adhesion process in real time, it was felt that the entire process (not just the failure point) needed to be monitored. To cover the widely variable time scale spanning from minutes to weeks, a flexible, reliable recording scheme was needed. In addition, mechanistic studies correlatable with wide temperature range viscoelastic data required precise temperature control over similar ranges.

With these requirements an automated multistation tester was built and controlled by an inexpensive microcomputer. Five samples were run simultaneously with linear variable differ- ential transformers (LVDT's) monitoring the shear adhesion process to a resolution of better than 10^{-6} inches. Testing surface temperatures were measured and controlled to $\pm 0.05°C$. Via a data compression scheme, the entire event regardless of length of time can be represented by thirty data points or less. In this study, smooth stainless steel surfaces were used. The schematic diagram for the apparatus is presented in Fig. 2.

C. Automated probe tack tester: Again, to gain insight into the transient nature of the probe tack process, a standard Poly-Ken probe tack tester was modified with a environmental chamber and the load cell output connected to a transient recorder. The loading waveform resulting from the debonding forces was digitized and recorded by a transient recorder at rates of up to 20,000 points per second. The stored waveform was then reproduced on a CRT or strip chart recorder. Option- ally, the reconstructed waveform can be transferred to the microcomputer via a digitizing tablet for further analysis. The overall block diagram for the tester is shown on Fig. 3.

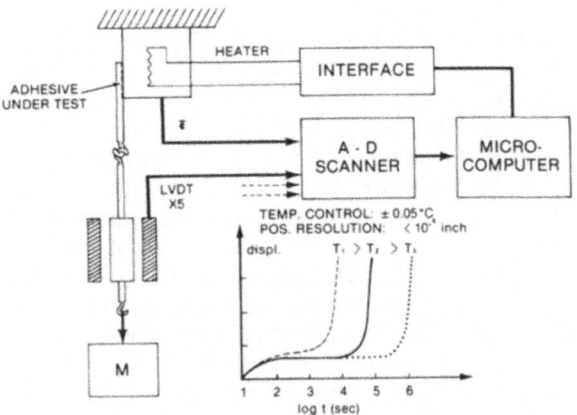

Figure 2.   Automated Multichannel Shear Adhesion Tester.

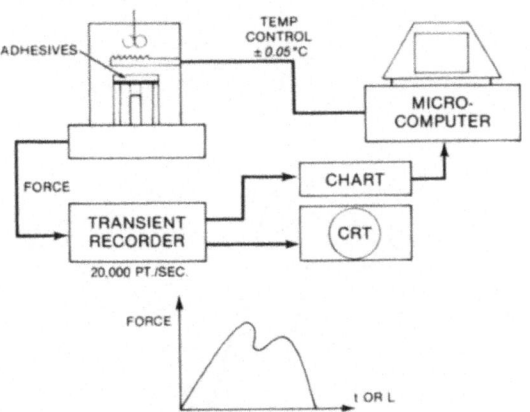

Figure 3.   Automated Probe Tack Tester.

RESULTS AND DISCUSSION

Probe Tack

A typical reconstructed real-time stress displacement curve
is shown in Fig. 4. As can be seen, the peak force developed on
the sample is a weak function of testing speed. This peak force
is near identical to the displayed tack values on the tack
tester, a reflection of the accuracy of calibration and data
conversion process. However, Fig. 4 also indicated a very strong
dependence of total tack energy on testing speed. Data for
sample B is presented in Table I. It is seen that for this
sample, the tack energy at high testing speeds is nearly constant
with temperature. A possible explanation is that the high
frequency (short time) viscous moduli involved at the testing
speed is near invariant with temperature. On the other hand, a
very sharp drop off of tack energy at each temperature with
decreasing testing speed indicates the relative donimant contribution of loss modulus over that of the elastic component. The
total tack energy as a new performance dimension in addition to
the traditional peak force measurements is expected to be
important for many applications such as high speed unwinding of
industrial tapes.

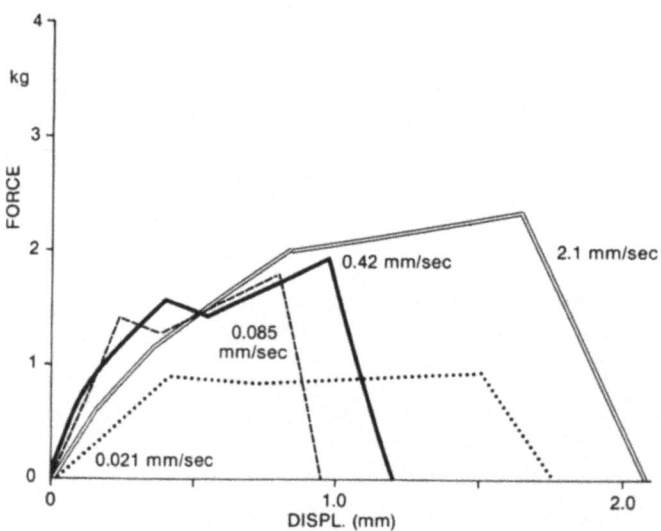

Figure 4.   Sample B Real Time Probe Tack Debonding Force vs.
Displacment.

TABLE I

PROBE TACK TESTING

SAMPLE B

Peak Force (kg) TEMPERATURE	PROBE SPEED (mm/sec)		
	2.1	0.42	0.085
25°C	2.95	2.90	2.0
40°C	2.33	1.95	1.82
60°C	1.90	1.50	1.0
Total Energy (Joules)			
25°C	1.24	1.15	.44
40°C	1.30	.60	.46
60°C	1.10	.16	.21

5 Sec. Dwell Time

SHEAR ADHESION

Real time shear adhesion data on sample A consists of a series of very reproducible, near parallel wave shapes (Figure 5). Because of the near constancy of wave shape, for kinetic purposes they can be consolidated into a single curve and use another parameter, the shift factor to express the temperature dependency. Indeed, when all the failure times for all temperatures and loadings are plotted, a series of parallel lines resulted (Figure 6). This indicated that most likely a single mechanism is operating for the entire temperature and stress levels. Empirically, all data can be represented by:

$$t_f = [3.2 *10^{11} * w ** (-2.96)] * \text{Shift factor} \quad \ldots\ldots(2)$$

Where

t_f: failure time in seconds

w = loading weight in grams

and

TEMPERATURE	SHIFT FACTOR (Shear Adhesion)
25°C	1.0 (Ref. Temp)
30°C	0.59
40°C	0.19
50°C	0.046
60°C	0.019
70°C	0.0067

The shift factors can be used for several purposes. For example, they can be used to shorten the testing processes by extrapolating from higher temperatures to predict with reasonable accuracy lower temperature performances which may take days or weeks to obtain. Secondly, they allow the kinetics of the shear adhesion process to be compared with data from other sources and thereby gaining molecular insight to the performance and facilitate product development.

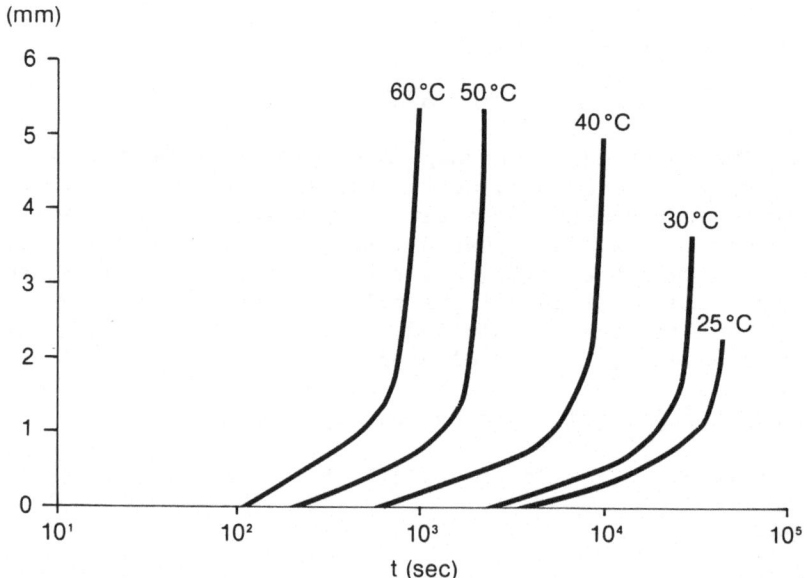

Figure 5. Sample A Shear Adhesion Displacement vs. Log t at 200g. Load.

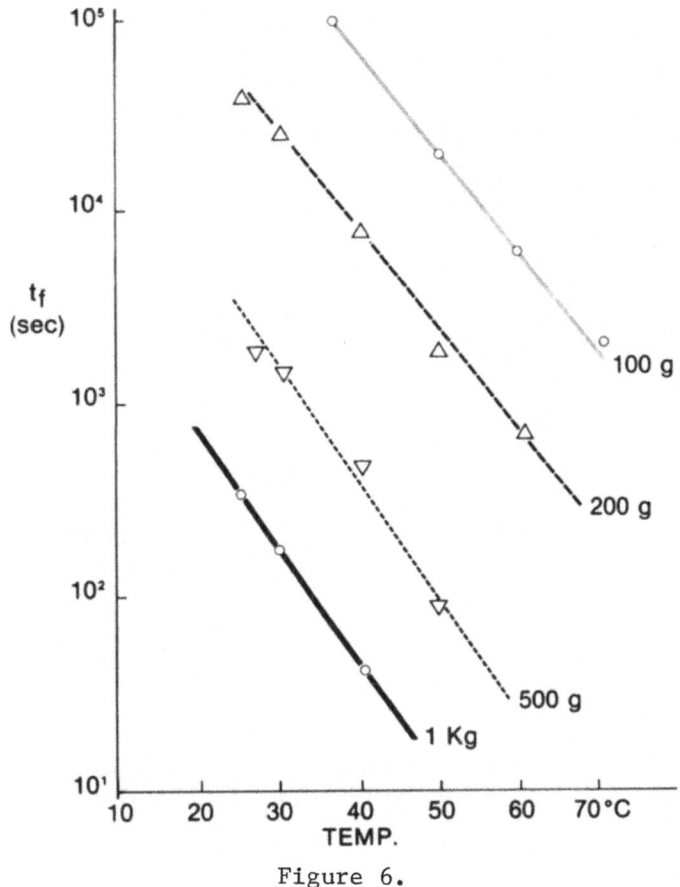

Figure 6.

On sample A, very interesting data was also determined from
scanning dynamic shear experiments. By application of standard
shifting procedures for viscoelastic functions (8), a master
curve was generated. Since for the range of temperature
covered, amount of vertical shift was negligble, it was
ignored. The master curve for G' and G" thus constructed is
shown in Figure 7, with shift factors tabulated in Table 2.
These strikingly near identical shift factors as compared with
actual shear adhesion data was immediately recognized.

TABLE 2

SAMPLE A

Viscoelastic Shift Factors

TEMPERATURE	A_T
25°C	1.0
30°C	0.48
40°C	0.175
50°C	0.051
60°C	1.017
70°C	0.0063

Using 25°C as our reference temperature again, a William Landel Ferry (WLF) equation is fitted to our shift factors as

$$\log a_t = \frac{-C_1(T - T_o)}{C_2 + (T - T_o)}$$

Where C_1, C_2 = constants

T = measurements temperture

T_o = reference temperature

For sample A, C_1 was found to be 5.5 and C_2, 80. The degree of fit is shown in Figure 8.

The significance of the WLF fit is that both the shear adhesion and viscoelastic functions are governed by the same molecular process, namely, the free volume of the system.

With the excellent agreement between shift factors from
viscoelastic measurements and shear adhesion, it will be
interesting to check the predictive ability of this concept on
sample B. First, a master curve is constructed from viscoelastic data and the shift factors tabulated. A careful
measurement of shear adhesion is then made at 60°C. Failure

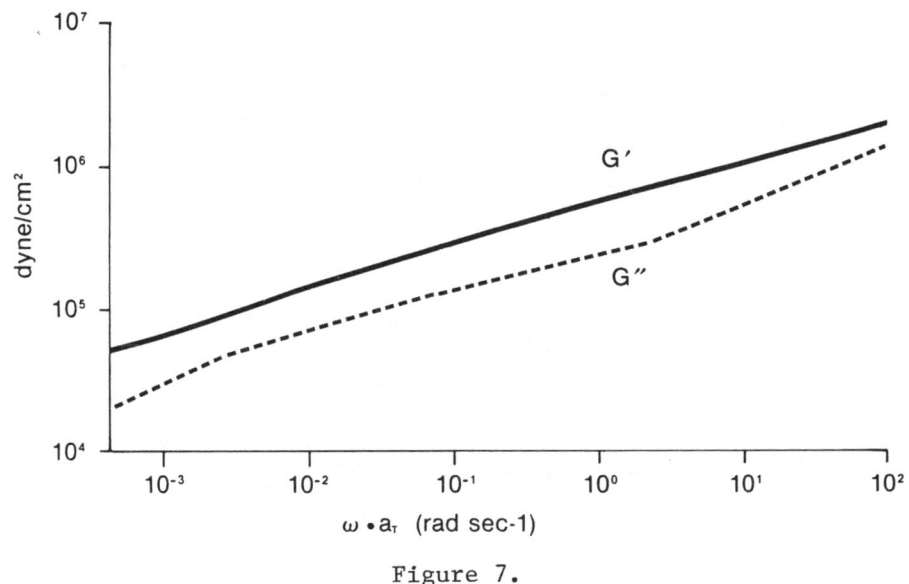

Figure 7.

time of 20 ± 1 sec was obtained. Failure times for other
temperatures are then calculated from viscoelastic shift factors
and the 60°C data. A comparison between predicted performance
and actual measurements are made in Table 3. The agreement is
quite good, especially considering that over two decades of
performance time was covered.

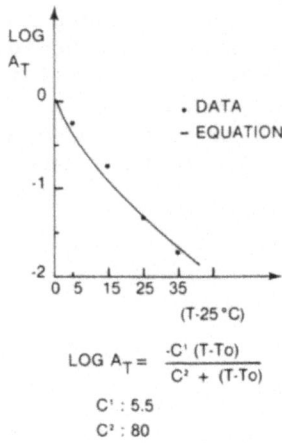

LOG A_T = $\dfrac{-C' (T \cdot To)}{C^2 + (T \cdot To)}$

C' : 5.5
C² : 80

Figure 8.   WLF Reduction, Sample A, Data Comparison.

TABLE 3

SAMPLE B

Shear Adhesion

| | | Failure Times (Sec.) | |
Temp	V/E Shift Factor	Predicted	Measured
25°C	1.0 (Ref.T)	3900	5300
30°C	0.533	2050	2800
40°C	0.107	410	450
50°C	0.0233	89	110
60°C	0.0052	--	20

From the two cases studies, we can conclude there exist a class (or classes) of pressure sensitive adhesives that readily obeys the WLF relationship. Also, it was noted that the modulus of both samples are relatively low, being of the order of 5 x 10^5

dyne cm^{-2} or less at low frequencies. Optical microscopy of the failure morphologies indicated both samples fails in a viscous mode. Smooth failure surfaces dominate and the adhesives are nearly equally partitioned between the substrate and the adherend surfaces. In much stiffer formulations, elastic or adhesive failure modes have been observed, and a modification to our procedures is obviously needed.

CONCLUSION

By automating the standard PSA testing methods for probe tack and shear adhesion, it was clearly demonstrated that one obtains statistically more significant data very efficiently. For probe tack, the newly measured debonding energies constitute a new dimension for short duration performance. In addition to the peak force, tack energy should be considered in many high speed applications. Since the peak force and tack energy appeared to have originated from different parts of the viscoelastic spectra, it offered an opportunity to adjust the formulation for optimum performance.

For both samples in this tudy, shear adhesion clearly obeys the WLF relationship and can be easily predicted from viscoelastic functions. This method of using fundamental molecular properties to predict long time performance should have significant advantages in product developments.

Since the compliance for both samples are relatively high, the shear adhesion result in this study may not be universally applicable to all systems. However, the method of using data at higher temperatures to accelerate a lengthy shear creep process should be quite useful regardless of the actual detail mechanism involved.

ACKNOWLEDGEMENT

This work was done while the author was at Arco Chemical Co. Philadelphia, Penna. Technical assistance from A.C. Jankowski, D. Maynard and J. Farrow and helpful discussions with Dr. J.A. Schlademan are all gratefully acknowledged.

REFERENCES

1. C.A. Dahlquist, in Treatise on Adhesion and Adhesives, R.L.
    Patrick, Editor (Marcel Dekker, Inc. N.Y. 1969). P.219.

2. G. Krauss, K.W. Rollmann & R.A. Gray, J. Adhesion Vol. 10,
    P.221, (1979).

3. D.H. Kaelble, Transactions of the Soc. of Rheology Vol. 4
    P.45, (1960).

4. A.N. Gent and G.R. Hamed, J. of Applied Polymer Science,
    Vol. 21, P.2817 (1977).

5. L. Woo and J.D. McGhee, U.S. Patent 4,034,602 (1978).

6. L. Woo, U.S. Patent 4,170,141 (1980).

7. R.L. Blaine, P.S. Gill, R.L. Hassel and L. Woo, J. Applied
    Polymer Science: Applied Polymer Symposium 34, p.157 (1978).

8. J.D. Ferry, Viscoelastic Properties of Polymers, 2nd Ed.,
    John Wiley, N.Y., Chapter 11.

EFFECT OF MOLECULAR STRUCTURE ON MESOMORPHISM. 15[1]. THERMAL AND X-RAY DIFFRACTION ANALYSIS OF A "SIAMESE TWIN" SERIES: 4,4'-DICARBOXY-α,ω-DIPHENOXYALKANES

Anselm C. Griffin[a], Sharon Gorman[a], and William E. Hughes[b]

Departments of Chemistry[a] and Physics and Astronomy[b]
University of Southern Mississippi
Box 5043 Southern Station
Hattiesburg, MS 39406

INTRODUCTION

The homologous series of 4,4'-dicarboxy-α,ω-diphenoxyalkanes shown below have been of considerable interest in our laboratories

$$HO_2C\text{—}\bigcirc\text{—}O(CH_2)_x\text{—}O\text{—}\bigcirc\text{—}CO_2H$$

[1]

X = 2-10,12

as intermediates in the synthesis of polyester liquid crystals[2] and other diester liquid crystalline compounds[3]. In the course of our work it ocurred to us that these diacids might themselves be likely candidates to exhibit liquid crystallinity (mesomorphism) since they are conceptually related (by a tail-to-tail coupling) to the well-known liquid crystalline family of 4-alkoxybenzoic acids (below). Members of series 2 show both nematic and/or smectic C

$$RO\text{—}\bigcirc\text{—}CO_2H$$

[2]

phases depending on the length of the carbon chain. The presence
of liquid crystalline behavior in these acids is attributed to the
existence of these compounds as dimers thereby obtaining the requi-
site geometric anisotropy (large length: breadth ratio) for meso-

$$RO-\text{\textcircled{O}}-\overset{O}{\underset{O-H\cdots O}{C}}\overset{\cdots H-O}{\underset{}{}}C-\text{\textcircled{O}}-OR$$

morphic behavior. We wish to report here investigations of the
thermal behavior and solid state structure of compounds in series 1,
related as Siamese Twins to the liquid crystalline family series 2.

EXPERIMENTAL

Samples of 4,4'-dicarboxy-α,ω-diphenoxyalkanes, 1, were prepared
according to the literature procedure.[2] Optical microscopy was
performed using a Reichert Thermovar polarizing light microscope in
conjunction with a Mettler FP5/52 microfurnace. A Du Pont 990
Thermal Analyzer equipped with a 910 DSC cell was used for calorimetric measurement. Indium standards were employed for calorimetry.
Peak areas were obtained at 20°/min heating rates and quantitated
by cutting and weighing photocopies of chart paper. The DSC sample
chamber was purged with nitrogen during all runs (flow rate 50
cc/min), Samples of 1 and copies of all resulting DSC peaks were
weighed on a Mettler AD-2 microbalance. X-ray powder diffraction
was performed on samples contained in 1.5 mm quartz capillaries using
nickel filtered copper Kα radiation. Scattered intensity was
recorded on Polariod Type 157 film using a flat plate camera. X-ray
spacings were calculated from the relation $n\lambda=2d \sin\theta$ where $n=1$. A
General Electric XRD 700 diffractometer was employed.

RESULTS AND DISCUSSION

Thermal

No evidence for liquid crystallinity of compounds 1 was found.
Hot stage polarizing light microscopy indicated no fluid birefringent
phases for these materials. Differential scanning calorimetry in
conjunction with microscopy revealed only solid-solid and/or solid-
isotropic liquid transitions for these compounds. A summary of
thermal data for series 1 is presented in Table 1. Calculation of
transition entropies was made using the relation $\Delta S=\Delta H/T$. Melting
points (solid-isotropic liquid temperatures) for series 1 are plotted
versus carbon chain length in Figure 1. The odd:even alternation in

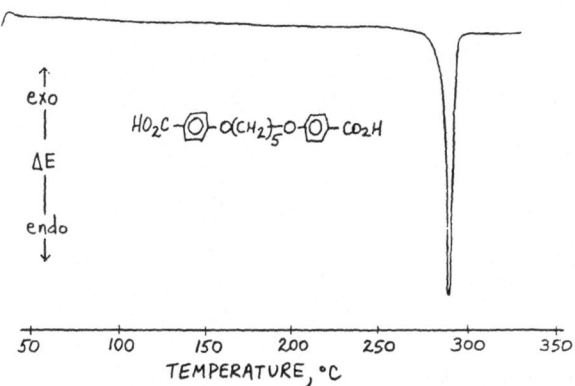

Fig. 1. Melting points (solid-isotropic liquid) for the homologous series of Twins 1.

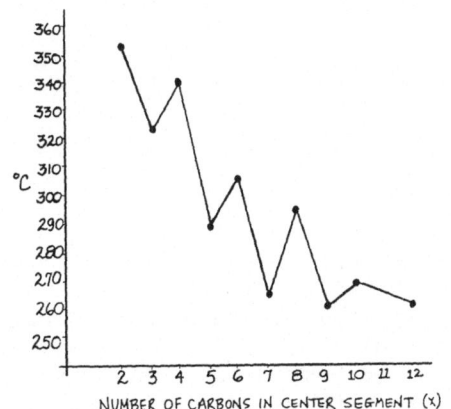

Fig. 2.   DSC curve for Twin 1 (x=5).

Table 1. A summary of thermal data for Twin series 1.

$$HO_2C-\boxed{O}-O(CH_2)_x-O-\boxed{O}-CO_2H$$

X	TRANSITION TEMP. (°C) (solid-isotropic)	ΔH (kcal/mole)	ΔS (e.u.)
2	354 (350 sh)	13.53	21.57
3	323.5	12.41	20.80
4	339.5	15.65	25.54
5	288.5	14.60	25.99
6	187 solid-solid 306	13.81	24.15
7	183 solid-solid 264	9.08	16.90
8	240 solid-solid 294	12.97	22.87
9	220 solid-solid 261	10.65	20.06
10	147 solid-solid 269	11.55	21.30
12	262	11.17	20.87

these points is readily apparent. Melting temperatures, solid-liquid crystal, for the half Twin series 2 show no such regular trend until the alkoxy chain becomes quite long. This behavior (series 2) has been explained by Bryan[4] as arising from changes in crystal structure as the homologous series is ascended. Figure 2 gives as an example the DSC curve of the x=5 member of series 1.

Plots of enthalpy and entropy changes for the melting transition for Twin compounds 1 are presented in Figures 3 and 4, respectively. These plotted points include enthalpic contributions from solid-solid transitions where applicable see (Table 1). It can be seen from these figures that there is no apparent simple relation (trend) between ΔH or ΔS and the number of carbons in the flexible segment joining the aromatic rings in 1. There is, however, a strong

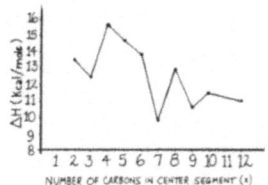

Fig. 3. Melting enthalpies for Twins 1.

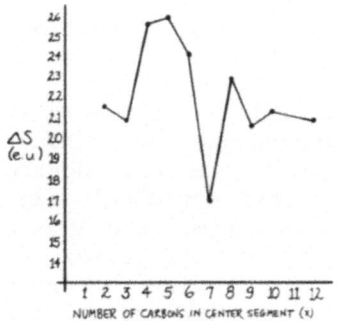

Fig. 4. Melting entropies for Twins 1.

correlation between the shapes of the ΔH line and the ΔS line with
the lone exception of x=5. By the way of comparison Figure 5
contains ΔS data for both series 1 (Twins) and series 2 (half-Twin)
compounds. Series 2 data points in Figure 5 represent summation of
all appropriate thermal transitions for these compounds, i.e. solid-
solid, solid-liquid crystal, liquid crystal-liquid crystal, liquid
crystal-isotropic liquid. Whereas points for the Twins 1 vary
irregularly with increasing carbon number, the ΔS values for half-
Twins 2 show a distinct odd:even alternation and increasing values
as the carbon number increases.

X-Ray

       In an effort to better understand the solid-state structure of
1 we undertook a room temperature x-ray powder diffraction study.
Results are presented in Table 2. The only diffraction ring of
significant intensity for all of the Twins was the wide angle ring
shown in the table indicating a relatively amorphous solid phase for
these samples having no previous thermal history. (For comparison
a series 2 half-Twin compound (R=C_7H_{15}) was examined. Several
diffraction rings were evident for this material. Results for both
thermally untreated and melt crystallized samples were nearly iden-
ticle for 2(R=C_7H_{15}). In the wide angle region 5.9, 4.9 and 4.4
Angstrom distances were found.) We feel the distances for Twins 1
are consistent with polymeric hydrogen-bonded structures. These
twins are difunctional and due to the known solid-state association
of carboxyl groups[4] in 2 one would expect polymeric associations to
obtain for the Twin diacids.

       The lack of crystallinity in these materials is ascribable in
part to the lack of independent head-to-head dimers of the type
found in series 2 mesogens. The large degree of "intermolecular"
hydrogen bonding in a polymer would help explain the high melting
points, somewhat broad endotherms, and loss of liquid crystalline
behavior in going from series 2 to series 1 compounds. Figure 6
depicts idealized schematic drawings of the two extreme hydrogen-
bonded polymeric forms of 1. The top structure, a, represents a
closed, 'linear' polymer; the bottom structure, b, represents an
open, 'network' polymer. We are hard put to choose between these
models and perhaps a dynamic equilibrium and/or a static combination
of both forms actually exists for these materials.[5] We prefer,
however, model b since it should be favored entropically and also
would lead to x-ray diffraction rings at large angles which are somewhat smaller than those found in the liquid crystalline phase of
series 2 compounds such as do the Twins 1 (x=2-9). For Twins 1
(x = 10, 12) this distance falls somewhat (see Table 2) to values
approaching what would be predicted for structures such as (a) in
Figure 6.

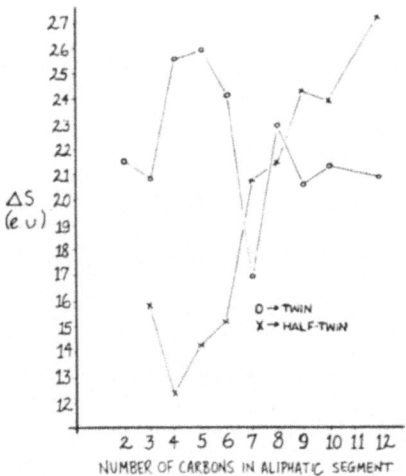

Fig. 5.   Comparison of total entropies for Twins 1 and half-Twins 2.

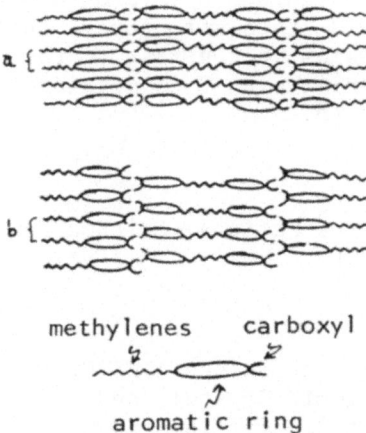

methylenes   carboxyl

aromatic ring

Fig. 6.   Schematic drawing for idealized hydrogen-bonded polymers of
          Twins 1; a) closed 'linear ploymer and b) open 'network'
          polymer.

Table 2.   X-ray powder diffraction data for series 1.

$$HO_2C\!-\!\boxed{O}\!-\!O(CH_2)_x\,O\!-\!\boxed{O}\!-\!CO_2H$$

X	Å	X	Å	
2	5.06	7	4.98	
3	5.12	8	5.00	Obtained at room temperature and calculated from the relation $n\lambda = 2d\sin\theta$.
4	4.95	9	5.00	
5	4.95	10	4.78	
6	5.05 4.86	12	4.3	

CONCLUSIONS

The series of compounds 1, the 4,4'-dicarboxy-α,ω-diphenoxy-alkanes, does not exhibit liquid crystallinity as might be predicted from structural considerations. These Twins do however show a strikingly regular odd:even alternation in solid-isotropic liquid melting points. Melting enthalpies and entropies do not vary so regularly but are intimately related to each other. X-ray diffraction from powder samples suggests that at least for 1(x=2-9) an open, 'network' polymer held together by hydrogen bonds is a significant contributor to the solid state structure. The lack of liquid crystallinity in the Twins in contrast to the ubiquitous liquid crystallinity in the half Twins 2 is ascribed to the polymeric nature of the solid state.

Acknowledgement

We wish to thank the National Science Foundation (DMR 8115703) for support of this work.

REFERENCES

1. For part 14 in this series see A.C. Griffin, G.A. Campbell and W.E. Hughes in "Liquid Crystals and Ordered Fluids, Vol. 4", J.F. Johnson and A.C. Griffin, eds., Plenum Press, NY, NY, in press.
2. A.C. Griffin and S.J. Havens, J. Polym. Sci. Polym. Phys. Ed., 19:951 (1981).
3. A.C. Griffin and S. Gorman, to be published.

4. R.F. Bryan, P. Hartley, R.W. Miller and M.-S. Shen, Mol. Cryst.
    Liq. Cryst., 62:281 (1980).
5. A dynamic equilibrium for 'closed' dimers and 'open' hydrogen-
    bonded structures has been proposed for the liquid crystalline
    phase of an 4-alkoxybenzoic acid. See E.F. Carr and L.S. Chou,
    J. Appl. Phys., 44:3365 (1973).

REVERSIBILITY AND CAPACITY OF DIFFERENT

ENERGY STORAGE REACTIONS BY CALORIMETRY

Pierre Le Parlouër
SETARAM
101-103 rue de Sèze
69541 Lyon Cedex 6
France

Solar and nuclear energies are the most interesting energies to be stored. Up-to-now, different types of energy storage have been tested, and some are practically used, especially for house heating. All the types of calories recuperation which are under study, are connected with the idea of reaction reversibility : melting – solidification for the latent heat storage, dehydration-hydration and absorption-desorption for the reaction heat storage, solid-solid transition for the transition heat storage... But the efficiency of the chemical materials used in energy storage is very variable. The heat stored by means of a reaction in a chemical system is much larger than a melting heat. A precise investigation of the different parameters which determine the reversibility and capacity of energy storage systems, is needed before an industrial application of such or such chemical system.

Calorimetry gives a powerful method for the simulation of storage reactions. The main thermodynamical parameters which are needed can be provided by the calorimetric test : latent heat, specific heat, reaction heat...

Compared to the DSC method, the calorimeter works on large samples at low scanning rates, which provide a better simulation of the thermal cycles occuring during an energy storage operation (1). For the investigation of absorption reactions, a vapour absorption device is connected to the calorimeter.

The use of calorimetry in the energy storage field is illustrated by means of different types of reaction :
- for low temperature storage : hydrated salts melting, reaction between two solids, absorption of a liquid on a solid
- for mean temperature storage : melting of salt mixtures having eutectics, solid-solid transition.

CALORIMETER DESCRIPTION

The SETARAM C 80 calorimeter is particularly well adapted for the investigation of thermal storage reactions. It is a modern type of Calvet microcalorimeters with a large sample volume (Fig. 1) (2).

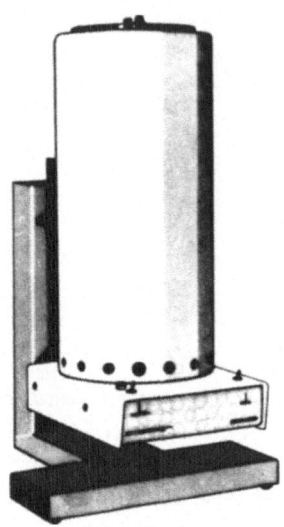 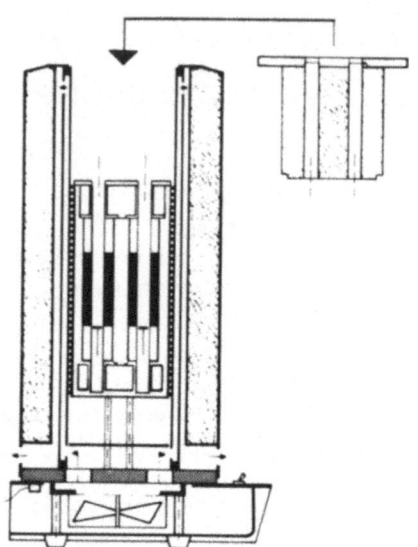

Fig. 1 - C 80 calorimeter with its   Fig. 2 - C 80 section
          reversing system.

The C 80 calorimeter is composed of two identical and symmetrical thermal fluxmeters which are set in a thermoregulated block (Fig. 2). The experimental cells which contain respectively the sample and an inert material, are completely surrounded by the fluxmeters.

The differential measurement is run in the isothermal way or in the scanning way ($2°C.mn^{-1}$ maximum), in a wide range of temperature (ambient up to $300°C$).

The instrument can be fitted on a reversing system (Fig.1) which is driven by an electric motor, in order to have a continuous reversing. This device is very helpful to perform a good homogeneity in solid mixtures used for energy storage.

Various experimental fittings can be adapted on the C 80 calorimeter, especially a vapour absorption device (Fig. 3) which allows to simulate the cycle of an absorption heat pump.

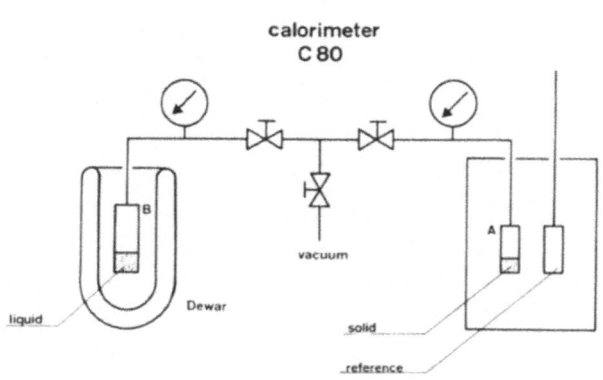

calorimeter
C 80

Fig. 3 - Vapour absorption device

Two vessels A and B are respectively containing an anhydrous solid and a liquid. The vessel A is initially isolated from the vessel B, in order to regenerate the solid under vacuum. On the other side, the liquid (in B) is set in a Dewar vessel containing liquid nitrogen. After liquid solidification, the Dewar vessel is replaced by a liquid bath at a constant temperature. It gives a constant vapour pressure of the solvent in vessel B. Then, the vapour is introduced on the anhydrous solid in vessel A, for the absorption.

EXAMPLES OF CALORIMETRIC INVESTIGATIONS

1 - Latent Heat storage at low temperature : hydrated salts

The latent heat is the energy evolved by a material when it undergoes a structure change. The effective storage of this heat requires the use of materials with large melting heat, that can be recovered during solidification. Both operations are practically done at constant temperature and the energy evolved heats or cools the surroundings. For latent heat storage below 130°C, hydrated salts are frequently used. Mostly, they are tightly encapsulated, in order to prevent evaporation or contamination of the material.

A calorimetric experimentation is run on a sample of $MgCl_2$-6 H_2O. The material is filled in a glass ampoule which is sealed under argon (sample mass : 1,715 g).

Then the sample is heated up in the calorimeter at a slow scanning rate (0.2°C.mn^{-1}). The hydrated salt begins to melt at about 90°C (Fig. 4). When cooling the calorimeter, a supercooling of the salt is detected. The solidification starts at 84°C.

The corresponding latent heats are 25.6 cal.g^{-1} for melting and 24.8 cal.g^{-1} for solidification.

From such a calorimetric test, it appears that the hydrated salt $MgCl_2$-6 H_2O does not considerably supercool. It is not necessary to add some other crystals to limit this supercooling. The little degradation observed in the enthalpy of transition is certainly due to a very slow rate of recombination.

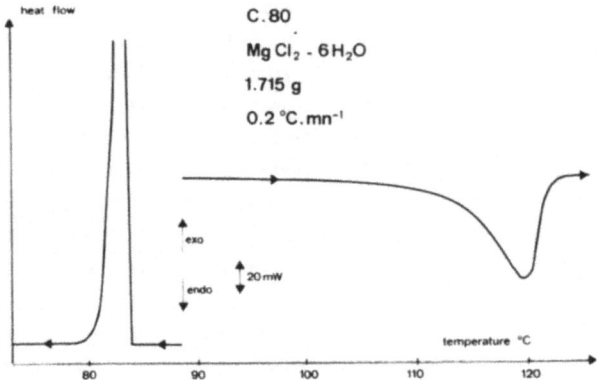

Fig. 4 – Melting and Solidification of
        $MgCl_2$ – 6 H_2O

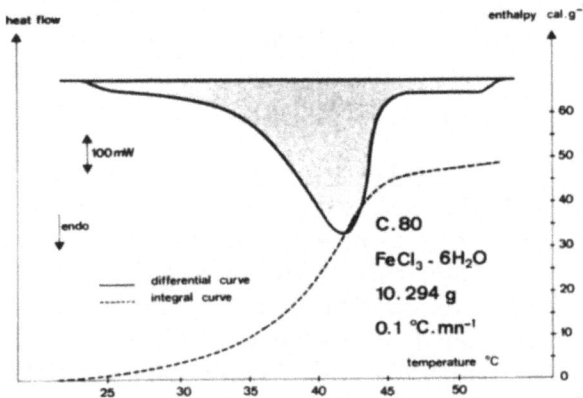

Fig. 5 – Differential and integral curves of
        $FeCl_3$ – 6 H_2O melting.

Whereas $MgCl_2$ – 6 H_2O works as a storage medium above 70°C, the hydrated
salt $FeCl_3$ – 6 H_2O melts in a lower temperature range (between 30°C and
47°C) (Fig. 5), that will be more suitable for some energy storage systems.

The calorimetric test run with a very large mass of sample (10.294 g), allows
to calculate the enthalpy by successive integration of the calorimetric curve.
This integral curve which takes in account the specific heat of the sample and
its heat of transition, gives the quantity of heat which can be stored for a defined range of temperature. In the case of $FeCl_3$ –6 H_2O, the amount of heat
stored between 20°C and 50°C is equal to 48 cal.g^{-1}.

Its limitation in energy storage application is due to its considerable supercooling.

2 – <u>Latent Heat storage at medium temperature : salts mixtures with eutectics.</u>

   Some salts mixtures, especially nitrates and nitrites of sodium and potassium, are investigated for heat storage (3,4). These relatively inexpensive salts have eutectics with low melting points, and are already used as heat transfer fluids in different manufacturing industries.

   The mixture nitrates of sodium and potassium shows only one eutectic at 222°C, for the following composition : 46 % $NaNO_3$ – 54 % KNO_3.

   A relatively large sample (850 mg) is investigated by calorimetry at 1°C.mn^{-1} (Fig. 6). The eutectic melts in a narrow range of temperature. When cooling, the mixture does not supercool. The corresponding transition heats are equal to 24.2 cal.g^{-1} for melting and solidification.

   The calorimetric test shows that such a system is interesting for medium temperature storage because it can work with a good reversibility, without heat degradation during the thermal cycle.

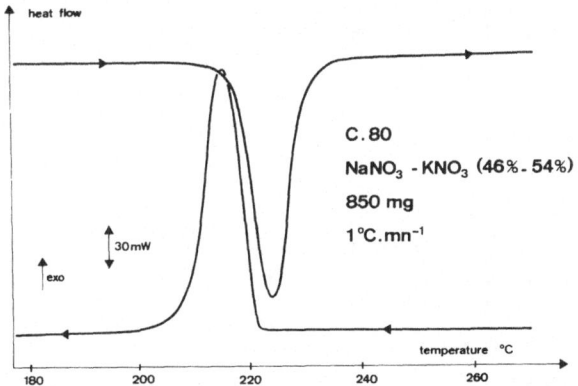

Fig. 6 – Melting and solidification of $NaNO_3$– KNO_3 eutectic.

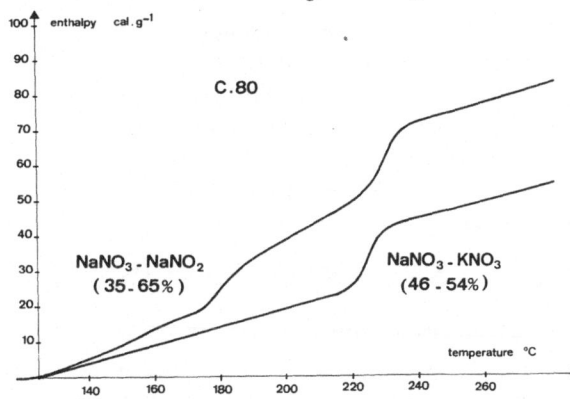

Fig. 7 – Enthalpy curves of $NaNO_3$ – KNO_3 and $NaNO_2$ eutectics.

Other salts mixtures having eutectics are also suitable for this type of energy storage. On Fig. 7, the enthalpy curves of $NaNO_3$ – KNO_3 eutectic are compared with a $NaNO_3$ – $NaNO_2$ mixture (35 % – 65 %).

In this case, it is useful to consider and compare the amount of heat stored over a temperature range encompassing melting temperatures. For example, between 200°C and 250°C, the enthalpies are respectively 28 cal.g^{-1} for $NaNO_3$ – KNO_3 eutectic and 36 cal.g^{-1} for $NaNO_3$ – $NaNO_2$ mixture.

When comparing the different materials, the energy storage capabilities of salts mixtures and hydrated salts are very similar, but in different temperature ranges of use.

3 – Reaction Heat storage : solid – solid materials

The storage of chemical heat consists in using the reversible endothermic and exothermic reactions of an chemical equilibrium. The principle of such a storage is given on Fig. 8 (5).

The heat coming from a source is used to displace the equilibrium on the endothermic side : that is the storage level. Then the different materials are cooled and separately stored.

The realisation of the reversible exothermic reaction allows to produce the heat that has been stored and at the same time, to regenerate the materials of the first reaction.

An example of reaction heat storage is simulated through the following calorimetric investigation. Two inorganic solid compounds, hydrated baryta and potassium nitrate, are mixed, and a large quantity of sample (17.40 g) is introduced in a calorimetric vessel. The expected reaction is the following :

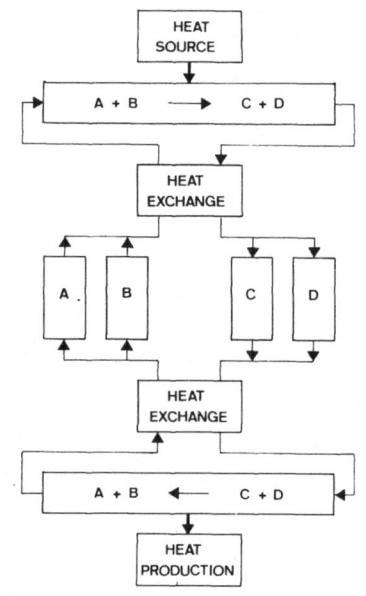

Fig. 8 – Reaction heat storage principle.

$$Ba(OH)_2 . 8 H_2O + 2 KNO_3 \rightleftharpoons Ba(NO_3)_2 + 2 KOH + 8 H_2O$$

So, the test has to be run in a closed vessel.

When the mixture is heated at a low scanning rate (0.1°C.mn^{-1}) the endothermic reaction starts at 68°C (Fig.9). The corresponding heat is equal to 54 cal.g^{-1}. The exothermic reaction, when cooling, is detected at a lower temperature (63°C) and the quantity of heat released (33 cal.g^{-1}) is much smaller.

If the mixture is heated again, it appears that there is a degradation in the quantity of heat which is evolved (36 cal.g^{-1}).

This calorimetric test shows that the reaction is reversible, but only 61 % of the energy which has been stored is recovered during cooling. But it will be also necessary to work with a slower scanning rate, to know its influence on the efficiency of the reaction.

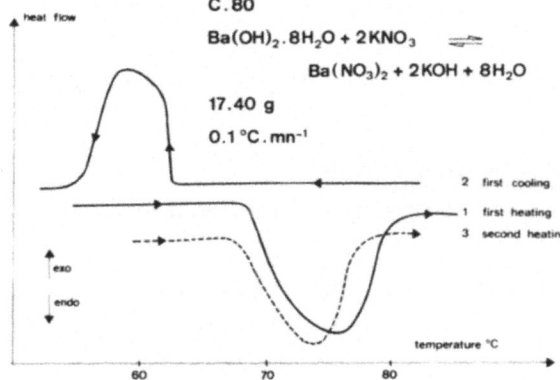

Fig. 9 - Hydrated baryta + potassium nitrate : reversibility of the reaction.

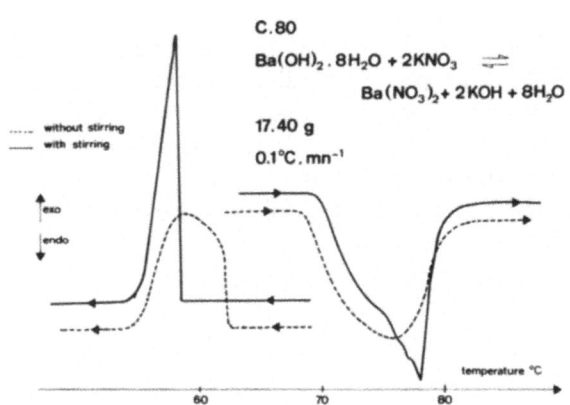

Fig. 10 - Hydrated baryta + potassium nitrate : effect of stirring.

In order to have a better homogeneity of the solid mixture during the reaction, a stirring is applied by means of the reversing mechanism of the calorimeter.

The shape of the thermal peaks are quite different (Fig. 10). The endothermic reaction occurs in the same range of temperature, but the exothermic reaction is shifted at a lower temperature.

In the following table, the values of heat stored or released during the cycle are compared, with or without stirring.

	Heat (cal.g^{-1})	
	without stirring	with stirring
1st Heating	54	55
1st Cooling	33	26
2nd Heating	36	30

For such a reaction, the stirring does not increase the efficiency of the system.

4 – Reaction Heat storage : solid – liquid materials

The principle of the reaction heat storage using a solid and a liquid is based on the cycle absorption-desorption.

This operation follows the cycle of an absorption heat pump, which is composed of two vessels under vacuum, separated by a valve (Fig. 11) (5).

In such a device, the liquid must have a large enthalpy of vaporization and the solid must be able to absorb large quantities of liquid.

The reversibility and efficiency of this heat pump can be calorimetrically simulated, when using the absorption vapour device which is described in the "calorimeter description" chapter. Its use is illustrated by the following example :

System Sodium sulfite/water.

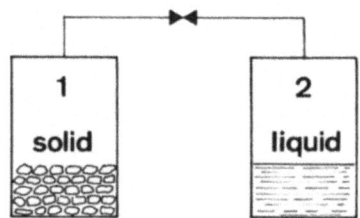

$$Na_2S . 5H_2O \rightleftharpoons Na_2S + 5H_2O$$

$$H_2O \text{ liq} \rightleftharpoons H_2O \text{ vap}$$

Fig. 11 – Absorption heat pump principle.

The system sodium sulfite – water has been developed by a Swedish company TEPIDUS and is industrially used for house heating. It works as follows :

(1) $Na_2S, 5 H_2O \rightleftharpoons Na_2S + 5 H_2O$ (2) $H_2O \text{ liq} \rightleftharpoons H_2O \text{ vap}$

The reaction of $Na_2S, 5 H_2O$ dissociation corresponds to the storage of heat in vessel (1). At the same time, the water vapour condensates, in vessel (II), giving its condensation heat to the external surroundings.

When the formation of Na_2S, 5 H_2O occurs, the heat is released in vessel (I), and the vessel (II) needs an external source to vaporize the water.

To simulate these reactions, a sample of anhydrous sodium sulfite (116.5 mg) is initially put under vacuum at 150°C during a night.

In the other vessel, water is initially frozen, and then placed in a thermostated bath at 13°C. The corresponding vapour pressure is 11 torr. The temperature of the calorimeter is fixed at 30.5°C.

The absorption of water on sodium sulfite is performed by opening the valve between the two vessels. The exothermic reaction is very energetic (Fig. 12) and corresponds to 238 cal.g^{-1}. By weighing the sample before and after absorption, the mole number of water fixed on sodium sulfite is determined. For the present test, it is equal to 4.4 moles of water for one mole of Na_2S.

The stable form of Na_2S, 5 H_2O is not reached, in this case, due probably to an insufficient regeneration of Na_2S.

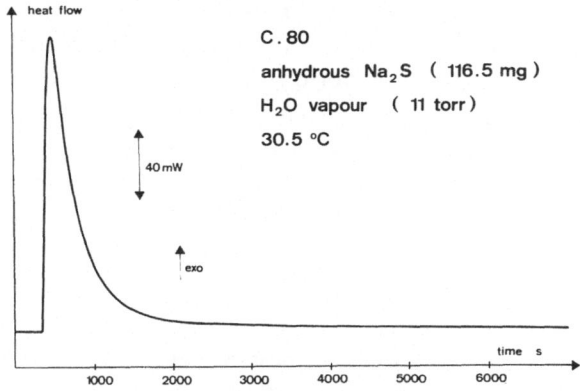

Fig. 12 – Absorption of water vapour on Na_2S.

Then, the valve being always opened, the sample is heated at 0.5°C.mn^{-1} up to 120°C. So the desorption occurs under a reduced water vapour pressure. The vessel II is put into ice, in order to condensate the desorbed water vapour. The thermogram (Fig. 13) shows a large endothermic peak between 60°C and 120°C.

Compared with the other types, the reaction heat storage allows to store very large quantities of heat, especially with a system like Na_2S - H_2O which works in an absorption range of 0 to 5 moles.

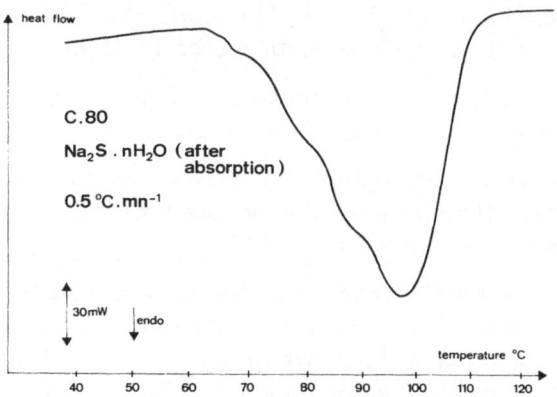

Fig. 13 – Thermal desorption of Na_2S, n H_2O.

5 – Transition Heat storage : pentaerythritol

The transition heat storage uses the reversibility of a solid-solid transition in typical materials. This type storage is less developed, but it can be interesting, because the material remains solid during the whole thermal cycle. There is no more problem of viscosity or corrosion.

Pentaerythritol undergoes such a transition at about 180°C.

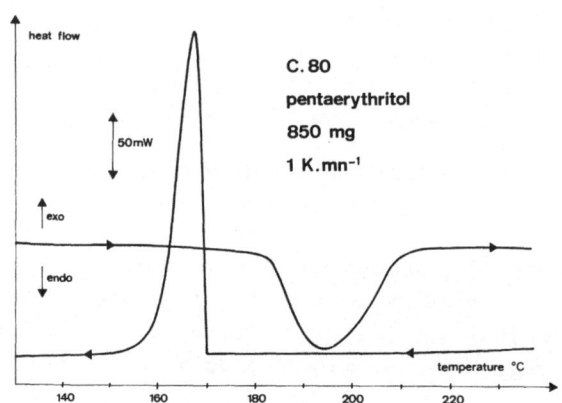

Fig. 14 – Solid-solid transition of pentaerythritol

A sample of this organic material (850 mg) is investigated at 1°C.mn^{-1} in the temperature range 140°C - 230°C (Fig. 14). The transition is reversible with a little temperature shift for the exothermic effect. The quantities of heat evolved are 70 cal.g^{-1} during heating and 65 cal.g^{-1} during cooling. In spite of a little heat degradation during the cycle, the amount of heat to be stored is similar to the quantity obtained with salts mixtures.

CONCLUSION

Through the different examples illustrating this paper, the calorimeter appears to be a powerful tool in the energy storage research. The reversibility and efficiency of various systems are rapidly investigated. The different thermodynamical parameters, useful for the definition of an energy storage unit, are also easily measured.

BIBLIOGRAPHY

1. S. Cantor, Thermochimica Acta, 26, 39 (1978)
2. P. Le Parlouër, "Thermal Analysis", Vol.1, Proceedings of the 7th ICTA conference (B. Miller Ed) p. 190-195, John Wiley and Sons (1982)
3. C.M. Kramer, Z.A. Munir, J.V. Volponi, Thermochimica Acta, 55, 11 (1982)
4. E.A. Dancy, Thermochimica Acta, 58, 53 (1982)
5. S. Elberg, P. Mathonnet, Conference " Le Génie Chimique et le stockage de l'énergie", Paris, Décembre 1980.

POWER SCANNING CALORIMETRY (PSC)

Joeseph Hakl

Sandoz AG
4002 Basel
Switzerland

INTRODUCTION

The heat of reaction and the rate of heat production in a reaction mixture as a function of temperature are important quantities for the design of reactors in chemical industry. Presently, several methods for the determination of these quantities are available, such as Differential Scanning Calorimetry, Differential Thermal Analysis, Bench Scale Calorimetry /1/ and adiabatic calorimetric methods.

The principal advantage of thermoanalytical micromethods is the prompt availability of the required data and small quantities of the educts necessary for the experiments. These methods, however, have many serious limitations and disadvantages: e.g. a strictly limited choice of the reaction mixture receptacles, lack of stirring, and the impossibility of addition of chemicals during the measurement; generally they can rarely take into consideration the macrokinetics.

On the other hand, bench scale calorimetric methods achieve a high level of plant conformability, but they have various disadvantages, such as limited flexibility in the choice of reactors and sometimes large volumes necessary to obtain reasonably accurate results; the large volumes comprise, e.g. an increased hazard level with operations under pressure. Moreover, they are usually time consumming and the instruments are expensive.

Among the best known disadvantages of the adiabatic methods is the impossibility to control the temperature, resulting sometimes in an uncontrollable run-away reaction and indirect availability of the required data /2/, thus entailing reduced accuracy and rapidity.

The scope of the PSC method is to eliminate the disadvantages of the methods listed above, and to provide reasonably accurate data on the heat production, under plant conformable conditions, to allow a safe design of production reactors.

PRINCIPLE

The reaction mixture is introduced in a suitable container fitted with a stirrer and an electric heating coil fed by a controlled power source. The ambient temperature is kept at a selected constant temperature lower than the temperature of the reaction mixture under investigation. The task of the controller is to keep the first derivative of the reaction mixture temperature with respect to time constant, i.e. to increase the sample temperature in a linear fashion. If there are no processes with heat effects taking place in this mixture, the electric power to achieve the conditions described above, remains constant. This electric power is to be measured and recorded on a suitable device.

At the temperature at which the reaction starts producing its own heat, the electric power necessary to control the linear rise of the mixture temperature is reduced, since the reaction contributes to the supply of necessary energy.

The difference between the original (constant) electric power and the reduced power is the value of the rate of heat production of the investigated reaction measured directly, i.e. obtained instantaneously and therefore exhibiting the least error. The values of the power differences (which equals the rate of heat production of the investigated reaction) plotted logarithmically versus reciprocal temperature in Kelvins, forms the Arrhenius plot, and can be directly used for the design of production reactors.

By a simple integration of the power difference curve, the overall heat of the studied reaction can easily be obtained.

EXPERIMENTAL EXAMPLE

The practical applicability and accuracy of the PSC method can be demonstrated by the examples of the thermal decomposition of aromatic diazonium salts in sulfuric acid, (important intermediates for the production of dyestuffs). The diazotation of both the amines was carried out by nitrosylsulfuric acid.

Example 1

123.8g of the reaction mixture after the diazotation were introduced in a beaker and placed in a SEDEX oven /3/. The temperature of this mixture was set at 50°C, the ambient (oven) temperature at 42°C, thus resulting in a temperature difference of 8°C. The stirrer speed 600 RPM and the heating rate selected was 20°C/h. Under the conditions described above, the electric power feeding the heating coil levels out at 5.1 W. The temperature increases in a linear fashion with the preset heating rate. Below 115°C, the power of the heating coil remains constant, thus indicating no reaction with heat effects is taking place in the reaction mixture. At 115°C, the electric power starts to decline – the diazonium salt mixture starts to decompose exothermically at a measurable rate. With the increasing temperature, the electric power continues to decrease as a result of the increasing contribution of the heat of decomposition. After reaching a minimum at 150°C, the power starts increasing as the reaction begins to abate. At about 170°C, the heating coil power reaches its original value, as the contribution of the reaction to the energy supply sinks to zero. The reaction is over. The difference between the origianal electric power and the reduced power at a selected temperature is the experimentally detected value of the rate of heat production of the studied reaction at that temperature (see Table 1, column dQ/dt(exp.)). Considering the decreasing concentration of the educt (see Table 1, C(norm.)), the true rate of heat production of this reaction can easily be obtained (Table 1, dQ/dt corr.). The values dQ/dt(corr.) plotted logarithmically vs. reciprocal temperature in Kelvins forms a straight line as required by the theory (see Fig. 1). The analogous data obtained by DSC are shown for comparison in the same diagram.

Table 1. Rate of heat Production Data

Temperature (°C)	dQ/dt(exp.) (W)	C(norm.)	dQ/dt(corr.) (mW/g)
120	-0.17	0.994	-1.38
125	-0.32	0.983	-2.63
130	-0.64	0.950	-5.82
135	-1.14	0.888	-11.74
140	-1.78	0.784	-22.41
145	-2.12	0.642	-37.59
165	-0.60	0.014	-335.47

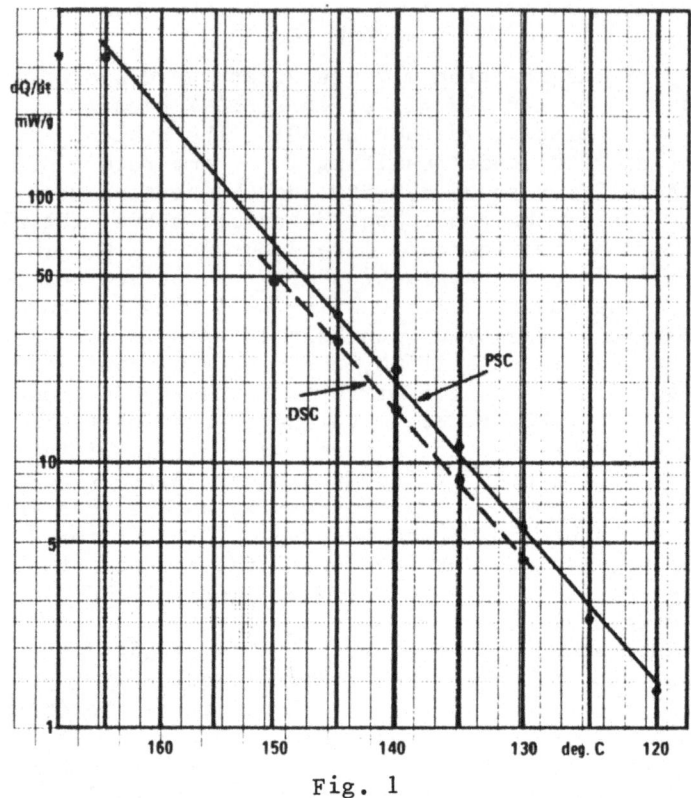

Fig. 1

Example 2

121.6g of another diazonium mixture was investigated under
the same conditions as described above. The initial temperature
of the scan was selected at 0°C, as this salt starts decomposing
at approximately 30°C. The evalutiaon of the PSC curve yielded
-74.8 J/g. Considering the specific heat of this mixture, estimated
to be (1.46 J/g.K) such heat would inflict, under adiabatic conditions, a temperature increase of 51°C. The adiabatic temperature increase observed with this reaction mixture in a Dewar
flask yielded 53°C.

DISCUSSION

The examples described above are seemingly not ideal to
demonstrate the usefulness of the presented method. However, they

were chosen intentionally to enable the verification of its accuracy by conventional methods under the conditions at which these
methods are supposed to supply reasonably correct results. Both
the studied reaction mixtures are homogeneous solutions and as
decompositions they may be supposed to obey first order kinetics.
The slight difference between the data found by DSC and PSC, as
seen in Fig. 1, are evidently due to a higher sensitivity of the
PSC, which also allows to detect the onset of exothermic reactions
at lower temperatures. Example 2 was chosen to make possible a
reasonable comparison with such a simple method as an adiabatic
measurement in a Dewar flask in the temperature range where such
measurements involve the least error.

An interesting variant of the PSC method is isothermal configuration. In this case, the reaction mixture temperature as
well as the ambient temperature, however several degrees lower,
are kept constant. One educt is placed in the receptacle and a
second one is (possible continuously) added. If the intended reaction is exothermic, the electric power feeding the heating coil
starts declining and, again from the power curve, can easily be
seen the rate of heat production of the studied reaction (e.g. as
a function of educt addition rate at a selected temperature). By
an integration of the power curve, the heat of this reaction is
easily obtained.

REFERENCES

1. F. Brogli et al.: "Assessment of Reaction Hazards by
    means of a Bench Scale Heat Flow Calorimeter",
    Proc. of the Symposium Runaway Reactions, Unstable
    Products and Combustible Powders, Chester, England,
    March (1981).

2. P. Hugo, W. Schaper: "Bestimmung kinetischer Daten von
    Flussigphase-Reaktionen mit dem adiabatischen
    Messverfahren", Chem. -Ing. -Tech. 51(1979), No. 8,
    p. 805.

3. J. Hakl: 11th NATAS Conference, New Orleans, Louisiana,
    U.S.A., October (1981).

A COMBINED TG-GC-MS SYSTEM FOR MATERIALS CHARACTERIZATION

Jen Chiu

Polymer Products Department, Experimental Station
E. I. du Pont de Nemours & Company, Inc.
Wilmington, Delaware 19898

ABSTRACT

Thermogravimetry is a powerful technique for studying
stability and determining composition of a variety of materials.
However, it derives information based on monitoring weight losses
as a function of temperature and does not identify the volatilized
components. Its capability for materials characterization is
greatly enhanced if other analytical techniques are combined with
it to analyze the off-gases at various weight loss steps. This
presentation describes a coupled thermogravimetry-gas chroma-
tography-mass spectrometry (TG-GC-MS) system based on the Du Pont
951 thermogravimetric analyzer and a Hewlett-Packard computerized
GC-MS instrument. Examples will be given to illustrate the
features of such a combination.

INTRODUCTION

Thermogravimetry (TG) is a powerful technique for materials
characterization based on continuous measurement of weight changes
upon controlled heating in a controlled environment. Its application on stability studies and compositional analyses are well
known. However, this technique provides mainly weight change
information and does not identify the nature of volatiles lost or
the residues remaining. It also does not detect any structural
transformations if there is no weight change. Attempts have been
made to couple other analytical techniques to TG in order to
enhance its capability, most notably infrared spectroscopy (IR)
(1,2), mass spectrometry (MS) (3-6) and titrimetry (7-9). These
coupled systems have shown great promise in analysis of volatile

197

substances evolved from TG, resulting in a wide variety of applications. Unfortunately, in many practical problems, the volatiles are complex mixtures not easily handled by these analytical methods. One logical solution is to couple a separations tool such as gas chromatography (GC) to TG (10-11). To go one step further, an ideal system would include a TG apparatus as a sample conditioner or reactor, a GC unit to perform separations and then an identification instrument. Such a combination has been demonstrated by combined TG-GC-IR (12) and TG-GC-MS (13), and is expected to gain wider use since coupled GC-IR and GC-MS are now commercially available. This presentation describes our combined TG-GC-MS system which can be alternately used for TG-MS, and illustrates its features with several examples.

EXPERIMENTAL

A schematic diagram for the instrumentation is shown in Figure 1. A Du Pont 990 or 1090 thermal analyzer equipped with a 951 thermogravimetric analysis module (Analytical Instruments Division, Du Pont Company, Wilmington, DE) is coupled with a Hewlett-Packard 5710A gas chromatograph and a 5982A mass spectrometer (Hewlett-Packard, Palo Alto, CA). A dedicated Hewlett-Packard 21MX computer system combined with a Tektronix 4012 CRT and a Tektronix 4610 copier is used for GC-MS data acquisition and computation.

The GC-MS interface is a standard jet separator provided by Hewlett-Packard. The TG-GC interface is a U-shaped condensation trap made of 1/8-inch O.D. stainless steel tubing and connected to a 6-port microvalve (Cat. No. 5521, Carle Instruments, Fullerton, CA) previously used for coupling TG to MS (3) and shown in Figure 2. This interface is connected to the GC injection port through a 1/16-inch stainless steel tubing silver-soldered to a syringe needle. In the TG-GC-MS operation mode, the injection port is connected through a GC column and the jet separator to the MS inlet A. The volatiles from TG are condensed in the trap by liquid nitrogen during sampling along the weight loss curve. The collected sample is injected onto the GC column upon rotating the microvalve by a compressed air actuator (11), and heating the trap with a hot flame. This method does not provide an injection as fast as that previously reported (11), thus reducing the column efficiency, but is simpler to operate, and adequate for most applications. A 6 ft x 1/8-inch diameter stainless steel column packed with 80/100-mesh Porapak Q was used for the brown speck work, and a similar column packed with 80/120-mesh Polypak #1 (Hewlett-Packard) for the cyclohexanol desorption work. Helium was used as carrier gas at a flow rate of 20 ml/minute. The injection port temperature was 250°C and the column was programmed from 50° to 250°C at 8°C/minute for the brown speck problem. For

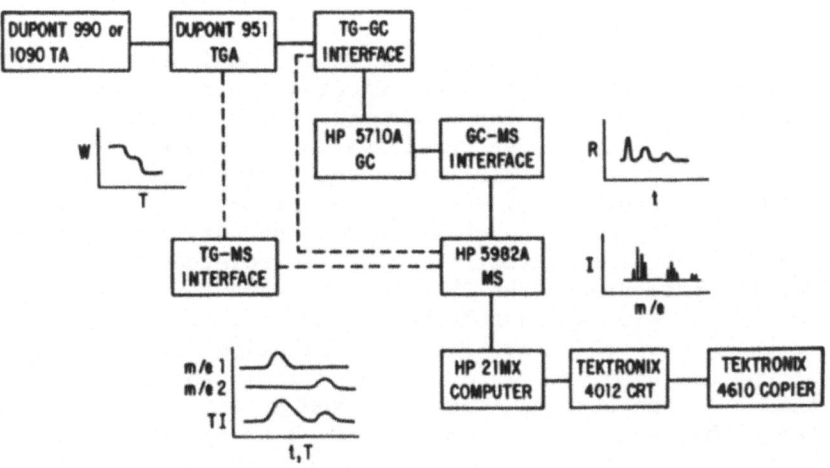

Fig. 1. Schematic Diagram of Combined TG-GC-MS System.

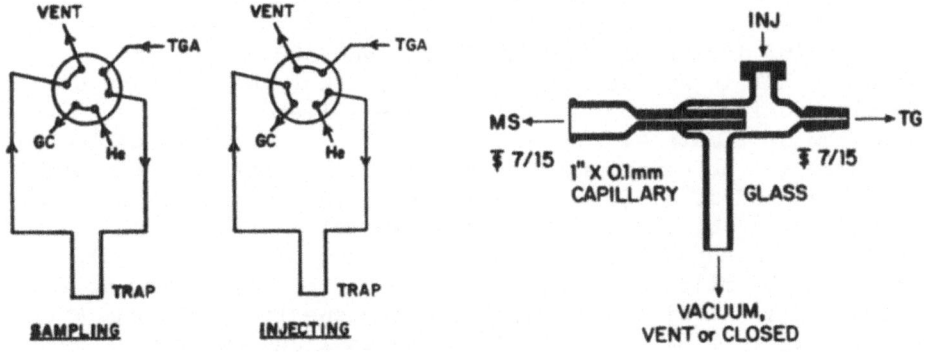

Fig. 2. TG-GC Interface.   Fig. 3. TG-MS Interface.

the cyclohexanol problem, the column was heated at 8°C/minute from
100° to 240°C, and then maintained at 240°C for another 5 minutes.
The GC scan was plotted as total ion intensity vs time as detected
by the MS. TG was operated in a standard fashion. The sample
size ranged from less than 1 mg to above 10 mg dependent on the
problem. The heating rate was typically 10°/minute.

In the TG-MS operation mode, the GC column is replaced with
an empty tubing which connects the injection port to the MS inlet
B directly without a molecular separator. We have found the condensation trap still the most sensitive and reliable interface for
TG-MS coupling. However, this method does not provide continuous
monitoring of evolved gases. One of the more successful interfaces
for continuous monitoring under various atmospheric conditions is a
leak valve (5) or similar restrictions (see references in Ref. 4)
to reduce the pressure of the TG effluence to a level compatible
with that of the MS. We have found the glass interface shown in
Figure 3 to be very effective for TG-MS coupling. It is simple
to construct and easy to maintain. The 0.1-mm capillary leaks a
small amount of the evolved gases into the MS while the main part
is vented. The capillary is easily cleaned by a suitable solvent
or by hydrofluoric acid. An additional port is added for direct
injection of a substance for calibration and evaluation of the MS.

RESULTS AND DISCUSSION

Combined TG-GC-MS

This system was employed to analyze a brown speck removed from
a molded resin part. The total sample available was less than 1
mg. As shown in Figure 4, the TG scan showed two weight loss steps
of 6% and 76% in the vicinities of 300° and 500°C, leaving a
residue of 18% in an inert helium atmosphere. The GC scans of
volatiles collected from weight loss steps 1 and 2 are shown at
the bottom of Figure 4. Since all the mass spectra are collected
and stored in the computer, we can recall any mass spectrum along
the GC scan and identify the volatiles at that point. Thus, the
major peak for cut #1 at a retention time of 4 minutes was identified as formaldehyde from spectrum #134 shown in Figure 5(A).
Similarly, the major peak in cut #2 with a retention time of 2.5
minutes was identified as tetrafluoroethylene from spectrum #81
shown in Figure 5(B). Since both polyacetal and polytetrafluoro-
ethylene depolymerize essentially completely into their respective
monomers, and the weight loss temperatures are consistent with
those of the decomposition temperatures of these two resins, we
conclude that the brown speck consists of 6% polyacetal, 76% PTFE
and 18% inorganic residue. Probably the brown speck was a
partially degraded resin from overheating.

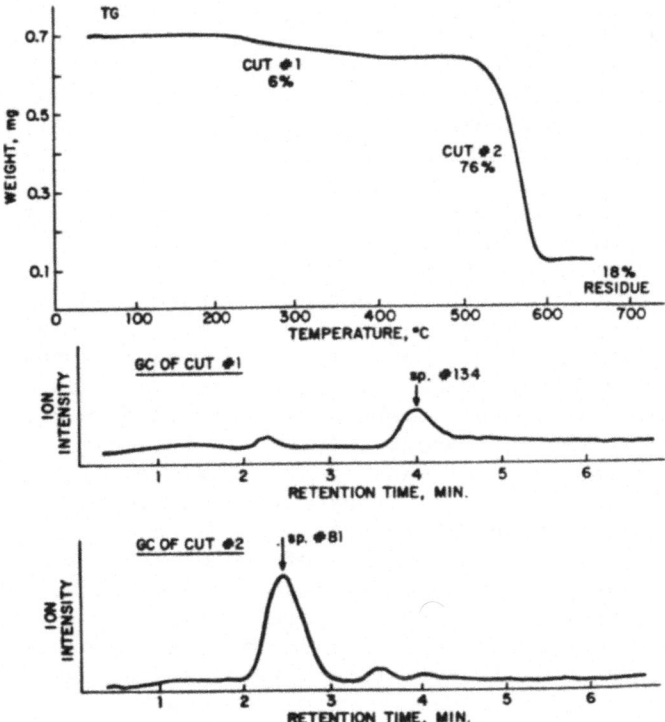

Fig. 4. TG-GC-MS Analysis of a Brown Speck from a Resin Part

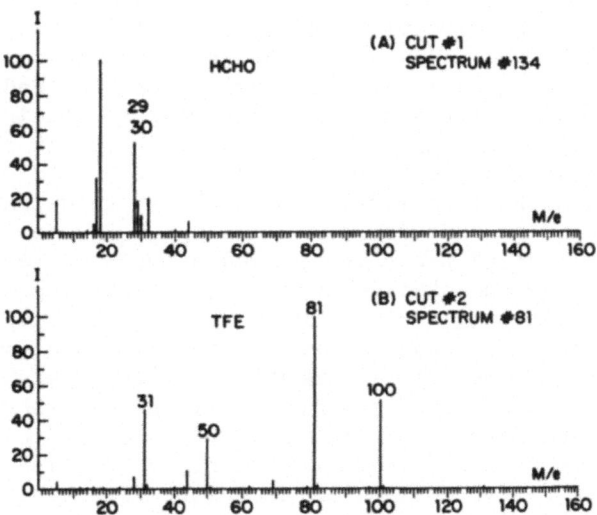

Fig. 5. Mass Spectra of Volatiles from a Brown Speck from
a Resin Part

The combined TG-GC-MS system is ideally suited for study of adsorption-desorption and catalytic reactions under either iso-thermal or programmed heating conditions. To illustrate, the desorption of cyclohexanol (Reagent grade, Fisher Scientific, Fair Lawn, NJ) from alumina (Activity grade 1, neutral, M. Woelm, Eschwege, Germany) or silica gel (Grade 12, Fisher Scientific) was briefly studied. A small amount of cyclohexanol was added to alumina or silica gel placed in the sample pan in the thermo-balance, evacuated to remove excess cyclohexanol, and heated at 10°C/minute in a helium flow. As shown in Figure 6, the TG scan provided information on the amount of desorbed gases at various temperature ranges. GC separated the desorbed gases and deter-mined quantitatively the amount of each component, whereas MS identified each GC component. In this case, the main GC peak in cut #1 was identified as cyclohexanol desorbed from alumina, while cyclohexene was the main product in cut #2 in a high temperature weight loss step. Typical mass spectra corresponding to the GC peaks are shown in Figure 7. Dehydration of alcohols on catalysts is well known but the products depend on the catalyst used. Figure 8 shows the results for desorption of cyclohexanol from silica gel. Cyclohexanol was the main product observed along the single step weight loss curve. It has been reported (14) that a large fraction of the material desorbed from alumina was cyclohexene, whereas off-gas from silica gel almost entirely recovered as cyclohexanol. The present work clearly demonstrates the capability and con-venience of the combined technique for such studies.

Coupled TG-MS

By using the interface shown in Figure 3, the TG effluence can be led to the MS directly and continuously as the TG experi-ment progresses. The main feature of the technique is to allow selective monitoring of individual components independent of each other. It was used effectively to distinguish certain polymer blends and copolymers. As shown in Figure 9, the TG scans showed a physical blend of 60:40 polymethyl methacrylate:polystyrene and a 60:40 methyl methacrylate:styrene copolymer to decompose approximately in the same temperature range. However, the total ion intensity curve of the blend showed a two-step decomposition different from the one-step decomposition of the copolymer. By simultaneously plotting the ion intensity curves of masses 104 and 41 to represent styrene and methyl methacrylate, we readily concluded that the less stable PMMA in the blend decomposed first and the decomposition of polystyrene followed. On the other hand, both methyl methacrylate and styrene were generated evenly during the entire range as would be expected from bond scission of a random copolymer. Evidently this technique is potentially useful in elucidation of polymer structures.

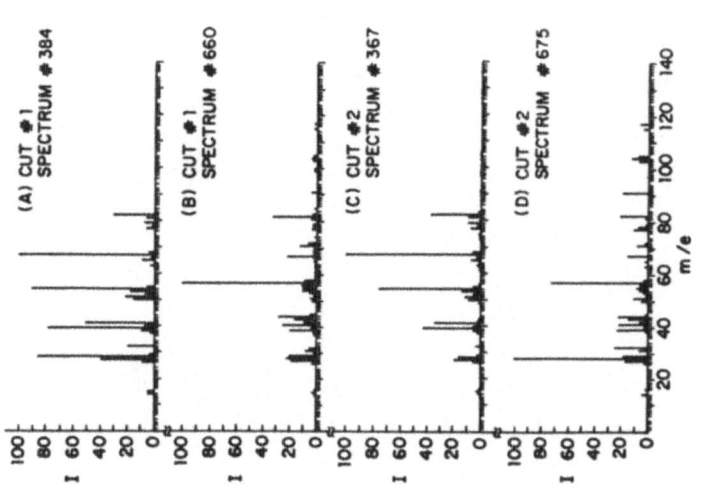

Fig. 7.   Mass Spectra of Volatiles From
          Cyclohexanol Desorption from
          Alumina

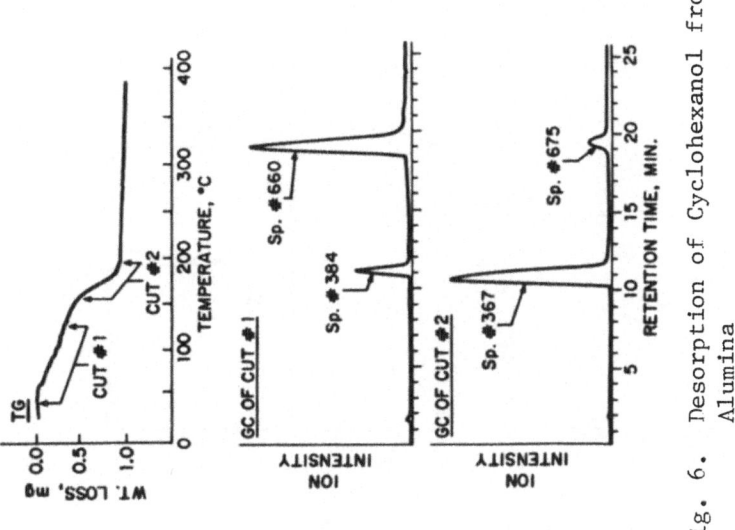

Fig. 6.   Desorption of Cyclohexanol from
          Alumina

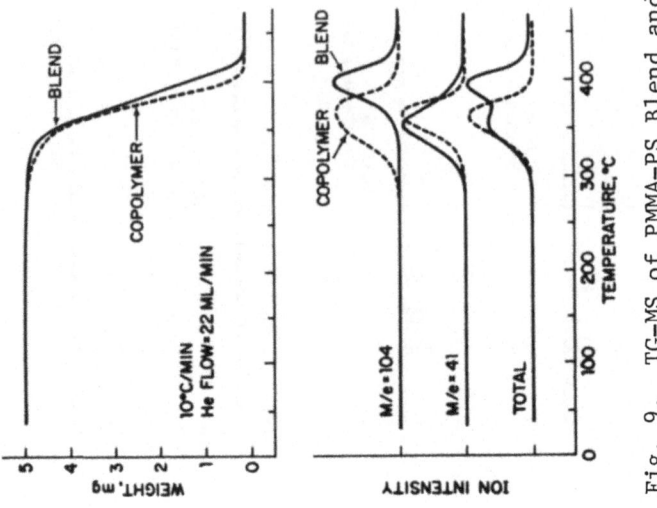

Fig. 9.   TG-MS of PMMA-PS Blend and
Copolymer

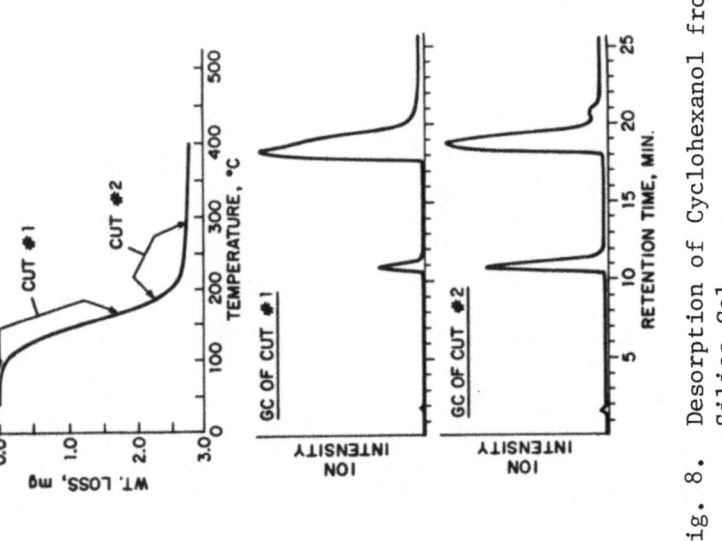

Fig. 8.   Desorption of Cyclohexanol from
Silica Gel

Toxic fumes generated from polymer degradation are of increasing concern in recent years, and hence techniques for better
defining the nature and amount of selected volatiles from a
material are highly desirable. The combined TG-MS system appears
to be suitable for obtaining the evolution profile of selected
toxic volatiles. Figure 10 shows simultaneous TG and MS scans
for thermal degradation of a developmental N-containing polymer.
The toxic component of concern here was hydrogen cyanide, which
was selectively monitored from the ion intensity of mass 27. Our
results showed HCN to form not at the beginning of the polymer
degradation, but during the maximum degradation stage and last
till the end of the degradation process.

To obtain the amount of HCN produced in relation to the
original sample weight, we applied the total condensation method
using the interface shown in Figure 2. Thus, the entirety of the
volatiles was condensed in the trap and later introduced into the
MS. The amount of HCN was measured from the peak area of the ion
intensity vs. time curve. The peak area was calibrated with known
amounts of HCN generated from potassium cyanide. The reactor, as
shown in Figure 11, is inserted between the TG furnace exit and
the TG-MS interface. A weighed amount of KCN (certified, Fisher
Scientific) was placed in a glass dish in the reactor, and an
excess of aqueous HCl solution (50%) was added through a syringe.
The HCN generated was condensed in the trap and then introduced
into the MS to produce an ion intensity vs. time curve. The
calibration curve is shown in Figure 12. The amount of HCN
generated from the polymer under present conditions was determined to be 1% by weight of the original sample.

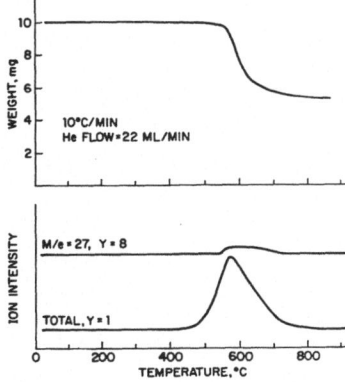

Fig. 10.   TG-MS Measurement of HCN Evolution

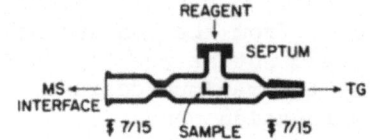

Fig. 11.   Sample Reactor for TG—MS.

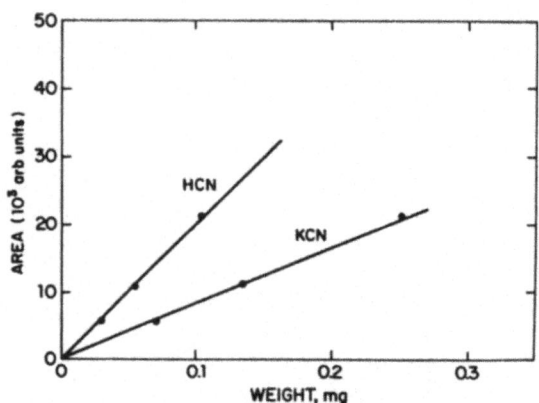

Fig. 12.   Calibration Curve for Quantitative Determination of
           HCN (M/e = 27).

CONCLUSION

I have described some techniques for combining TG, GC and MS and also some unique applications. Such a combined system is now in routine use in our laboratory for problem solving. I believe similar systems are also in use in many other laboratories. I can easily visualize the wide use of such combined techniques in the future. Another complementary and comparable technique, combined TG-GC-FTIR, is also expected to grow rapidly and gain popularity.

ACKNOWLEDGEMENT

The author wishes to thank Richard A. Parkinson for his experimental assistance and many helpful discussions.

REFERENCES

1. D. E. Smith, Thermochim.Acta, 14, 370 (1976).
2. C. A. Cody, L. DiCarlo and B. K. Faulseit, Proc. 10th North American Thermal Analysis Society Conference, October 26, 1980, Boston, MA; Am.Lab., 13, (1), 93 (1981).
3. Jen Chiu and A. J. Beattie, Thermochim.Acta, 40, 251 (1980).
4. Jen Chiu and A. J. Beattie, Thermochim.Acta, 50, 49 (1981).
5. H. K. Yuen, G. W. Mappes and W. A. Grote, Thermochim.Acta, 52, 143 (1982).
6. T. Szendrei and P. C. Van Berge, Thermochim.Acta, 44, 11 (1981).
7. J. Paulik, F. Paulik and L. Erdey, Mikrochim.Acta, 4-5, 886 (1966).
8. G. T. Kerr and A. W. Chester, Thermochim.Acta, 3, 113 (1971).
9. S. G. Fischer and Jen Chiu, Proc. 7th International Conference Therm.Ana., August 22, 1982, Kingston, Canada.
10. Jen Chiu, Anal.Chem., 40, 1516 (1968).
11. Jen Chiu, Thermochim.Acta, 1, 231 (1970).
12. P. Cukor and E. W. Lanning, J.Chromatogr.Sci., 9, 487 (1971).
13. T. L. Chang and T. E. Mead, Anal.Chem., 43, 534 (1971).
14. B. R. Smith and J. M. Thorp, J.Phys.Chem., 67, 2617 (1963).

A NEW APPROACH FOR DETERMINING
MECHANICAL PROPERTIES OF THERMOSET RESINS

Charles Gramelt

Research and Development
Owens-Corning Fiberglas Corporation
Granville, Ohio

ABSTRACT

A new approach for using Dynamic Mechanical Analysis (DMA), to
characterize the mechanical properties of thermoset resins has been
developed. This new thermoanalytical method requires that the existing clamping mechanism of the DuPont 981 Dynamic Mechanical
Analyzer be modified with semi-half rounds on each face of the clamps.
This modification allows the clamping of thermoplastic tubing to be
attached to the vibrating arms of the instrument. Thermosetting
resin is loaded into the tubing and sealed with thermoplastic plugs.
The resin is polymerized upon heating and the cure cycle is obtained.
In addition, a rod specimen that has dimensional stability is formed
and can be analyzed for the mechanical properties of the resin only.

Since no mechanical, or interfacial bonding, occurs between the
resin and thermoplastic tubing, the cured specimen can be removed
from the tubing and attached to the arms of the instrument. The
characterization of the mechanical properties of the cured rod can
be obtained from the usual testing procedure. The purpose of this
paper is to show how this modification can be used to obtain curing
data and mechanical information on thermoset resins.

INTRODUCTION

Dynamic Mechanical testing has become one of the most important
and sensitive indicators of mechanical properties of thermoset
resins. There are just as many techniques as there are instruments
available for the study of thermosets. This method eliminates the

209

need to prepare a batch resin and then form a test specimen from
sheet molding or injection molding. The usual method is to prepare
a molded sheet or injection molded object from the resin and cut
rectangular specimens for testing on some type of dynamic mechanical
instrument to obtain the end-use properties. In the investigation
discussed here the DuPont 981 Dynamic Mechanical Analyzer and 1090
Microprocessor was used for the ease at which it could be modified
for liquid resin characterization. The test clamps attached to the
DMA vibrating arms were modified by machining semi-rounds on the
surface of the sample holder and clamp jaws. This modification
formed a circular cylinder so that thermoplastic tubing with plugs
at each end could be clamped into the DMA instrument. These pressure
tubes were loaded with resin and cured on the instrument and the
cure cycle information obtained. The rod specimen formed is removed from the tubing and attached to the instrument and its
mechanical properties determined.

DMA INSTRUMENT AND MACHINED CLAMPS

The behavior of a material from DMA analysis is based on the
dual concept of viscoelasticity (1,2). The DMA technique provides
a measure of the real and imaginary components of the modulus of
viscoelastic material. The real part, which is the elastic modulus,
is the coefficient for storage of mechanical energy. The imaginary
part is the viscous component or damping and represents dissipation
of mechanical energy by converting it to heat. The DMA instrument
will measure the ability of a specimen to dissipate and store mechanical energy upon deformation (3). During deformation of the
specimen, a portion of the energy required to deform the test
specimen is dissipated as heat. The energy being released is at a
rate that is a characteristic of that test material (4). The
mechanical configuration of the instrument is shown in figure 1.
The Dynamic Mechanical spectra of a material will show the modulus,
damping (loss modulus) and loss tangent (tanδ). The modulus is a
quantitative measure of the stiffness or rigidity of a material.
The loss modulus is the ability of the material to dissipate mechanical energy into heat. Tan δ is a ratio of loss modulus to
elastic modulus. It is a useful index of the viscoelasticity of the
test material.

The sample clamps in figure 1 were machined with 3.2 mm diameter
semi-rounds. Each semi-round was machined on the surface of the
sample jaw and clamp by drilling into the center of the assembled
clamp. Figure 2 shows this modification. The tightening of the
machined sample clamp screws will bring the two semi-rounds together
and form a cylinder in the center of each clamp. Figure 3 shows the
machined clamp assembly. This modification allows tubing to be
attached to both arms of the instrument.

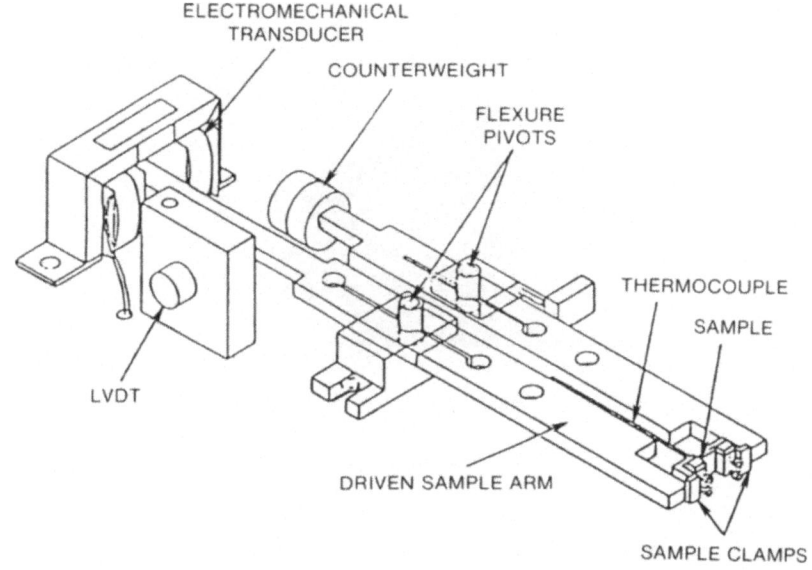

ELECTROMECHANICAL
TRANSDUCER
COUNTERWEIGHT
FLEXURE
PIVOTS
THERMOCOUPLE
SAMPLE
LVDT
DRIVEN SAMPLE ARM
SAMPLE CLAMPS

Figure 1. Mechanical configuration.

Figure 2.

Figure 3

Experimental

The pressure tubes were cut to a length of 25.0 mm with a 3.3 mm
outside diameter and a 3.0 mm inside diameter. Thermoplastic plugs
were machined to a length of 7.0 mm and a diameter of 3.0 mm. These
plugs are used to seal each end of the tubing containing the resin.
Figure 4 shows the assembled pressure tube without resin. The
pressure tube is attached to the DMA instrument by removing the
clamping jaws with clamp inserts from the driven and passive arms.
The pressure tube is inserted into each end of the two modified
clamps and the semi-rounds are tightened around each end of the
sample tube. The attachment of the tube with resin to the instrument
is illustrated in figure 5. The resin used for this technique is a
styrenated polyester resin with one percent benzoyl peroxide initiator. The curing of the thermoset resin was obtained by loading
120 mg of resin into a pressure tube with an eye dropper. The plugs
are attached to each end of the tube and any air trapped is expelled.
The plugs are inserted into each end of the tubing so that a length
of 20 mm is obtained which is the largest gap setting for the instrument. Upon curing the resin, a rod is formed with a length of
approximately 7 mm. The rod may be attached to the instrument at
its lowest gap setting for maximum sensitivity. Using this procedure
a rod with an average length-to-diameter of 9:1 was obtained. The
thermoset inside the pressure tube was attached to the arms of the
instrument and cured using a temperature program from

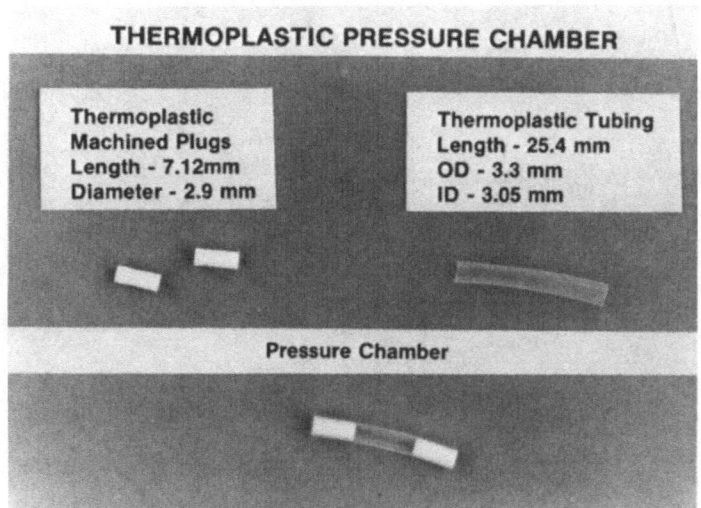

Figure 4.

Figure 5.

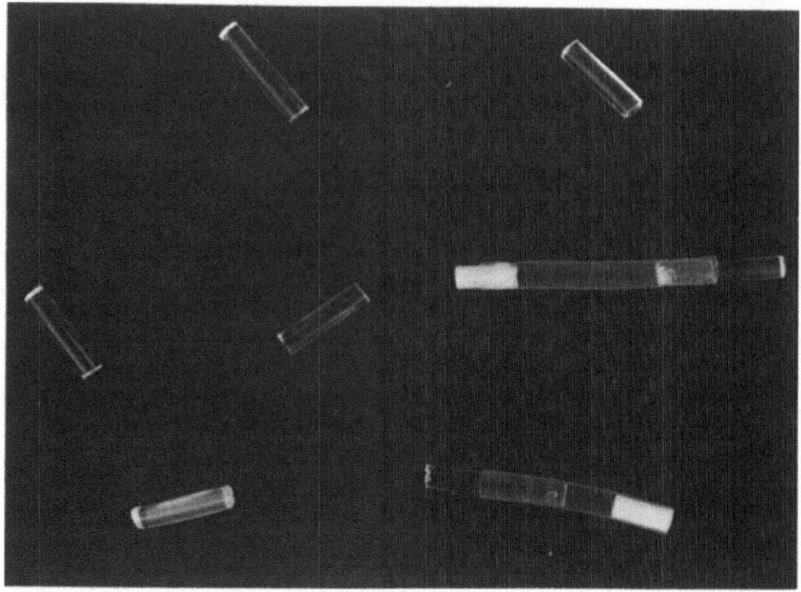

Figure 6.

70°C to 170°C at 5°C per minute and kept at its upper limit for thirty minutes. Since no mechanical, or interfacial bonding, occurs between the resin and thermoplastic tubing the cured rods can be removed from the tubing and attached to the arms of the instrument. The cured rod specimen are shown in figure 6. For the analysis of the cured rods the clamp jaw is turned so that the flat surface is holding the rod in only one of the semi-rounds of the clamp. This prevents the sliding of the rod in and out of the clamps during testing.

Results and Discussion

The DMA responses of several blank pressure tubes were added and normalized. The normalized spectra of the pressure tubes are shown in figure 7. The modulus decreases as a function of increasing temperature. Also, the damping curve decreases in intensity and at 140°C it reaches a negative value. This is from the thermoplastic tubing becoming soft and flexible and can no longer contribute any restoring energy to the natural resonant vibration at a constant amplitude. Spectra of the cured resin and thermoplastic tube are shown in figure 8. This is an expanded plot of the curing of the thermoset. As the chemical reaction begins at 106°C the frequency begins to increase. At 131°C, the frequency and damping both decrease. Spectra of three cured resins and tubes of the damping plots are shown in figure 9. The data from these three analyses

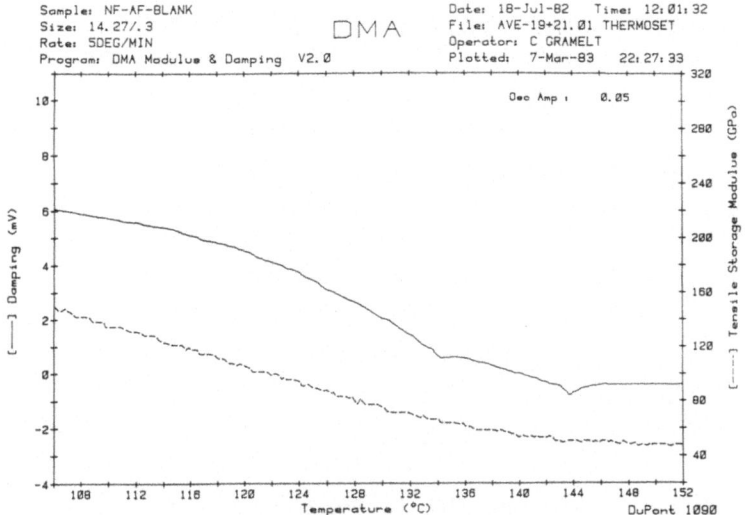

Figure 7. Blank pressure tube.

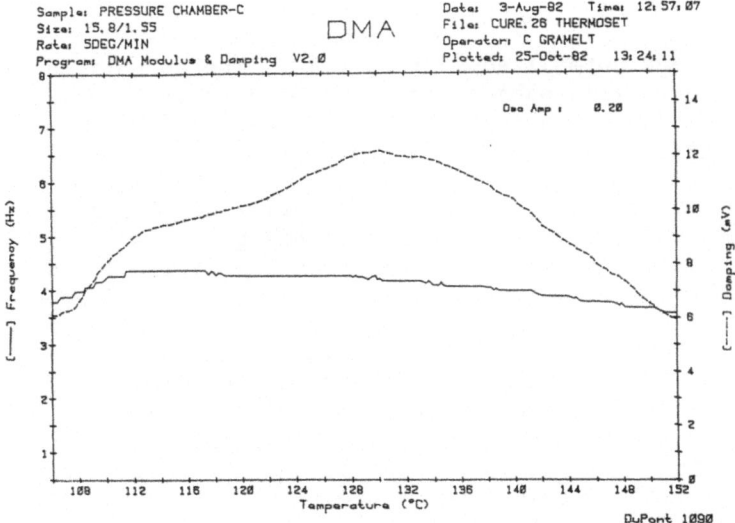

Figure 8. Cured resin and thermoplastic tube.

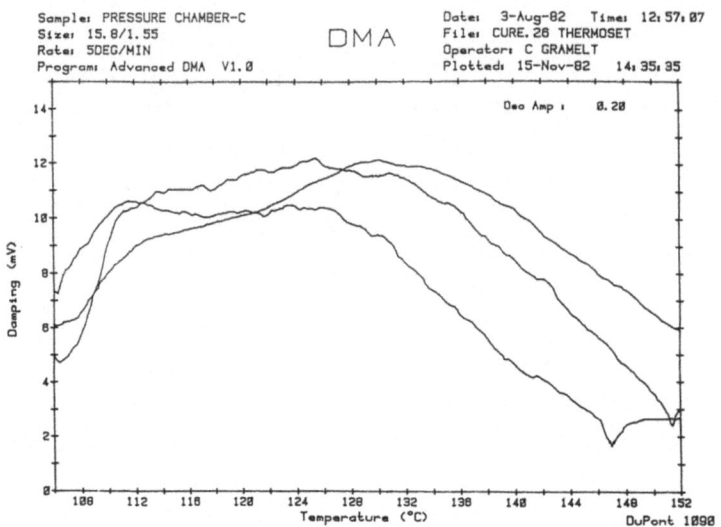

Figure 9. Reproducibility of cured resin and tubes.

were added and normalized in figure 10. The normalized blank tube
data in figure 7 is subtracted from the data in figure 10 with the
aid of a microprocessor to obtain the qualitative effects of curing
the thermoset.

The cure process begins by initiation and linear growth of the
chain backbone and ends with cross-linking. This change is an
irreversible transformation from a viscous liquid to an elastic gel,
which is the beginning of the network, and is called the gel point
(6). Gelation is an important parameter in characterizing thermosets,
since this represents the point at which the polymer will not flow
and is therefore critical in processing the thermosets. Another
important step in the curing process is vitrification. Vitrification
occurs when the viscous liquid or elastic gel changes to a glass.
The cure cycle will begin with gelation and continue through vitrification, until a cross-linked network forms. Figure 11 shows frequency and damping versus temperature for the resin alone, having
subtracted the contribution of the pressure tubes. The gelation and
vitrification of the resin are observed at distinct temperatures
during the cure cycle. Gelation is noted on the frequency curve as
a discontinuous frequency increase at 106°C, caused by crosslinking,
producing a stiffness change. A similar discontinuity is also
observed in the damping

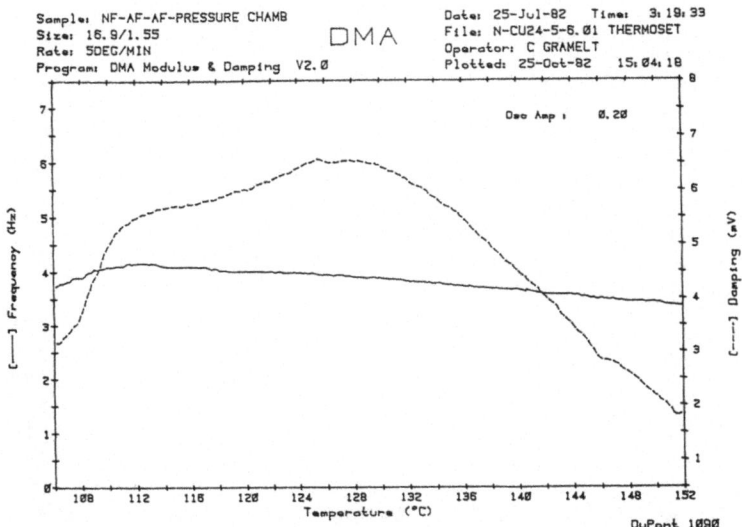

Figure 10. Normalized damping and frequency spectra.

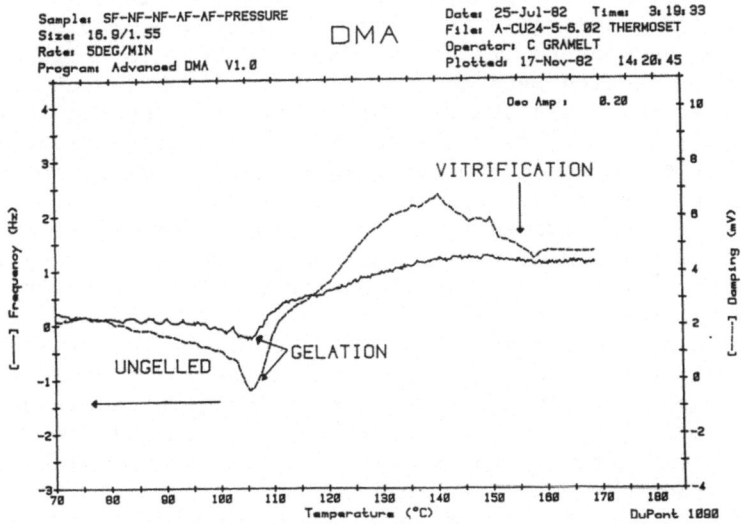

Figure 11. Cure cycle of thermoset.

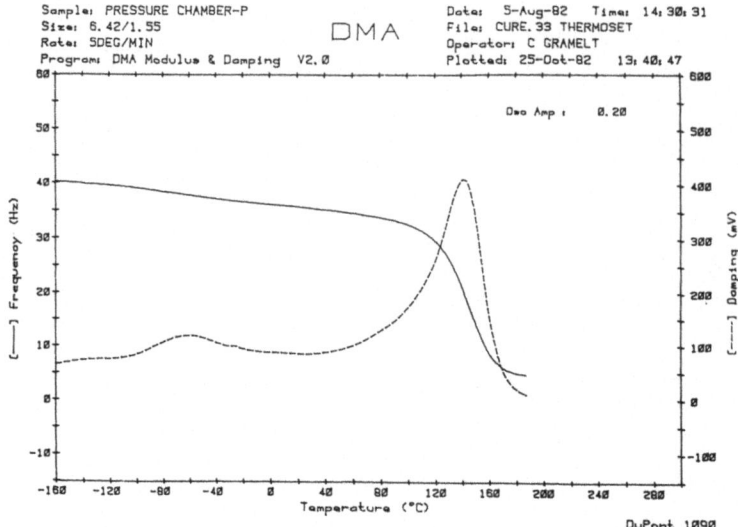

Figure 12. Cured rod: end use properties.

curve at 106°C. Vitrification is also observed in figure 11.
Vitrification occurs as a result of the increased crosslinking
causing the samples to transform from a rubbery to a glassy state,
causing the reaction rate to diminish significantly due to slow
diffusion in the solid state. The viscoelastic changes at vitrification are observed as a plateau in the frequency trace and a flat
valley in the damping curve, both indicative of cessation of the cure
process. Figure 12 is the damping and frequency trace of the data
for the cured rod removed from the pressure tube and attached to the
instrument for analysis of the end-use properties. The glass transition temperature is observed to be 142°C. This is the temperature
at which the thermoset loses its glasslike properties and becomes
semi-liquid and more flexible in nature. The spectra of the cure
cycle in figure 12 has a peak temperature of 137°C which suggests
that the cure reaction and glass transition temperature occurred
at the same time. Also, a beta or secondary transition occurred
at -60°C. This transition is related to different molecular motion
within the polymer. The tan δ and mechanical contributions are shown
in figures 13 and 14.   In figure 13 the tan δ peak temperature
occurred at 160°C. This is a ratio of the loss modulus to elastic
modulus. It is a useful index of viscoelasticity since it is a

measure of the ratio of viscous and elastic moduli. This index is
tabulated in figure 14. High numbers are indicative of a liquid-
like material and low numbers represent an elastic material.

CONCLUSION

    The pressure tube method for Dynamic Mechanical Analysis,
eliminates the need of preparing a large batch resin and molding

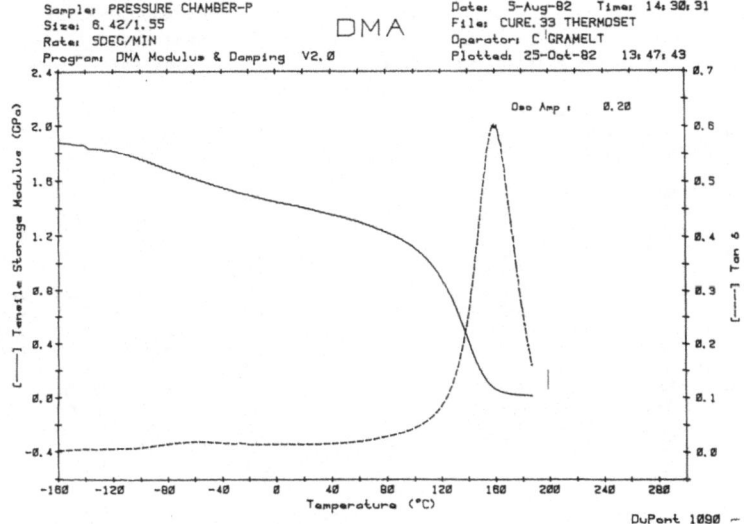

Figure 13. Tan δ spectra of cured rod.

the resin by sheet or injection molding to obtain a test specimen
for end-use properties. Using this approach the experimentalist
can cure a thermoset into a rod specimen for analyzing the cure
cycle and end-use mechanical properties. During the curing of the
thermoset the gelation, vitrification, cure temperatures and the
modulus during the cure are obtained. Analyzing the cured rod
specimen that was formed in the instrument will give the end-use
properties such as tan δ, modulus and glass transition temperature
of the cured thermoset.

Sample: PRESSURE CHAMBER-P   Date: 5-Aug-82   Time: 14:30:31
Size: 8.42/1.55   File: CURE.33 THERMOSET
Rate: 5DEG/MIN   Operator: C GRAMELT
Program: DMA Modulus & Damping V2.0   Plotted: 7-Sep-83   13:00:20

Temp [°C]	Freq [Hz]	Tensile Store [GPa]	Tan δ		Temp [°C]	Freq [Hz]	Tensile Store [GPa]	Tan δ
-100.0	39.3	4.25	0.0018		40.1	35.4	3.06	0.0126
-80.0	39.0	4.17	0.0048		50.1	35.2	3.01	0.0146
-70.0	38.6	4.04	0.0095		60.1	34.9	2.93	0.0177
-60.0	38.3	3.92	0.0134		70.1	34.5	2.84	0.0224
-50.0	38.0	3.83	0.0150		80.0	34.0	2.71	0.0293
-40.0	37.7	3.73	0.0147		90.1	33.4	2.58	0.0370
-30.0	37.4	3.64	0.0192		100.1	32.7	2.41	0.0485
-20.0	37.1	3.55	0.0120		110.0	31.6	2.19	0.0654
-10.0	36.9	3.48	0.0110		120.1	29.7	1.86	0.0943
-0.0	36.8	3.40	0.0104		130.2	26.4	1.38	0.155
0.0	36.4	3.34	0.0102		140.0	21.2	0.804	0.273
10.1	36.2	3.28	0.0104		150.1	14.1	0.325	0.489
20.0	36.0	3.22	0.0104		160.2	8.49	0.109	0.608
30.0	35.7	3.14	0.0111		170.0	5.96	0.0491	0.460
					180.0	4.97	0.0315	0.265

Figure]4. Mechanical contributions of cured rod.

REFERENCES

1. DuPont Company (Scientific and Process Instruments Division)
    DuPont 982 Dynamic Mechanical Analysis System, Product Bulletin.
2. DuPont Company (Scientific and Process Instruments Division)
    981 Dynamic Mechanical Analysis System (DMA), Product Bulletin.
3. M. G. Lofthouse and P. Burroughs, J. Thermal Anal. 13, 439–453
    (1978).
4. T. Murayama, T., Dynamic Mechanical Analysis of Polymeric
    Material, Elsevier Scientific Publishing Co., 36–57 (1978).
5. Creedon, J. P., Analytical Colorimetry, Vol. 2 Pelenum Press,
    185–199 (1970).
6. R. B. Prime, "Thermosets", Chapter 6, Thermal Characterization of
    Polymeric Materials (E. A. Turi, ed.) Academic Press, (1981).

THE MELTING TEMPERATURE OF POLYMERS: THEORETICAL AND EXPERIMENTAL

L. Mandelkern, G.M. Stack and P.J.M. Mathieu

Department of Chemistry and Institute of Molecular
   Biophysics
Florida State University, Tallahassee, Florida 32306

INTRODUCTION

The equilibrium melting temperature, T_m°, of a crystalline polymer, the melting temperature of a perfect crystal of infinite molecular weight, is a very important parameter. It reflects not only the molecular and conformational characteristics of a chain but when used in analyzing crystallization kinetics yields the interfacial free energy for nucleation. A difference of only a few degrees in T_m° can be significant in establishing basic crystallization mechanisms. Despite the importance of this melting temperature its direct experimental determination has proven to be very elusive. Consequently, recourse has been made to theory and to the development of extrapolative procedures of experimental data.

The basic theory is due to Flory and Vrij (1) who analyzed the melting temperature of molecular crystals as a function of chain length. For a given molecular weight all molecules in the system are required to be exactly the same length (2). Otherwise molecular crystals cannot be formed. Although not applicable to polyethylene of finite molecular weight, no matter how well fractionated, by analyzing the fusion of the n-hydrocarbons extrapolation to T_m° for infinite molecular weight polyethylene can be satisfactorily made (1). The requirement of perfectly uniform chain length has not always been recognized. Thus the theory, or minor modifications thereof, has been applied to polymer fractions (3)-(5). Unfortunately, such systems, no matter how well fractionated, do not satisfy the requirements (1)(2).

Two main extrapolative experimental procedures have been developed for polymeric systems. One of these involves extrapolation

223

of the depencence of the observed melting temperature on the crystallization temperature (6)-(8). It involves a variation of the
Gibbs-Thomson equation for the melting of crystals of finite size
and is not unique to polymers or interfacial structures. Unfortunately it is very difficult to satisfy the basic premises of this
method (8)(9). Experimentally it is found that the results, at a
given crystallization temperature, depend on the level of crystallinity at which the measurements are made (7)(8)(10). The reason
for this is the extensive increases in the crystallite size distribution that takes place during isothermal crystallization (11)-
(13).

The other method involves measuring the melting temperature as
a function of crystallite thickness. T_m° is then determined by
extrapolating the results, in terms of a reciprocal thickness plot,
to infinite crystallite thickness, again utilizing the general theory
for the melting of crystals of finite size. Here an absolute
melting temperature must be measured and related to a definite
crystallite thickness. Since the larger crystallite sizes are most
often associated with very broad distributions there is an obvious
difficulty in deciding what thickness is to be assigned to the
observed melting temperature. In addition, as has been amply
demonstrated for solution formed crystals of linear polyethylene,
for the thinner crystals melting-recrystallization problems, which
would vitiate the melting temperature analysis, can be anticipated
(14)-(19). Of further concern is the possibility that the interfacial free energy associated with the basal plane, a basic assumption in the analysis, may not be constant with crystallization
temperature. The change in the crystallization temperature, in
practice, gives the variation in crystallite thickness. Because
of structural differences on the surface, resulting from differences
in chain tilt, this possibility must be given serious consideration
(20).

We list in Table I the extrapolated values of T_m° that have
been deduced by these methods. The inherent difficulties that have
been described have not always been recognized. For reference and
comparison we list in Table II a set of the highest directly determined melting temperatures.

Examination of the directly determined melting temperatures
makes abundantly clear that proposed equilibrium melting temperatures of 138.1°C and 138.9°C (24)(25) are incompatible with
experiment. They obviously should be dismissed. The know experimental facts were apparently overlooked by the investigators. The
remainder of the extrapolated T_m°'s fall into two groups, one around
142°C and the other about 146°C. On more detailed analysis, the
original Flory-Vrij conclusions are found to be strongly favored
(32).

TABLE I

Extrapolated T_m° for Linear Polyethylene

Authors	T_m° °C	Method	References
Flory and Vrij (1963)	145.1 ± 1	Extrapolation of n-alkanes	(1)
Hay (1976)	146.0	Extrapolation of n-alkanes and polyethylene	(5)
Weeks (1963)	145.5	$T_m - T_c$ analysis	(10)
Mandelkern and Gopalan (1967)	146.0	$T_m - T_c$ analysis	(8)
Bair et al. (1968)	145.8	$T_m - 1/L$ (solution crystals	(18)
Wunderlich and Czornyj (1977)	141.6	Specific heat of n-alkanes	(3)
Bassett et al. (1981)	142.0	$T_m - 1/L$ (bulk crystallized)	(21)
Brown and Eby (1964)	143.5	$T_m - 1/L$ (bulk crystallized)	(22)
Illers and Hendus (1968)	141.0	$T_m - 1/L$ (bulk and solution)	(23)
Varnell, Harrison, Wang (1982)	138.9	$T_m - 1/L$ (bulk crystallized)	(24)
Runt, Harrison, Dobson (1980)	138.1	$T_m - 1/L$ (solution crystals)	(25)

TABLE II

Directly Determined Melting Temperature for Linear Polyethylene

Authors	Sample	Crystallization Method	T_m °C	References
Chiang & Flory (1961)	M = 4.9x10^5	131.3°C, atmospheric	138.5[a]	(26)
Mandelkern et al. (1966)	M = 5.7x10^5	130.0°C, atmospheric	139.0[a]	(27)
Gopalan & Mandelkern (1967)	M = 4.7x10^5	130.0°C, atmospheric	138.5[a]	(8)
Ergöz (1970)	M = 6.6x10^5	132.1°C, atmospheric	138.6[a]	(28)
Rijke & Mandelkern (1970)	Unfractionated	Solution stirring	146.0[a]	(29)
Hoffman (1979)	Unfractionated	High pressure	145.0[b]	(30)
Arakawa & Wunderlich (1965)	Polymethylene	High pressure	141.5	(31)
Brown & Eby (1964)	Unfractionated	High pressure	139.2[c]	(22)

(a) Dilatometry; (b) DTA extrapolated to zero heating rate; (c) Rapid heating on hot stage

We have already pointed out the difficulties of analyzing the dependence of the melting temperatures on crystallization temperature because of the size increases that take place. To properly carry out these experiments it is necessary to extrapolate the observed melting temperature, at a given crystallization temperature, to zero levels of crystallinity.

In the present work we address ourselves to two problems directly concerned with determining the physical quantities required for the different extrapolations. The first of these is concerned with the relation of the melting temperature determined by differential calorimetry to the true melting temperature. The other involves an analysis of melting-recrystallization during a DSC scan and its influence on the observed melting temperature.

EXPERIMENTAL

The source and characterization of the linear polyethylene molecular weight fractions have already been described in previous publications from this laboratory (33). A Perkin-Elmer DSC-2 differential scanning calorimeter was used. The heating rates and sample sizes are indicated in the specific experiments. Special thermal history procedures that were adopted are described in the text.

The fusion process was also studied by conventional dilatometry (34). After isothermal crystallization the dilatometers were rapidly heated to 133°C, after which the temperature was increased in intervals of about 0.3°C. The temperature was held fixed after each interval until stable readings were obtained. This procedure resulted in heating rates as low as 0.6°C/day for the highest molecular weight sample.

The relatively large crystallite thicknesses that were deliberately developed for these dilatometric experiments (12) assured that melting-recrystallization processes did not complicate this particular study of the fusion process. Consequently, none of the well-known effects of melt-recrystallization reported previously (35) were observed during these fusion experiments.

The crystallite size distribution was determined from the low frequency Raman acoustical mode using methods previously described (36)-(38).

RESULTS AND DISCUSSION

In the very early use of differential calorimetry, a very strong dependence of the melting temperature on heating rate was

reported (39). The ease with which polymer samples could be super-
heated was pointed out and questions raised as to the quantitative
validity of melting temperatures determined by this technique. The
effects of heating rate, sample mass and thickness have been studied
further (40). There are, however, two distinctly different concerns.
One of these is establishing a melting temperature (endotherm peak)
which is independent of the heating rate, mass and sample configuration. The other is, once having established such a melting temperature to undetstand its relation to the true melting temperature, the
one of thermodynamic significance. We have analyzed the first of
these problems, by studying the influence of molecular weight, using
fractions, on the corrections that are necessary to establish the
endothermic peak. This corrected peak melting temperature, as well
as the temperature of the onset of melting are then compared with
dilatometric studies of identical samples. The DSC determined onset
of melting, is defined as the temperature corresponding to the
intersection of the base line with the leading edge of the endotherm.

By treating the n-hydrocarbons, as well as other low molecular
weight substances, as pure compounds it is possible to obtain melting temperatures which are in agreement with more classical static
type measurements (41). We have recently confirmed this conclusion
with $C_{94}H_{190}$, tetranonacontane; $C_{70}H_{142}$, heptacontane; and $C_{50}H_{102}$,
pentacontane.

The results for two low molecular weight fractions of linear
polyethylene are given in Table III. The main body of the data are

TABLE III

Endothermic Peak Location T_m, for Low Molecular Weight

Heating Rate	M_w/M_N	T_m
1.25°/min.	1756/1586	123.2
2.5°/min.	"	123.4
5.0°/min.	"	123.5 (122.6)
Step Wise	"	123.6
1.25°/min.	5800/5600	133.0
2.5°/min.	"	133.1
5.0°/min.	"	133.6 (133.0)
Step Wise	"	133.3

1 mg. samp.; ($\leq$ 0.1 mg.)

for 1 mg. samples. It is clear that in this molecular weight range
there is only a very small effect, if any at all, of heating rate
on the location of the endothermic peak. The temperature in paren-
theses are for sample masses less than about 0.1 mg. Again there
is no significant influence of the mass.

Interrupted scan experiments were also performed. In this
type of experiment a sample is rapidly heated to a predetermined
temperature and held there for several minutes. The amount of
crystallinity remaining can then be determined by subsequent heating from this temperature. By repeating this process at progressively higher temperatures, the temperature at which the last trace
of crystallinity disappears can be determined. This temperature
was taken as the melting temperature. The melting temperature
determined in this manner was found to be 123.6°C and 133.3°C
respectively for the two samples. These results are in very good
agreement with the values obtained by scanning. Clearly with low
molecular weight polymers reliable melting temperatures can be
obtained by the scanning technique.

This situation does not hold for the higher molecular weight
fractions that we have studied, namely $M_w = 1.54 \times 10^4$, 5.3×10^4 and
1.62×10^5. We find first of all a very definite influence of heating rate and sample mass on the position of the endothermic peak.
The results for $M = 1.62 \times 10^5$, which are typical of these higher
molecular weight samples, are given in Fig. 1. At a given heating
rate the temperature of the observed endothermic peak decreases
with decreasing sample mass for samples of the order of 1.0 mg. or
less. For samples of the order of 0.5 mg. this temperature is
essentially invariant with decreasing mass at a given heating rate.
There is, however, a significant influence of heating rate for any
given mass. These data are easily extrapolated to zero heating
rate. This extrapolated melting temperature is usually about 2°C
lower than that directly observed with the conventional heating
rates of 5°/min. or 10°/min. with 1 mg. samples. A similar magnitude of correction has been inferred previously (24)(40).

We can now compare the endothermic peak temperatures extrapolated to small mass and zero heating rate with the results of the
dilatometric experiments. The dilatometrically determined melting
temperatures represent the true melting temperatures (9). This
comparison is made in Fig. 2. The temperature for the onset of
melting, as determined by the DSC experiments is also indicated
in this figure.*

---

*The temperature of the onset of melting as previously defined for
the DSC experiment, T_i, is, except for the highest molecular weight
sample, independent of both the heating rate and sample weight.
This behavior stands in strong contrast to the endothermic peak
temperature.

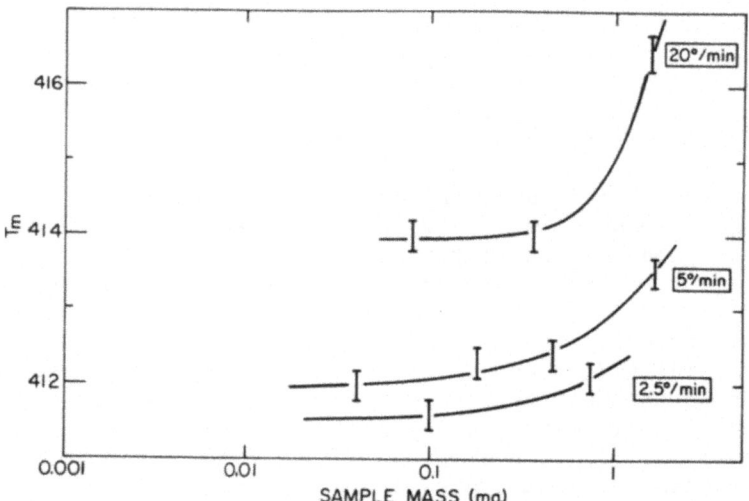

Fig. 1. Temperature of endothermic peak, T_m, in °K as a function
of sample mass at indicated heating rates. Linear polyethylene
fraction $M_W = 1.62 \times 10^5$ crystallized at 130°C for 21000 min.

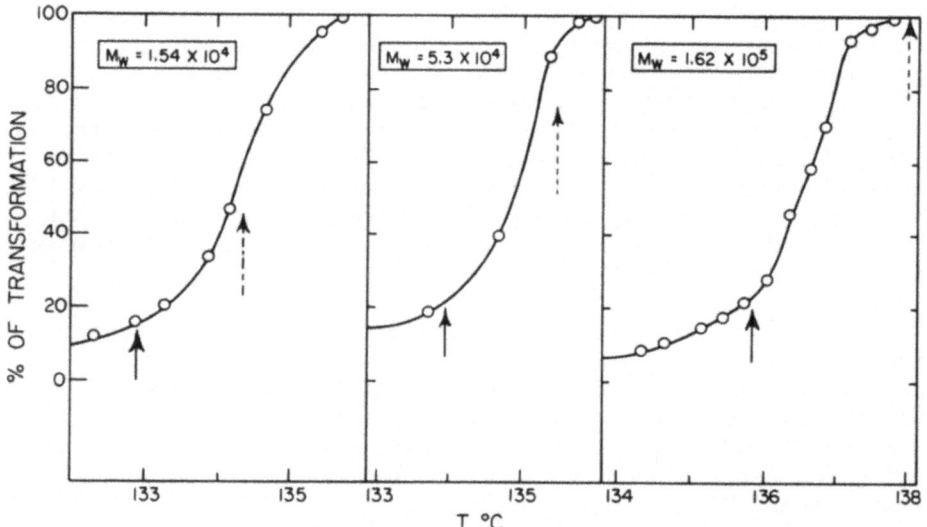

Fig. 2. Dilatometric study of the course of fusion. Plot of extent
of transformation, relative to initial degree of crystallinity,
against temperature for three indicated molecular weight fractions
of linear polyethylene. ↑ onset of melting from DSC; ↑ extrapolated endothermic peak temperature from DSC.

An examination of Fig. 2 shows that the differential scanning calorimeter can give a very good indication of the onset of melting.
T_i corresponds to the temperature where about 10% to 15% of the
transformation has already taken place. However, it is also clear
from the figure that the endothermic peak temperature, <u>extrapolated</u>
for heating rate and mass, is a poor representation of the true
melting temperature. For the lowest molecular weight fraction,
$M_w = 1.54 \times 10^4$, the peak temperature is about 2° too low. For
$M_w = 5.3 \times 10^4$ it is only 1° low. However, for the highest molecular
weight sample studied, $M_w = 1.62 \times 10^5$ the peak temperature is 1° too
high. Presumably this discrepancy will become even greater with
higher molecular weights. Thus, the extrapolated melting temperatures, obtained by scanning, do not represent absolute values and
cannot be used in proper thermodynamic analysis. Moreover, they
are not correct on a relative basis, of one to another, since there
is a very strong effect of molecular weight on the location of the
extrapolated endothermic peak. We must, therefore, conclude that
despite extensive concern, precaution and corrections for heating
rate, mass and sample configuration the resulting endothermic peak
temperature does not represent a true melting temperature for
moderate and high molecular weight chains.

Interrupted scan experiments, similar to those carried out
with the very low molecular weight fractions, were also performed
with the three higher molecular weights. For $M = 1.54 \times 10^4$ and
5.3×10^4 the melting temperatures determined by this method were
now found to be in good agreement with the dilatometric values.
Only a one minute time interval was necessary to obtain sample
equilibrium in these cases, when the scan was stopped at a given
temperature. For the highest molecular weight sample, $M_w = 1.62 \times 10^5$,
a one minute interval resulted in a melting temperature still about
1°C higher than the dilatometric value. However, if the time
interval is increased to 10 minutes the two melting temperatures
could be brought into coincidence.

There are, thus, methods by which the DSC can be used to
determine absolute melting temperatures. However, the scanning
mode cannot be used and a very extensive calibration will have
to be undertaken for a given polymer which definitely involves
molecular weight and very possibly crystallite thickness and
supermolecular structure. It appears that in the scanning mode
the rapidly changing crystalline/non-crystalline ratio between
T_i and melting preclude the determination of a true melting
temperature.

As was pointed out earlier, the determination of the melting
temperature of solution formed crystals of linear polyethylene by
differential scanning calorimetry is complicated by melting, or
partial melting followed by recrystallization. In essence, what
happens is that, because of the very small crystallite thicknesses

that are usually formed in this crystallization mode ($\approx100-150\text{Å}$),
melting takes place in a temperature region $\approx120°C$ where recrystallization is very rapid (17). In favorable cases the melting of the
crystallites initially formed, their recrystallization and subsequent melting at a higher temperature, can be resolved by differential scanning calorimetry (15). Most often, however, only the
endothermic peak observed at the higher temperature, representative
of the recrystallized material is detected (15)(16)(18)(23). This
melting temperature, even when corrected for the various effects
that we have discussed in detail, clearly does not reflect the
original crystallite size nor its structure. It can be off anywhere
from 10° to 15°C. Clearly the use of such a temperature in thermodynamic analysis would be fraught with error. Both direct and
indirect methods of reliably determining the appropriate temperature
for solution formed crystals under these circumstances have been
described (16)(17).

Linear polyethylene samples rapidly crystallized from the melt,
at low temperatures, yield similar ranges in the crystallite thicknesses (42)(43). Thus, even when possible differences in the interfacial free energies are recognized, similar problems can be anticipated when the melting temperatures of such crystallites are
determined by differential scanning calorimetry. This concern is
further borne out by the scans illustrated in Fig. 3. Here a
fraction $M = 3.7\text{x}10^4$ was quenched at -130°C. The peak in the
crystallite size distribution, from Raman longitudinal acoustic
mode (LAM), is at 110Å for a relatively narrow distribution. The
endothermic peak is found to be at 130°C for all heating rates.
Theoretically, this value is substantially greater than anticipated
for the melting temperature and is similar to what is usually
observed with solution formed crystals. In fact the dashed curve
in the figure represents the results for a solution formed crystal
of comparable thickness, having almost the same melting temperature
as the bulk system. We are apparently observing the same phenomenon,
with the same problems, in both cases. The extremes in apparent
heating rates that have been used still do not allow the melting-
recrystallization-melting process to be resolved. If these problems
were not recognized then the melting temperature would be taken to
be 130°C.

The serious error that is involved in accepting the aforementioned conclusion is vividly demonstrated in the following set of
experiments. In independent experiments aliquots of the sample
were heated (annealed) for 10 minutes at a predetermined temperature
below 130°C. The samples were then rapidly cooled to room temperature and the enthalpy of fusion and crystallite size distribution
determined. The results of these experiments are illustrated in
Fig. 4.* Up to annealing temperatures of 80°C the crystallite size
distribution is essentially the same as found in the original sample,
indicating that no changes have occurred. Minor changes begin to

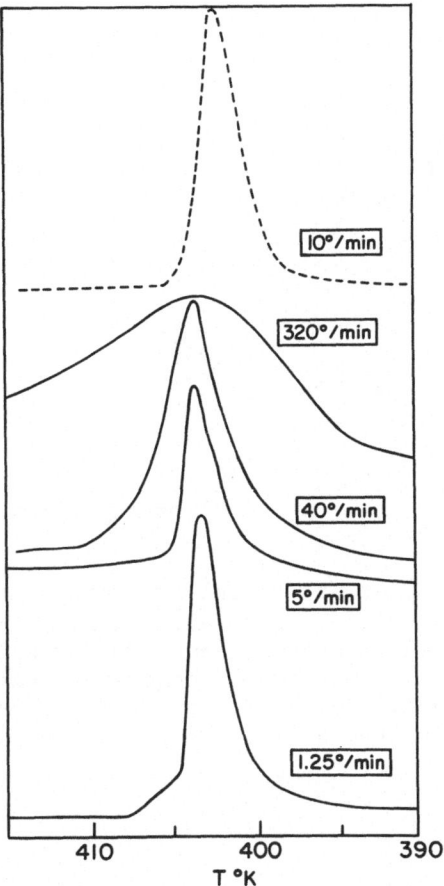

Fig. 3. DSC scan of bulk and solution
crystallized polymers with very thin crystallite sizes. Solid line bulk crystallized;
Dashed line solution crystallized.

be observed at about 90°C. Major changes in size distribution begin
to be observed after annealings just above 100°C. At 120°C a completely different size distribution has developed. The changes at
this temperature can only have taken place by complete melting.

*In studying these results it should be borne in mind that the data
plotted do not represent a continuous heating type fusion curve.
The results for each temperature is for a sample annealed at that
temperature and then cooled back to room temperature.

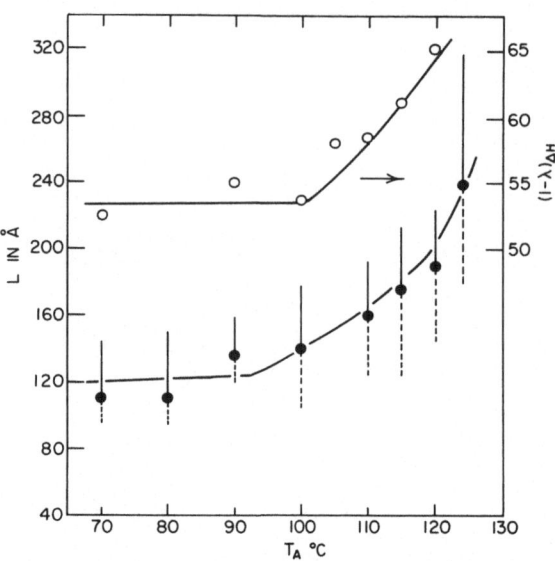

Fig. 4. Plot of crystallite size distribution
(width at half-height) and degree of crystallinity, $(1-\lambda)_{\Delta H}$ in percent from enthalpy of
fusion as a function of annealing temperature.
Linear polyethylene fraction, M = 3.7x10^4,
initially crystallized by rapid cooling to
-130°C.

The process from 90° to 120°C is a continuous one and highly co-
operative. A logical conclusion, therefore, is that the major
changes in size distribution at the lower temperatures, in this
range, are a consequence of partial melting and recrystallization.
This procedure cannot give a true melting temperature. However,
it can clearly set limits thereon. For this sample the melting
temperature cannot be greater than 120°C and it is obviously above
80°C. A more exact delineation of the limits is complicated by the
fusion range being the order of 20°C. The levels of crystallinity
measurements determined from the enthalpies of fusion are in complete accord with the changes in the crystallite size distribution.

    In previous electron microscope studies of the annealing of
solution crystals, of similar thickness, Takayanagi (44) explained
the results at high temperature by some actual melting taking place.

His micrographs showed that after high temperature annealing the original morphology is replaced by one that is similar to that of a bulk crystallized sample. However, he attributes the results at lower temperatures, where in agreement with our studies major changes were not observed, to some type of motion along the chain axis.

These kinds of results are not limited to this particular molecular weight range but are also found in much higher molecular weight fractions. These effects are not observed in a sample with a relatively narrow crystallite size distribution with a peak height at about 240Å. Thus for these sizes, and greater, the endothermic peak temperature, when properly corrected, does properly represent the melting temperature of these crystallites. The size range between 115Å and 200Å for bulk crystallized polymers would have to be studied in more detail in order to discuss the validity of the directly determined melting temperature.

CONCLUSION

We have seen that the experimental determination of the melting temperature of polymeric systems, suitable for use in thermodynamic analysis, possesses several inherent difficulties concerned with both concept and technique. Some of the main problems have been pointed out in this paper and the procedures by which they could be overcome has been indicated. Hence there are obvious difficulties in determining the equilibrium melting temperature from polymer data, even admitting the extrapolative procedure. A more detailed discussion of this aspect of the problem will be taken up in a subsequent publication (32).

Acknowledgment

Support of this work by the Exxon Chemical Corporation is gratefully acknowledged.

REFERENCES

1. P.J. Flory and A. Vrij, J. Am. Chem. Soc., 85 3548(1963)
2. J.G. Fatou and L. Mandelkern, J. Phys. Chem., 69 417(1965)
3. B. Wunderlich and G. Czornyj, Macromolecules, 10 960(1977)
4. C.P. Buckley and A.J. Kovacs, Progr. Colloid Polym. Sci., 58 44(1975)
5. J.N. Hay, J. Polym. Sci., Polym. Chem. Ed., 14 2845(1976)
6. J.D. Hoffman and J.J. Weeks, J. Res. Natl. Bur. Stand., A66 13(1962)
7. D.R. Beech and C. Booth, J. Polym. Sci., B8 731(1970)

8. M.R. Gopalan and L. Mandelkern, J. Phys. Chem., 71 3833(1967)
9. L. Mandelkern, "Crystallization of Polymers", McGraw-Hill, New York, N.Y. (1964)
10. J.J. Weeks, J. Res. Natl. Bur. Stand., A67 441(1963)
11. R.A. Chivers, P.J. Barham, J. Martinez-Salazar and A. Keller, J. Polym. Sci., Polym. Phys. Ed., 20 1717(1982)
12. G.M. Stack, L. Mandelkern and I.G. Voigt-Martin, Polym. Bull., 8 421(1982)
13. I.G. Voigt-Martin and L. Mandelkern, J. Polym. Sci., Polym. Phys. Ed., 19 1769(1981)
14. L. Mandelkern, A.S. Posner, A.F. Dioro and D.E. Roberts, J. Appl. Phys., 32 1509(1961)
15. L. Mandelkern and A.L. Allou Jr., J. Polym. Sci., B4 447(1966)
16. B. Wunderlich, P. Sullivan, T. Arakawa, A.B. Dicyan and J.F. Flood, J. Polym. Sci., A1 3581(1963)
17. L. Mandelkern, R.K. Sharma and J.F. Jackson, Macromolecules, 2 644(1969)
18. H.E. Bair, T.W. Huseby and R. Salovey, Amer. Chem. Soc. Polym. Preprints, 9 795(1968)
19. E.W. Fischer and G.F. Schmidt, Angew Chem., 1 488(1962)
20. G.M. Stack, L. Mandelkern and I.G. Voigt-Martin (to be published)
21. D.C. Bassett, A.M. Hodge and R.H. Olley, Proc. Roy. Soc. London, A377 39(1981)
22. R.G. Brown and R.K. Eby, J. Appl. Phys., 35 1156(1964)
23. K.H. Illers and H. Hendus, Makromol. Chem., 113 1(1968)
24. W.D. Varnell, I.R. Harrison and J.I. Wang, J. Polym. Sci., Polym. Phys. Ed., 19 1577(1981)
25. J. Runt, I.R. Harrison and S. Dobson, J. Macromol. Sci. Phys., B17 99(1980)
26. R. Chiang and P.J. Flory, J. Amer. Chem. Soc., 83 2857(1961)
27. L. Mandelkern, J.M. Price, M. Gopalan and J.G. Fatou, J. Polym. Sci., A2 4 385(1966)
28. E. Ergöz, Ph.D. Dissertation, Florida State University, Dec. 1970
29. A.M. Rijke and L. Mandelkern, J. Polym. Sci., A2 8 225(1970)
30. J.D. Hoffman, Disc. Farad. Soc. 68 470(1979)
31. T. Arakawa and B. Wunderlich, J. Polym. Sci., C16 653(1965)
32. L. Mandelkern and G.M. Stack (to be published)
33. L. Mandelkern, M. Glotin and R.A. Benson, Macromolecules, 14 22(1981)
34. E. Ergöz, J.G. Fatou and L. Mandelkern, Macromolecules, 5 147 (1972)
35. P.J. Flory, L. Mandelkern and H.K. Hall, J. Amer. Chem. Soc., 73 2532(1951)
36. R.G. Snyder, S.J. Krause and J.R. Scherer, J. Polym. Sci., Polym. Phys. Ed., 16 1593(1978)
37. R.G. Snyder and J.R. Scherer, J. Polym. Sci., Polym. Phys. Ed., 18 1421(1980)
38. M. Glotin and L. Mandelkern, J. Polym. Sci., Polym. Phys. Ed., 21 29(1983)

39. L. Mandelkern, J.G. Fatou, R. Denison and J. Justin, J. Polym.
    Sci., 3 803(1965)
40. I.R. Harrison and W.D. Varnell, J. Thermal Anal. (in press)
41. E. Pella and M. Nebuloni, J. Thermal. Anal., 3 229(1971)
42. D.E. Axelson, L. Mandelkern, R. Popli and P. Mathieu, J. Polym.
    Sci., Polym. Phys. Ed. (submitted)
43. R. Popli, M. Glotin, L. Mandelkern and R.S. Benson, J. Polym.
    Sci., Polym. Phys. Ed. (submitted)
44. M. Takayanagi and F. Nagatashi, Mem. Fac. Eng., Kyushu Univ.,
    24 33(1965)

NONEQUILIBRIUM THERMODYNAMICS AND MODERN

DYNAMIC THERMAL ANALYSIS TECHNIQUES

Paul H. Lindenmeyer   Genia Paul

Dynamic Materials, Inc.   Mettler Instrument Corporation
165 Lee Street   Hightstown, NJ 08520
Seattle, WA 98109

INTRODUCTION

The development of various thermal analysis techniques over the past several decades has had a profound effect in accelerating the use of thermodynamic, kinetic and pseudo thermodynamic measurements in the analysis and characterization of materials. All of these methods have in common the use of very small samples and a means of causing the temperature to change linearly with time. The measured quantity can be weight loss (TGA), a mechanical quantity (TMA) or a comparison between the behavior of two specimens (DTA) which, when properly calibrated, yields thermodynamic quantities that compare favorably with older, more conventional calorimetric techniques. Since the measurement time and equipment costs for scanning calorimetry (DSC) are orders of magnitude lower, the DSC has essentially replaced the conventional adiabatic calorimeter and finds a place in nearly every modern analytical laboratory.

More recently, the rapid advancement in microprocessor and data reduction technology has allowed the automation of these techniques and permits the rapid and simultaneous accumulation of kinetic, thermodynamic and pseudo thermodynamic data. Unfortunately these experimental techniques have in some cases outpaced a sound theoretical understanding of what is involved in the data reduction. Since these calculations are accomplished by software that is not generally accessible to the experimenter, the recognition and correction of problems is not easily accomplished. The situation is further complicated by the fact that thermodynamics and kinetics are traditionally taught as separate and distinct sciences, whereas the problems of concern here clearly belong to a nonequilibrium thermodynamics that encompasses both kinetics and (equilibrium) thermodynamics.

It is the purpose of this paper to discuss and illustrate some
of the problems that arise in defining and separating the kinetic
and thermodynamic parameters. A linear nonequilibrium thermodynamics,
applicable to systems sufficiently close to equilibrium so that the
time rate of change is a linear function of the deviation from
equilibrium has been well established[1]. Attempts to extend this
theory into regions far removed from equilibrium[2] have led to
considerable controversy[3,4,5] which has yet to be resolved. For
a recent review of this situation see Lavenda[5]. We shall not
attempt to take either side in this polemic but instead we propose
a novel experimental approach[6] that allows the experimenter to
operationally define the state of the system by his selection of
time and distance scales and by controlling the power level (i.e.,
rate of energy flow).

THE DIFFERENTIAL SCANNING CALORIMETER (DSC)

We shall discuss here only the DSC although these arguments
can be extended to any of the scanning thermal analysis techniques
mentioned above. Two different kinds of DSC's are presently on the
market. Both provide two essentially identical channels for heat
flow, one of which contains the sample of interest and the other
only some inert reference material. In both cases the temperature
in the reference channel is caused to change linearly with time.
In one type of instrument the temperature in the sample channel is
also caused to change linearly with time and the difference between
the power required to drive the two channels is recorded. In the
other type of instrument the same driving force is applied to both
channels and the difference in temperature is measured. In the
first case the difference in power, dH/dt, is divided by the scanning rate, dT/dt, to obtain the total derivative of heat content
with respect to temperature, dH/dT. In the second type of instrument a measure of dH/dT is obtained by suitable instrument calibration. In both cases careful calibration with known reference materials is required to obtain absolute numerical data.

Our concern is not with the difference between these two
methods of measurement but rather with the interpretation that is
usually made of dH/dT. The heat capacity of a thermodynamic system
is defined as the _partial_ derivative of enthalpy with respect to
temperature with all other independent variables held constant.
In equilibrium thermodynamics the energy of a system may be considered to be a homogeneous bilinear function of pairs of intensive
and extensive variables, either of which can be considered as the
independent variable. For example, either pressure or volume may
be considered as an independent variable depending upon the environment. The difference between the heat capacity at constant volume
and at constant pressure is well known in equilibrium thermodynamics.
Thus, in a single component equilibrium system where temperature,

pressure and the number of moles of a single molecular species are
considered to be the independent variables, the total derivative of
enthalpy with respect to temperature becomes

$$\frac{dH}{dT} = \left(\frac{\partial H}{\partial T}\right)_{p,N} + \left(\frac{\partial H}{\partial P}\right)_{T,N}\frac{dp}{dT} + \left(\frac{\partial H}{\partial N}\right)_{p,T}\frac{dN}{dT} \qquad (1)$$

If the system is closed and the pressure held constant this re-
duces to

$$\frac{dH}{dT} = \left(\frac{\partial H}{\partial T}\right)_{p,N} \equiv C_p \qquad (2)$$

However, if the system is not at equilibrium, time must also be
considered as an independent variable so that

$$\frac{dH}{dT} = \left(\frac{\partial H}{\partial T}\right)_{p,t,N} + \left(\frac{\partial H}{\partial t}\right)_{p,T,N}\frac{dt}{dT} \qquad (3)$$

Now only the first term on the right hand side is the thermodynamic heat capacity. The second term is a kinetic contribution that
will only vanish if the system is at equilibrium (i.e., independent
of time). But this can never happen when the DSC is operating in
the scanning mode. Furthermore, since the kinetic term is inversely
proportional to the scanning rate, dT/dt, the error in assuming
that dH/dT is the true heat capacity, <u>increases</u> as the scanning
rate is reduced! Why, then, has the DSC yielded such excellent results in comparison with the adiabatic calorimeter?

The answer lies in the fact that, almost invariably, the comparisons have been made for integrated values of enthalpy rather than
differential values of heat capacity. It is well known to operators
of the DSC that in order to measure the ΔH of some process or transition, it is necessary to start the temperature scan well below
the transition temperature, to extend it to well above, and then
integrate over the total area under the curve in order to obtain a
measure of the total enthalpy change for the process. The basic
reason why this integration of an inaccurate estimate of heat capacity yields an accurate estimate of total enthalpy is not generally understood or easily explained. The simple explanation that
integration over time is equivalent to integration over temperature
since temperature is always changed linearly with time is not adequate since it fails to provide any constraint upon the kinetic
term.

The relationship between kinetics and thermodynamics is given by the phenomonological equations - so named because they can only be established from experimentally measured phenomena rather than derived theoretically. The linear nonequilibrium thermodynamics[1] is limited to a region sufficiently close to equilibrium so that the rate of change of thermodynamic quantities are linear functions of the deviation of the system from equilibrium. Practical experience has shown that the kinetics of many phenomena can be restricted to this linear region as a first approximation. When this is the case, the simple explanation in the previous paragraph is adequate. Before discussing the extension to nonlinear phenomonological equations we consider the operation of a DSC in the "isothermal mode".

THE ISOTHERMAL MODE[*]

Operation of the DSC in the isothermal mode appears to be straightforward, with the measured power required to maintain a constant temperature being equated to the change in enthalpy with time. Expanding the total derivative in partial derivatives of the independent variables yields:

$$\frac{dH}{dt} = \left(\frac{\partial H}{\partial t}\right)_{p,T,N} + \left(\frac{\partial H}{\partial T}\right)_{p,t,N} \frac{dT}{dt} \qquad (4)$$

Now the first term is the kinetic term that is desired and the second term, which contains the thermodynamic heat capacity, must vanish since $dT/dt = 0$. Thus, at first sight, it would appear that the DSC, operating in the isothermal mode provides an excellent, convenient, and unequivocal means of measuring the kinetics of the approach of a sample to equilibrium under constant thermodynamic conditions. However, let us examine more carefully how a specimen is maintained at a constant temperature in a DSC. Using the same procedure of expanding the total derivative in partials we have:

$$\frac{dT}{dt} = \left(\frac{\partial T}{\partial t}\right)_{H,p,N} + \left(\frac{\partial T}{\partial H}\right)_{p,t,N} \frac{dH}{dt} = 0 \qquad (5)$$

---

[*] This section discusses only the DSC with controlled power management. Similar conclusions can be arrived at using the other type of DSC but the discussion is not as straightforward.

Now in order for dT/dt to vanish either both terms must vanish or the first must be the negative of the second. However, the second term cannot vanish because it is composed of the reciprocal of the thermodynamic heat capacity and the measured differential power. The significance of this observation is that a specimen held at a constant temperature in a DSC is not the same as a specimen held in a constant temperature bath. In order to maintain a stable operation in a DSC, both the sample and the reference pan are immersed in a heat sink and maintained at the predetermined temperature by some unknown level of power applied independently to each channel. Thus the sample in a DSC may approach a time independent steady state but it can never become homogeneous since it must always have a temperature gradient sufficient to cause a flow of energy (power) through the specimen equivalent to the flow through the reference channel.

THE POWER LEVEL

The existence of this power that passes through the system is the essential difference between equilibrium and nonequilibrium thermodynamics. A little thought will show that this same unknown level of power is also passing through the sample in the scanning mode, where it is not only unknown, but it is varying in time with the difference between the operating temperature and the ambient temperature of the surrounding heat sink.

The power passing through the system must be sufficiently high so that the instrument can detect the difference between the behavior of the two channels but not so high that it influences the results. That is to say, not so high that it drives the system sufficiently far from equilibrium so that linear phenomonological equations are no longer a good approximation to the relationship between kinetics and thermodynamics. Again this is all very well known to operators of the DSC who long ago discovered empirically that one must adjust both the sample size and the scanning rate to the "proper" range in order to obtain reasonable results from their data reduction techniques. Unfortunately the "proper" range of sample size and scanning rate differs from one sample or reaction to another and as yet no clear cut directions have been available for determining this proper combination of scanning rate and sample size. In fact there may very well exist reactions for which no combination of sample size and scanning rate is available within the operating capability of present DSC's. These empirically observed results are illustrated in the experimental section of this paper.

EXPERIMENTAL SECTION

Two different materials were used to illustrate these ideas experimentally. The first of these was a relatively simple comm - ercial epoxy system, ARALDITE KU600 in which the curing reaction of the neat resins was studied. The second material was a graph- ite prepreg that meets the BMS-8-212 performance specification. Here the resin is already embedded in a graphite fiber tow and B- staged. The detailed composition is generally propriatary to the manufacturer subject only to the BMS-8-212 performance specifi- cations. The measurements were performed with a Mettler TA3000 microprocessor control system using a Mettler DCS20 cell and the programs supplied with this system.

A number of scans were made at scanning rates ranging from 1 degree per minute to as high as 50 degrees per minute and with sample size ranging from 2-3 mg up to as high as 26 mg. The results were analyzed using the programs supplied with the Mettler TA3000 systems and the data summarized in table I for the KU600 resins and in Table II for the 8-212 prepreg. The seperation of the kinetic and the thermodynamic data from a DSC is based upon the following equations:

$$\frac{dH}{dt} \cdot \frac{1}{\Delta H_{TOT}} = \frac{d\alpha}{dt} = k(T) \cdot f(\alpha) \qquad (6)$$

Where α is the degree of conversion, $k(T)$ is the temperature depend- ance of the reaction rate, $d\alpha/dt$ and $f(\alpha)$ is some function of the degree of conversion that may vary with the particular kinetic model. The most commonly used data reduction model assumes that $k(T)$ is given by the Arrhenius equation and $f(\alpha)$ is some power of $(1-\alpha)$ so that:

$$d\alpha/dt = k_o e^{-\frac{EA}{RT}} (1-\alpha)^n \qquad (7)$$

If it is assumed (as is the case in the Mettler software and we be- lieve most other systems) that $d\alpha/dt$ is experimentally given by the left half of equation (6) then the three constants k_o, E_A and n can be obtained by solving equation (2) with $d\alpha/dt$ measured at a minimum of three values of α. But, according to the first half of equation (6) the DSC obtains a measurement of $d\alpha/dt$ as a continuous function of time so that the microprocessor can do a regression analysis on the entire curve and obtain a best fit value for ko, E_A and n along with calculated 95% confidence limits that indicate the goodness of the fit.

Using this goodness of fit as a criterion and the data in Table I one can easily select a scanning rate of 5 degrees per minute as an optimum rate and observe the sample size is not very critical

TABLE I

EFFECT OF SAMPLE SIZE AND HEATING RATE ON THE MEASURED
VALUES OF KINETIC AND THERMODYNAMIC PARAMETERS OF ARALDITE
KU 600 EPOXY

Heating Rate °C/ min	Sample Wt. mg	ΔH J/g	n ± conf. limit	EA ± conf. limit	Ln Ko ± conf. limit	T.50% °C
50	10.57	64.4	1.43 ± .18	116 ± 14.5	24.1 ± 3.5	240.2
50	12.35	66.0	1.42 ± .17	115 ± 13.8	24.0 ± 3.4	246.1
50	13.92	65.3	1.41 ± .16	113 ± 13.2	23.0 ± 3.2	240.8
20	2.60	71.4	2.04 ± .16	192 ± 14.2	44 ± 3.6	212.0
20	8.34	66.4	1.97 ± .19	182 ± 16.0	41 ± 4.0	217.8
20	10.97	70.0	2.03 ± .17	186 ± 14.5	42 ± 3.7	217.9
20	11.85	69.5	2.06 ± .16	189 ± 14.0	43 ± 3.6	218.9
20	24.88	81.6	1.67 ± .12	148 ± 10.3	32 ± 2.6	214.0
5	9.70	69.4	1.83 ± .07	222 ± 7.0	52 ± 1.9	194.7
5	13.18	69.1	1.78 ± .06	219 ± 6.3	52 ± 1.7	194.7
5	16.71	70.6	1.73 ± .06	211 ± 5.3	50 ± 1.5	194.7
5	24.51	75.1	1.47 ± .04	182 ± 4.0	43 ± 1.1	188.5
1	3.23	84.3	.78 ± .09	129 ± 56.8	29 ± 16.1	165.3
1	16.30	72.2	1.06 ± .23	186 ± 24.8	43 ± 6.9	172.2
1	20.83	71.8	1.05 ± .22	180 ± 21.4	42 ± 6.0	172.2
1	21.24	67.0	.92 ± .20	167 ± 21.6	38 ± 6.0	172.2

TABLE II

EFFECTS OF SAMPLE SIZE AND HEATING RATE ON THE MEASURED
VALUES OF KINETIC AND THERMODYNAMIC PARAMETER OF BMS 8-212
GRAPHITE - EPOXY PREPREG

Heating Rate °C / min	Sample wt. mg	ΔH J/gm	n + conf. limit	E_A + conf. limit	Ln Ko + conf. limit	T.50% °C
20	5.72	161	1.23 + .37	69 + 14.5	13 + 3.9	205.3
20	23.63	163	1.48 + .36	82 + 15.11	16 + 4.0	214.2
10	24.48	190	1.43 + .22	96 + 10.26	19 + 2.8	203.7
5	5.79	114	1.26 + .12	111 + 6.7	24 + 1.9	181.2
5	23.90	164	1.76 + .05	119 + 2.7	26 + .8	194.0
1	5.87	66.1	2.17 + 1.05	182 + 86.8	44 + 24.8	161.8
1	20.38	153.0	1.96 + .14	122 + 8.6	27 + 2.5	166.2

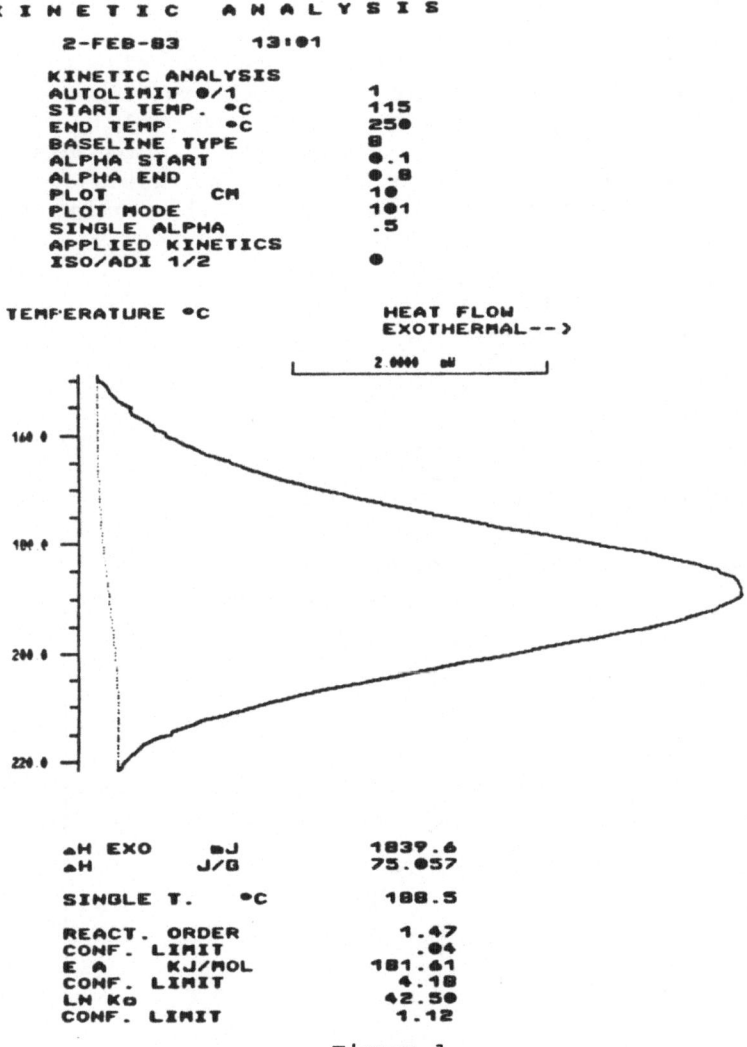

Figure 1.

in the usual range of 10-15 mg. Using the kinetic constants obtained in this manner the software will calculate an isothermal
reaction curve for any temperature and one can verify these calculations by making isothermal measurements at least for those
temperatures near the maximun in heat evolution. Figures 1 and 2
show typical plots of the initial scan and the calculated isothermal conversions.

As might be expected the more complex reactions occuring in the
BMS-8-212 material yielded results (Table II) that are not so readily interpreted. The typical DSC scan, Figure 3, clearly shows
evidence of more than one reaction so that one would not expect
the kinetic model used in the data reduction to yield valid kinetic data. Nevertheless a comparison of heat flow measured
isothermally at 175° C (see Table III) and integrated over the
times calculated for α=.25, .50, .75 and .90 show surprisingly

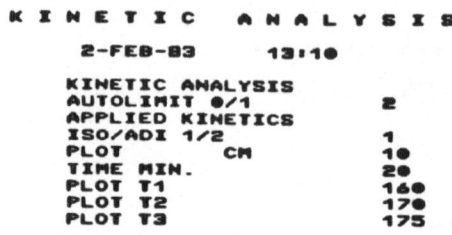

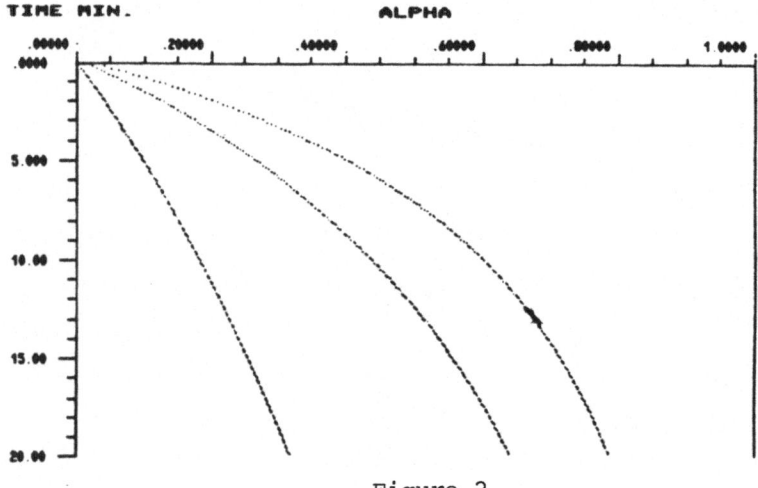

Figure 2.

good agreement with the actual degree of conversion as measured isothermally. One might reasonable ascribe those descrepancies that
do exist to the fact that a system involving multiple reactions
would require a more complex kinetic model than the one given in
equation (7).

We wish to suggest an alternate possibility. The validity of
the left side of equation (6) rests upon not one but two different
assumptions. In the first place in order that the total area under
the curve be proportional to the total ΔH for the reaction, the scan-

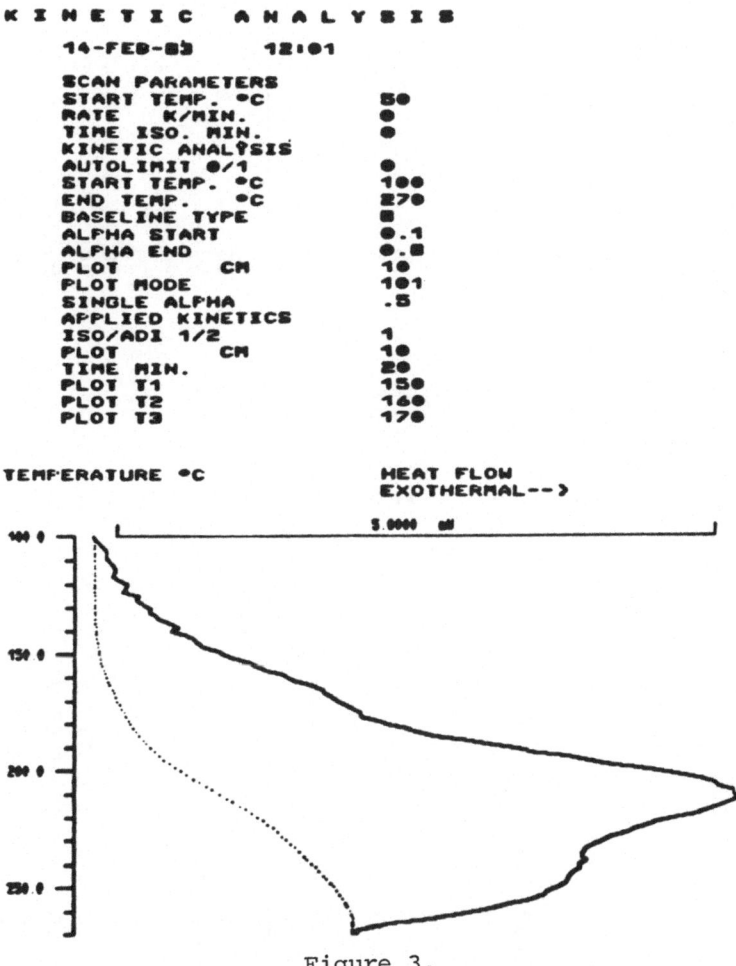

Figure 3.

ning rate must not be so great that the deviation from equilibrium exceeds the linear region so that integration over time is equivalent to integration over temperature. On-the-other-hand, equation (6) also requires that $dH/dt = Cp\ dT/dt$ <u>at all points on the curve</u>. This in turn requires that the scanning rate be sufficiently large so that the second term on the right hand side of equation (3) be negligible in comparison to the first term. Thus we have a requirement that the scanning rate be high enough to meet our criterion yet not so high as to exceed another. Clearly the sample size also enters into the determination of these criteria and there is no theoretical requirement that an optimum combination of sample size and scanning rate need exist within the capability of the instrumentation.

TABLE III

COMPARISON OF ISOTHERMAL CURING OF BMS-8-212 AT 175° C WITH

THAT CALCULATED FROM KENETIC PARAMETERS MEASURED BY DYNAMIC SCANNING

Sample Wt. mg	Total Area J/G	Integration Time— Min.	Area J/G	Estimate from Heating rate °C/ min	mg	Predicted α	Actual α
29.03	168.2	2.2	49.7	1	20.38	25	29.5
		6.3	94.1			50	55.9
		17.5	143.6			75	85.4
		46.4	165.3			90	98.3
		10.4	120.0	5	23.90	50	71.3
		20.8	149.5	10	24.48	75	89.1
		2.3	51.2	20	23.63	25	30.4
		6.2	93.2			50	55.4
		14.9	136.9			75	81.4
		31.9	160.9			90	95.7
5.95	173.2	0.6	18.9	1	5.87	25	10.9
		1.9	43.5			50	25.1
		6.3	91.3			75	52.7
		21.4	152.6			90	88.1
		1.8	42.0	5	5.79	25	24.2
		4.6	76.1			50	43.9
		10.0	115.8			75	66.9
		19.0	147.7			90	85.3
		1.6	38.9	20	5.72	25	22.5
		4.0	69.9			50	40.4
		8.7	108.3			75	62.5
		16.3	140.7			90	81.2

RELATIONSHIP BETWEEN SCANNING RATE AND SAMPLE SIZE

It is one of the objectives of this paper to explain the fundamental reasons why one must have the proper relationship between the scanning rate and the sample size in order to obtain both the kinetic and thermodynamic parameters from calorimetric measurements Two different factors are involved in this explanation, neither of which have previously been discussed in this connection. The first of these is the level of power (i.e., the rate of energy flow) passing through the system. This power is determined by the external forces acting on the system (i.e., the temperature difference between the source and the sink) and the thermal resistance to the flow of heat. This power level is a major factor in determining the deviation of the system from equilibrium. In other words, the driving force that operates a DSC also influences how far the sample is from equilibrium which in turn influences how rapidly the properties are changing. Thus the technique used to measure a property of the system actually has a direct and important influence upon the value of the property being measured.

More simply stated, in order to measure a change in a property of a system, one must supply sufficient power to cause that property to change by a measurable amount. But what constitutes a measurable amount? This brings us to the second important factor to be considered, the measurement of rates or velocities.

MEASUREMENT OF VELOCITIES

The concept of velocity has undergone a very significant change with the acceptance of Einstein's theories of relativity even though it has taken a very long time for the more subtle aspects of these ideas to become appreciated in everyday life. In essence a velocity is a dimensionless number that depends upon the reference frame of the observer and whether or not that reference frame is moving or fixed with respect to the system whose velocity is being measured. In other words, a velocity can be defined only with respect to a given point in space-time and relative to a particular observer. Since one cannot measure velocity at an infinitesimal point in time or space, any measurement of a rate or a velocity requires averaging the change in the quantity of interest at some point in space over a period of time as well as averaging it over a volume at some point in time. The period of time over which the quantity is averaged determines the time scale and the volume determines the space (or distance scales). We note that the measurement precision of the resulting rate or velocity is directly proportional to the time scale and inversely proportional to the distance scales. But since these time and distance scales may be arbitrarily determined (within certain limiting constraints), the observer can influence the measurement precision and thus influence the power required to measure a velocity. Conversely the power applied by the observer will influence not just the measured value but the actual value of the velocity or rate. Thus the Heizenbergs' uncertainty principle and Einstein's relativity theories must be applied not only to esoteric quantum mechanics and high speed particles but also to the everyday measurement of rates.

To make this rather unusual point more explicitly we write

$$\text{rate or velocity} = \frac{\text{change in quantity}}{\text{change in time}} \quad x \quad \frac{\text{time scale}}{\text{distance scales}}$$

If one measures the change in some quantity averaged over a short enough time scale (but not shorter than the measurement precision of time) or averaged over a large enough volume or distance (but no larger than the system being measured) the numerical value of the rate or velocity may be less than the measurement precision. Conversely, if one measures the change in the quantity over a long enough time scale (but no longer than the total available time) or over a small enough volume (but not smaller than the measurement precision), then one may obtain a measurable velocity. Thus the existence or nonexistence of a measurable rate can be influenced by the time and distance scales adopted by the observer.

This may appear to be a rather long and involved way to arrive at the obvious conclusion that an observer may change his observed results by changing his measurement precision. However, it was necessary to show how the time and distance scales can affect the measurement in order to describe how this occurs in the DSC as we shall be in the next section. In addition we hope that this analysis has emphasized the relative nature of a velocity or a rate. Reaction rates (or any rates or velocities) are not just "out there" waiting for someone to measure them with greater and greater precision. The act of measuring them can change their value depending upon how much external force must be applied.

OPERATION OF A DSC

Returning now to the operation of a DSC we observe that the operator can independently control the sample size (within the limits of the size of his pans). However, the time scale and the power level are not independently controllable since increasing the scanning rate increases the power (i.e., deviation from equilibrium) but decreases the time scale (i.e., measurement precision of velocity). Thus only if it should happen that there exists a sample size, compatible with the specimen cups, and a scanning rate, compatible with the instrument controls, for which a given reaction proceeds fast enough to yield a heat flow greater than the measurement precision yet not so fast as to cause a deviation from equilibrium sufficient to yield a heat flow in the nonlinear region, will it be possible to equate an integration over time to an integration over temperature and to obtain realistic kinetic and thermodynamic parameters.

Thus the experimentalists have arrived at a dilemma similar to that of the theorists who must assume that there exists a region of time and space small enough to be considered infinitesimal yet large enough in order to apply thermodynamics (i.e., assumption of "local or cellular" equilibrium).

However, the experimentalist has advantages not enjoyed by the theorist. First of all, he can tell when he is wrong and secondly he has the possibility of changing his apparatus to correct his problem once he realizes its nature. Thus once it is recognized that the scanning rate influences both the power level and the time scale in opposite directions, it should be possible to effect at least a partial separation and control by changing the difference in temperature between the source and the sink independent of the scanning rate. This would make it possible to increase the power level (driving force) without increasing the scanning rate. The reverse of this, that is, decreasing the power level for a given scanning rate, will generally not be possible.

Thus we see that the nonequilibrium thermodynamic parameters which encompass both the traditional thermodynamic and kinetic parameters are not simply material parameters sitting "out there" waiting for someone to devise methods of more accurately measuring them. Rather they are influenced by the power applied by the observer in his attempts to measure them.

SUMMARY AND CONCLUSIONS

The DSC can only measure a true heat capacity when there are no time dependent processes occurring within the specimen. That is to say, only when $(\partial H/\partial t)_{p,T,N} = 0$ so that $dH/dT = (\partial H/\partial T)_{p,N,t} = C_p$.

The DSC can only measure a true total enthalpy change for a chemical or physical process when the specimen size and the scanning rate are such that the deviation of the sample from equilibrium remains in the range where the assumption of linear phenomonological equations is valid and when integration is carried out over the total range of temperature where the reaction or process may occur.

There may occur reactions where the proper selection of sample size and scanning rate required to yield a true measure of total enthalpy change lie outside of the capability or present instruments. This situation may be corrected or at least substantially improved by providing a means of controlling the temperature of the source and the sink independently from the scanning rate.

However, the most important conclusion of this paper is that reaction rates (in common with all velocities) are relative to the reference frame of the observer and his measurement precision. His measurement precision in turn depends upon the amount of power he applies in order to make the measurement as well as the time and distance scales he uses to express his results. By independently controlling the power and the time and distance scales, it should be possible to restrict the reaction to the linear range and effect a separation of the traditional kinetic and (equilibrium) thermodynamic parameter.

An additional, still more important conclusion that goes well beyond what has been discussed here is that since the power that influences the reaction rates can be controlled independently from the scanning rate (i.e., the change in thermodynamic parameters), then why should we be interested only in restricting the system to the linear region in order to measure the traditional kinetic and thermodynamic constants? By increasing the power we can actually change the reaction rates independent of the average thermodynamic parameters. This offers the potential of opening a new era in materials processing which unfortunately cannot be discussed at the present time due to impending patent action.

REFERENCES

1.   S. R. DeGroot and P. Mazur, "Nonequilibrium Thermodynamics", North Holland, Amsterdam, (1962).
2.   P. Glausdorff and I. Prigogine, "Thermodynamic Theory of Structure, Stability and Fluctuations", Wiley-Interscience, New York, (1971).
3.   J. Keizer and R. F. Fox, Proc. Nat. Acad. Sci. USA, _71_, 192-196 (1974).
4.   R. Landauer, Phys. Rev. A12, 636 (1975).
5.   B. H. Lavenda, "Thermodynamics of Irreversible Processes", Wiley-Halsted, New York (1978).
6.   P. H. Lindenmeyer, "An Approach to the Theory of Polymer Solids", Interscience, in preparation.

INTEGRAL-DIFFERENTIAL RELATIONSHIPS IN THERMAL ANALYSIS

John P. Elder

I.M.M.R., University of Kentucky

Lexington, KY 40512

Introduction

Recent investigations of the pyrolysis of bituminous coals[1] have shown that the TG/DTG data characterizing the overall complex thermal degradation obeys the well-known Kissinger equation[2]. The resulting global kinetic parameters are very close in magnitude to those given by other investigators for similar coals, using a statistical approach to the analysis of thermogravimetric data. The Kissinger equation has been used empirically by other workers in the fossil fuel field[3]. Is the Kissinger equation generally applicable to thermal reactions, irrespective of the form of the mathematical relationship used to describe the kinetic model? This paper is the author's answer to this question. The caveats of Garn[4], regarding the use of the Arrhenius kinetic equation, incorporating analytical degree of reaction functions encompassed by the Sestak-Berggren general equation[5] in solid state reactions, are well taken, and should be borne in mind by all interested parties.

Theoretical Principles

The basic equation describing the rate of a physico-chemical process in a homogeneous system under non-isothermal conditions at constant heating rate, β, is generally expressed in the form (1) with $f(\alpha)$, the function of the degree of reaction, α, given by (1a)[5]. The Criado argument[6] justifies the use of the total differential.

$$\frac{d\alpha}{dT} = \frac{A}{\beta} . T.^m \exp\left(\frac{-E}{RT}\right). f(\alpha) \qquad (1)$$

$$f(\alpha) = (1-\alpha)^p. \alpha.^q[-\ln(1-\alpha)]^r \qquad (1a)$$

where p, q and r are empirical numbers obtained from a suitably designated closeness of fit analysis of experimental data, or are integer values obtained from a theoretical analysis of postulated rate-controlling physico-chemical processes. Although arguments have been presented advocating their non-usage (cf. Garn[4]), the prevailing view is to refer to E as the activation energy of the rate-determining step. A is generally referred to by the innocuous term, pre-exponential factor. The exponent, m, in (1) is generally set equal to zero, i.e. the Arrhenius assumption is followed. Transition State theory predicts m = 1, and in homogeneous gas phase reactions, collision theory predicts m = ½. In what follows, it is assumed that the basic equations are descriptive of heterogeneous, solid state processes, even if only from an empirical point of view.

The integrated form of (1) is

$$\int_0^\alpha \frac{d\alpha}{f(\alpha)} = F(\alpha) = \frac{A}{\beta} . \int_{T_0}^T T^m . \exp\left(\frac{-E}{RT}\right) dt = \frac{AI}{\beta} \qquad (2)$$

The evaluation of the temperature integral, I, has been discussed by numerous authors over several decades. Representing for simplicity the Arrhenius exponent, E/RT, by x, equation (2) may be expressed in the form (3)

$$F(\alpha) = \frac{A}{k\beta} . \left(\frac{E}{R}\right)^{m+1}. p_m (x) \qquad (3)$$

The $p_m (x)$ function is given by (4)

$$p_m(x) = \frac{\exp(-x)}{x^{m+2}} . \gamma_m (x) \qquad (4)$$

For m = ½, k = 1. For all integer m, k = (m+1)! The evaluation of the function $\gamma_m(x)$ will be discussed later.

The second temperature differential of (1) is given by (5), where $f'(\alpha) = df(\alpha)/d\alpha$.

$$\frac{d^2\alpha}{dT^2} = \frac{d\alpha}{dT} . \left[\frac{f'(\alpha)}{f(\alpha)} . \frac{d\alpha}{dT} + \frac{m}{T} + \frac{E}{RT^2}\right] \qquad (5)$$

At a certain temperature, T_{max}, $(d\alpha/dT)_{T_{max}}$ attains a maximum value, and hence the second differential equals zero. Then,

$$\frac{E}{RT^2_{max}} = -\frac{A}{\beta} \cdot f'(\alpha_{max}) \cdot T^m_{max} \cdot \exp\left(\frac{-E}{RT_{max}}\right) - \frac{m}{T_{max}} \qquad (6)$$

Combining this equation with the integral equation (3) at T_{max}, one has the simple relationship (7)

$$-f'(\alpha_{max}) \cdot F(\alpha_{max}) = \eta_m (E/RT_{max}) \qquad (7)$$

where:

$$\eta_m(E/RT_{max}) = (1 + mRT_{max}/E) \cdot \gamma_m(E/RT_{max}) \qquad (8)$$

Rearranging the logarithmic form of (6), one has the generalized Kissinger equation, (9)

$$\ln(\beta/T^{m+2}_{max}) = \ln(AR/E) + \ln \phi_m (\alpha_{max}) - E/RT_{max} \qquad (9)$$

where:

$$\phi_m(\alpha_{max}) = \frac{-f'(\alpha_{max})}{(1+mRT_{max}/E)} \qquad (10)$$

Kinetic Models

Rather than attempt to derive a general expression for α_{max} from equations (1a) and (7), a procedure fraught with difficulties, and of questionable utility, the values of $-f'(\alpha_{max})$, $F(\alpha_{max})$ and α_{max} have been calculated for a number of kinetic models, falling within the Šesták-Berggren general type, (1a).

$$n^{th} \text{ Order } (F,n) \qquad f(\alpha) = (1-\alpha)^n \qquad (11)$$

$$\underline{n = 1} \qquad \alpha_{max} = 1 - \exp(-\eta) \qquad (11a)$$

$$\underline{n \neq 1} \qquad \alpha_{max} = 1 - [1-(\tfrac{n-1}{n})-\eta]^{1/(n-1)} \qquad (11a')$$

$$F(\alpha_{max}) = \frac{\eta}{n-(n-1)\cdot\eta} \qquad (11b)$$

$$\underline{\text{All } n} \qquad -f'(\alpha_{max}) = n - (n-1)\cdot\eta \qquad (11c)$$

It should be noted that Balarin[7] has proposed the use of
equation (11a') for the case m = 0.

<u>Linear Growth of Nuclei</u> (E1, B1)

<u>Chain Model</u> (E1)   <u>Branching Model</u> (B1)

$f(\alpha) = \alpha$ (12) $f(\alpha) = \alpha - \alpha^2$ (13)

$\alpha_{max} = \exp(-\eta)$ (12a) $(2\alpha_{max} - 1) \cdot \ln\left(\dfrac{\alpha_{max}}{1 - \alpha_{max}}\right) = \eta$ (13a)

$F(\alpha_{max}) = \ln \alpha_{max}$ (12b) $F(\alpha_{max}) = \ln\left(\dfrac{\alpha_{max}}{1 - \alpha_{max}}\right)$ (13b)

$-f'(\alpha_{max}) = -1$ (12c) $-f'(\alpha_{max}) = 2\,\alpha_{max} - 1$ (13c)

The left hand side of (13a) is essentially a parabolic function
of α_{max}. However, in the range of most interest to the thermal
analyst, $0.9 \leqslant \eta \leqslant 1.0$, ($20 \leqslant E/RT \leqslant 100$), α_{max} is a linear function
of η, given by (13d)

$$\alpha_{max} = 0.6863 + 0.1378 \cdot \eta \qquad (13d)$$

<u>Random Nucleation: Avrami-Erofeev[8,9] Models</u> (A2, A3)

$$f(\alpha) = n \cdot (1-\alpha) \cdot [-\ln(1-\alpha)]^{\frac{n-1}{n}} \qquad (14)$$

$$\alpha_{max} = 1 - \exp\left(\frac{1-n-\eta}{n}\right) \qquad (14a)$$

$$F(\alpha_{max}) = \left(\frac{n + \eta - 1}{n}\right)^{1/n} \qquad (14b)$$

$$-f'(\alpha_{max}) = \left(\frac{n\,\eta^n}{n+\eta-1}\right)^{1/n} \qquad (14c)$$

where n = 2 (A2) or n = 3 (A3)

<u>Phase Boundary Reactions</u> (R2, R3)

$$f(\alpha) = n \cdot (1-\alpha)^{(n-1)/n} \qquad (15)$$

$$\alpha_{max} = 1 - \left(\frac{n - 1}{n + \eta - 1}\right)^{n} \qquad (15a)$$

$$F(\alpha_{max}) = \frac{\eta}{n + \eta - 1} \qquad (15b)$$

$$-f'(\alpha_{max}) = n + \eta - 1 \qquad\qquad (15c)$$

Contracting area (cylindrical symmetry, R2) n = 2
Contracting volume (spherical symmetry, R3) n = 3

Two-Dimensional Diffusion Control (D2)

$$f(\alpha) = - [\ln(1-\alpha)]^{-1} \qquad\qquad (16)$$

$$\frac{1}{\ln(1-\alpha_{max})} + \frac{\alpha_{max}}{(1-\alpha_{max})} \cdot \frac{1}{[\ln(1-\alpha_{max})]^2} = \eta \quad (16a)$$

$$F(\alpha_{max}) = (1-\alpha_{max}) \cdot \ln(1-\alpha_{max}) + \alpha_{max} \qquad (16b)$$

$$-f'(\alpha_{max}) = \frac{[\ln(1-\alpha_{max})]^{-2}}{(1-\alpha_{max})} \qquad\qquad (16c)$$

Equation (16a) is best solved graphically. Using the Sharp et al[10] data, one can calculate $-f'(\alpha) \cdot F(\alpha)$ as a function of α. α_{max} is the value of α at which the product equals η.

Three Dimensional Diffusion (D3, D4)

$$f(\alpha) = \frac{3h(\alpha)}{2 \cdot [(1-\alpha)^{-1/3} - 1]} \qquad\qquad (17)$$

In the Jander[11] model (D3), $h(\alpha) = (1-\alpha)^{2/3}$

$$\alpha_{max} = 1 - (3/2 - \eta)^3 \qquad\qquad (17a)$$

$$F(\alpha_{max}) = (\eta - 1/2)^2 \qquad\qquad (17b)$$

$$-f'(\alpha_{max}) = \frac{\eta}{(\eta - 1/2)^2} \qquad\qquad (17c)$$

In the Ginstling and Brounshlein[12] model, (D4), $h(\alpha) = 1$
In this case, the expressions for α_{max}, $F(\alpha_{max})$ and $-f'(\alpha_{max})$ are more complicated. The equation relating α_{max} as a function of η can be expressed in the form of a quartic, which has one imaginary root and three real roots, two of which are equal, yielding the trivial solution, $\alpha_{max} = 0$. The only root of significance is given by (18a). (18b) and (18c) follow.

$$\alpha_{max} = 1 - \{1/(6\eta) + [1/(36\eta^2) + 1/(6\eta)]^{1/2}\}^3 \qquad (18a)$$

$$F(\alpha_{max}) = 1 - \frac{1}{3}.(1 + \alpha_{max}) - (1 - \alpha_{max})^{2/3} \qquad (18b)$$

$$-f'(\alpha_{max}) = \frac{1}{2.(1-\alpha_{max})^{2/3}.[(1-\alpha_{max})^{1/3} - 1]^2} \qquad (18c)$$

Evaluation of p_m (E/RT) and γ_m (E/RT)

The general expression for $\gamma_m(E/RT)$ is (19)

$$\gamma_m(E/RT) = 1 + \sum_{j=1}^{j=\infty} (-RT/E)^j . \prod_{i=0}^{i=j} (m + i + 1) \qquad (19)$$

This is the most widely used series[13]. Vallet[14] has given
extensive tables of this function for m = ½ and m = 0 to
eight significant figures. γ_1 (E/RT) values may be calculated
from the γ_0 (E/RT) values using relationship (19a)

$$\gamma_1(E/RT) = E/RT . [1 - \gamma_0(E/RT)] \qquad (19a)$$

and from (4) and (19a), one has p_1(E/RT).

$$p_1(E/RT) = \exp (-E/RT).(RT/E)^2 - p_0(E/RT) \qquad (19b)$$

A number of different polynominal series have been proposed
in an attempt to develop the best, approximate, simple analytical
expression for p_m (E/RT). Blazejowski[15] has recently compared
and contrasted their relative efficiency and accuracy. He
recommends the use of continued fraction expansion series,
since a minimal number of polynomial terms are required to
obtain agreement with the exact Vallet values.

Certain aspects of kinetic analysis may require the
calculation of an E/RT value from a known p_m (E/RT) value.
For example, for a first order reaction with the Arrhenius
assumption p_0 (E/RT) = $-\beta R.\ln(1-\alpha)/AE$. Thus, one may calculate α - T relationships using known kinetic parameters.
The Doyle[16] equation does enable the calculation of p_0 (E/RT)
for known E/RT and the inverse. Unfortunately, using this
equation in comparison with the Vallet[14] values, one finds
coefficients of variation, (CV), ranging from -24% (E/RT = 20)
through 0.4% (E/RT = 45) to -19% (E/RT = 60).

One may generalize the Doyle expression in the form:

$$\ln p_m(E/RT) = \sum_{i=0}^{i=j} a_{m,i} . (E/RT)^i \qquad (20)$$

Table 1. Generalized Doyle Equation Parameters

$$\ln\ P_m(x) = a_{m,0} + a_{m,1} \cdot x + a_{m,2} \cdot x^2 + a_{m,3} \cdot x^3$$

(x = E/RT)

Parameter	m	$5 \leqslant x < 25$	$25 \leqslant x < 50$	$50 \leqslant x < 75$	$75 \leqslant x < 100$
$a_{m,0}$	0	-1.64064	-3.65794	-4.68230	-5.36830
	½	-2.02759	-4.56827	-5.84971	-6.70350
	1	-2.40733	-5.47847	-7.01782	-8.04624
$a_{m,1}$	0	-1.45328	-1.16035	-1.09482	-1.06682
	½	-1.56998	-1.20068	-1.11867	-1.08377
	1	-1.68766	-1.24097	-1.14246	-1.10043
$a_{m,2} \cdot 10^4$	0	165.971	21.685	7.589	3.762
	½	209.205	27.158	9.510	4.731
	1	252.917	32.616	11.419	5.664
$a_{m,3} \cdot 10^7$	0	-2505.8	-128.7	-26.9	-9.4
	½	-3162.3	-161.3	-33.8	-11.9
	1	-3826.6	-193.7	-40.6	-14.2

If one limits j = 3, $p_m(E/RT)$ may be calculated for $5 \leqslant E/RT \leqslant 100$
with relative errors (CV) no greater than 1.5% for $5 \leqslant E/RT \leqslant 30$,
and no greater than 0.1% for $30 \leqslant E/RT \leqslant 100$. The values of
the coefficients, $a_{m,i}$ are dependent upon the E/RT value,
as well as m, as shown in table 1. For convenience, k
(cf. equation 3) has been incorporated into the tabulated
coefficents. Some examples of p_m (E/RT) calculations are
given in table 2, with the coefficients of variation from
the Vallet[14] values. With j=3, (20) is a cubic equation
in E/RT, which is solvable. Thus, one can calculate E/RT
for a given p_m (E/RT) value. A Fortran program, used routinely
by the author for such calculations is available upon request.

Table 2. Typical p_m (E/RT) Values

E/RT	m	Equation 20	Exact Value[14]	cv(%)
10	0	$3.87199.10^{-7}$	$3.83024.10^{-7}$	+1.1
20	½	$1.04314.10^{-12}$	$1.02891.10^{-12}$	+1.4
30	1	$3.16314.10^{-18}$	$3.15897.10^{-18}$	+0.1
40	0	$2.53034.10^{-21}$	$2.53153.10^{-21}$	-0.05
50	½	$1.03994.10^{-26}$	$1.04002.10^{-26}$	<-0.01
60	1	$3.86328.10^{-32}$	$3.86373.10^{-32}$	-0.01
70	0	$7.89104.10^{-35}$	$7.89076.10^{-35}$	<0.01
80	½	$3.05870.10^{-40}$	$3.05850.10^{-40}$	<0.01
90	1	$1.08811.10^{-45}$	$1.08812.10^{-45}$	<0.001
100	0	$3.64784.10^{-48}$	$3.64782.10^{-48}$	<0.001

The Generalized Kissinger Equation

The γ_m (E/RT) and η_m (E/RT) functions can now be easily
calculated for any E/RT value one is likely to encounter
in thermal analysis. For any of the kinetic models considered,
one can calculate α_{max}, $-f'(\alpha_{max})$ and hence $\ln \phi_m (\alpha_{max})$.
Table 3 shows the results of such calculations, using typical
kinetic parameter values and making the Arrhenius assumption.
In this case, $\phi_0 (\alpha_{max}) = -f'(\alpha_{max})$ and $\eta_0 (E/RT_{max}) = \gamma_0 (E/RT_{max})$.
As can be seen, with the exception of the R3, D3 and D4 models,
the exclusion of the $\ln \phi_m (\alpha_{max})$ term in the generalized
Kissinger equation (9) has less than a 3% effect, and for
the simple n^{th} order and Avrami-Erofeev models, A2, A3, less
than a 1% effect. Similar figures result if one sets m = ½
or m = 1 in the rate equation (1). As the Arrhenius exponent
E/RT_{max} increases, there is a gradual small increase in
α_{max} for all models. For all n^{th} order and diffusion controlled
models, $\phi_m (\alpha_{max})$ decreases slightly with increase in E/RT_{max}.
For all other models, there is a slight increase. However,
over the entire range $25 \leqslant E/RT \leqslant 100$, the orders of magnitude
of the effects of the $\phi_m (\alpha_{max})$ term are as given in table 3.

Table 3. $\phi_O(\alpha_{max})$ Values for Various Kinetic Models

$E = 220 \text{ kJ.mole}^{-1}$ $A = 10^{15} \text{ min}^{-1}$ $T_{max} = 442°C$

η_O (37) = 0.949129

Model	α_{max}	$\phi_O(\alpha_{max})$	$\ln \phi_O(\alpha_{max})$	$\dfrac{\ln \phi_O(\alpha_{max})}{\ln(AR/E)}$
F,1	0.6129	1	0	0
F,2	0.4746	1.0509	0.0496	0.002
F,5	0.2996	1.2036	0.1853	0.008
E1	0.3871	-1	0	0
B1	0.8171	0.6342	-0.4554	-0.019
A2	0.6226	0.9614	-0.0394	-0.002
A3	0.6258	0.9545	-0.0465	-0.002
R2	0.7368	1.9491	0.6674	0.027
R3	0.6881	2.9491	1.0815	0.044
D2	0.8125	1.9033	0.6435	0.026
D3	0.8328	4.7057	1.5488	0.064
D4	0.7500	9.2013	2.2193	0.091

The Kissinger equation, when expressed in its modified form (9) is a general thermoanalytical expression, applicable to a wide range of kinetic models. In the absence of other criteria, the value of α_{max} will assist in indentifying the model. Unfortunately, one cannot distinguish between simple first order and Avrami-Erofeev models (F1, A2, A3). Using the DTG capability, available with most modern thermogravimetric instruments, one can easily calculate α_{max} for reactions involving weight changes, especially if moderate to high heating rates are used. In this case, the DTG signal is enchanced. However, one must suitably calibrate the instrument such that the sample temperature is known with accuracy and precision, [17] since the essence of the method is the establishment of a significant temperature. In the case of DSC measurements, it will be necessary to measure the partial reaction heat from the quiescent state to the maximum heat flow, as well as the total heat of reaction in, at least one experimental run, in order to evaluate α_{max} and thereby calculate the magnitude of the $\phi_m(\alpha_{max})$ correction term.

References

1. Elder, J.P. and Harris, M.B., paper submitted to Fuel
2. Kissinger, H.E., J. Res. NBS, 57, 217 (1956)
3. Hanaba, P., Jüntgen, H. and Peters, W. Brennstoffe-Chemie, 49, 368 (1968)
4. Garn, P.D., CRC Critical Reviews in Analytical Chemistry, 65, Sept. 1972
5. Šesták, J., and Berggren, G., Thermochim. Acta, 3, 1 (1971)
6. Criado, J.M. Thermochim. Acta, 43, 111 (1981)
7. Balarin, M., Thermochim. Acta, 24, 176 (1978)
8. Avrami, M., J. Chem. Phys., 9, 177 (1941)
9. Erofeev, B.V., CR (Doklady) Acad. Sci., URSS, 52(6), 511 (1946)
10. Sharp, J.H., Brindley, G.W. and Achar, B.N.N., J. Amer. Ceram. Soc., 47, 379 (1966)
11. Jander, W., Z. Anorg. Allgem. Chem., 163 (1-2), 1 (1927)
12. Ginstling, A.M. and Brounshtein, B.I., J. App. Chem. USSR, 23(12), 1327 (1950)
13. Šesták, J., Thermochim. Acta, 3, 150 (1971)
14. Vallet, P., Numerical Tables, Gauthier-Villars et Cie, Paris, 1961
15. Blazejowski, J., Thermochim. Acta, 48, 125 (1981)
16. Doyle, C.D., J. App. Pol. Sci., 6(24), 639 (1962)
17. Elder, J.P., Thermochim. Acta, 52, 235 (1982)

THE CHARACTERIZATION OF WATER IN TREED CROSSLINKED POLYETHYLENE
FILMS

R. A. Weiss, S. H. Shaw and M. T. Shaw

Institute of Materials Science
University of Connecticut
Storrs, CT   06268

INTRODUCTION

A particularly troublesome problem to the electric utility
industry is the occurrence of tree-like defects in polyethylene-
insulated transmission cables that have been energized for a
period of time in the presence of water (1). These defects, water
trees, are associated with decreased dielectric strength (2) and
may be responsible for premature failure of underground cables
(3). Although it is generally agreed that both water and electrical stress are necessary for the initiation and growth of water
trees, no single mechanism for the formation of these defects is
universally accepted (4).

Despite the intensive effort over the past decade to determine the origin and structure of water trees in polyolefin insulation, their morphology is still not known with any certainty.
Careful electron microscopy by several groups (3,5) has produced
somewhat contradictory results: Kalkner et al (3) report that
water trees are channel-like and grow between spherulites, while
Bamji et al (5) claim that a water tree consists of a cloud of
voids with no apparent connection between them. When the water is
removed from a treed polyethylene the defect disappears, but the
water tree reappears upon re-immersion of the specimen (6). This
result not only indicates that the tree-related damage to the
polymer is permanent, but also that the tree contains water.

The purpose of the research described herein was to study the
content and transport of water in electrically stressed crosslinked polyethylene (XLPE). Of particular interest was the effect
of water-tree damage on the diffusion of water and on the

265

structure of water (i.e., bound versus free water) in XLPE. Water that fills the voids in the trees is expected to be free water, and the quantitative measurement of this water may provide an alternative to microscopy as a method for detecting and sizing trees. The non-destructive coulometric and calorimetric techniques used for water characterization are simpler and more convenient than the tedious and artifact-prone microscopic techniques normally used to evaluate treeing damage in polyethylene insulation. Furthermore, these methods are also applicable to filled, non-transparent insulation, for which microscopy is especially difficult.

Mass transport in the direction of the tree propagation may possibly resolve the morphological dispute concerning whether water trees are channels or discrete voids. If channels are present, one would expect unusually high water diffusivities in treed polymer relative to the untreed material. Therefore, one objective of this work was to study the diffusion constant of water in treed XLPE and to relate the results to the morphology of the trees.

EXPERIMENTAL

Crosslinked polyethylene (XLPE) samples 1 mm thick, were prepared from a commercial low density polyethylene, USI-NA143, by adding 2% (wt) of dicumyl peroxide on a heated two-roll mill at 115°C and curing between Mylar film and steel plates at 177°C and a pressure of 41 MPa for 24 min. Residual breakdown products of the curing reaction, e.g., acetophenone, were removed by placing the samples overnight in a vacuum oven at 85°C.

Sites for tree-growth were introduced by scratching the surface of XLPE films in a controlled and uniform manner with 120-grit sandpaper using a specially designed fixture attached to a Rheometrics Mechanical Spectrometer (RMS). The scratching was done in a 1-cm diameter area using eccentric rotating disks at 1.5 mm offset, 10 rad/s angular velocity, and a 0.5-kg normal force applied for 10s. Laboratory-treed XLPE samples were prepared in the apparatus described by Nunes et al (7) by immersing the scratched specimens in a solution of 0.5N NaOH and exposing them for various times to a voltage of 2.5 kV and a frequency of 1 kHz. Controls were treated identically, except that no voltage was applied.

Water concentration in XLPE film was determined coulometrically using a DuPont Model 903, Moisture Evolution Analyzer (MEA). The instrument was isolated in a dry box flushed with dry N_2 to minimize extraneous sources of moisture. Analyses were run for 30 min. at 70°C and the total amount of water evolved from a pre-weighed sample was read directly from the instrument.

Calorimetric measurements were made with a Perkin Elmer
Differential Scanning Calorimeter, Model 2 (DSC), that was equipped
with a liquid N_2 cooling accessory and a Perkin Elmer Data Station,
Model 6500. DSC thermograms were obtained between 180K and 300K
using cooling and heating rates of 40K/min. The concentration of
clustered water, C, in ppm was calculated from the thermograms
with the following equation:

$$C = \frac{\Delta H_{tr}}{\Delta H_f} \cdot 10^6 \qquad\qquad\qquad (1)$$

where ΔH_{tr} is the heat of transition and ΔH_f is the heat of fusion
or crystallization of water. Since ΔH_f is a function of temperature, different values were used in equation 1 depending upon
where the transition was observed. For the cooling experiments,
clustered water exhibited supercoolings of the order of 40°C and
a value of 55 cal/g was used for ΔH_f (8); upon heating, the normal
melting point of ice was observed, ca. 0°C, and ΔH_f was taken to
be 79.7 cal/g (9).

Microscopic examinations of the treed samples were made using
an optical microscope at a magnification of 100X. The samples
were stained with methylene blue following the method described
by Henkel et al (10). Cross sections were cut from the center of
treed sample and a representative field was selected and photo-
graphed. The percent treed area was calculated from the micrograph
and was used as an index of tree damage.

Water desorption and absorption measurements were carried
out on specimens treed for 5 and 15 days. Drying was accomplished
in a convection oven at 40°C. Samples were weighed periodically
until they were dry on a Perkin Elmer Autobalance, Model AD-2Z.
At the completion of the drying experiment, the samples were placed
in water at 40°C, and the resorption of water was measured as a
function of time by periodically removing the sample from the
water, drying the surface with a tissue, and weighing.

RESULTS AND DISCUSSION

Water Analysis and Characterization

A typical micrograph of a water tree grown in XLPE energized
for 5 days is shown in Fig. 1. The percent treed area is plotted
against treeing time in Fig. 2; with a few exceptions, each datum
point represents at least 4 determinations. Tree growth was
observed within the first 12 hours of the experiment, and after
15 days many of the trees extended nearly through the sample.
The unusually low treed area in the samples treed for 25 days

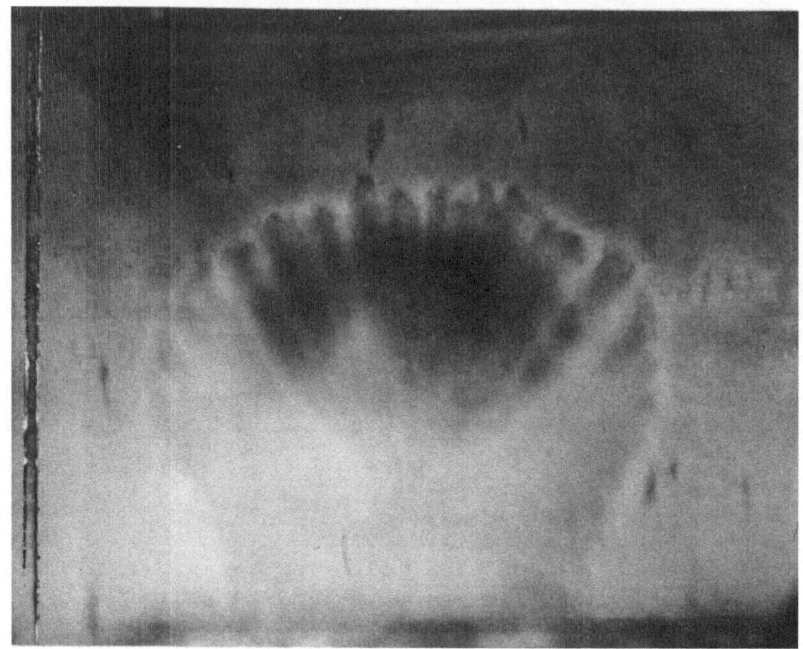

Fig. 1. Water tree in XLPE energized for 5 days.

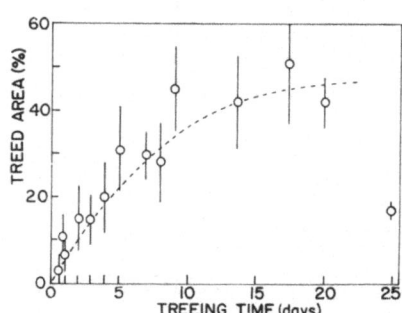

Fig. 2, Treeing damage versus treeing time for XLPE.

most likely can be attributed to a large number of samples in the treeing apparatus at that time that resulted in a low effective voltage.

The effect of water trees on the amount of water sorbed by XLPE as measured by the Moisture Evolution Analyzer (MEA) is shown in Fig. 3, and these data are replotted against the treed area, determined from Fig. 2, in Fig. 4. These data clearly demonstrate that water trees significantly increase the total amount of water in XLPE, and these results are consistent with the hypothesis that the trees are voids that can accommodate water.

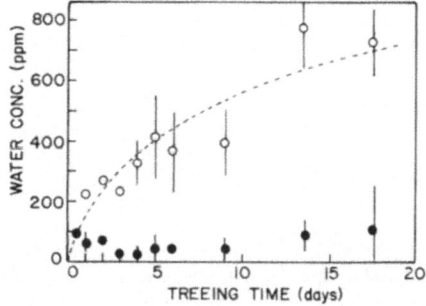

Fig. 3 Water concentration versus treeing time for XLPE: (O) treed samples, (●) control samples.

Bair and co-workers have considered the phenomenon of water clustering in polyethylene (11) and other polymers (12). In polyethylenes soaked in water at elevated temperatures, they observed a low temperature crystallization exotherm upon cooling the sample in a DSC. A melting endotherm near 0°C was found when the sample was reheated. They attributed these results to clustered water that resulted from supersaturation of the sample when the polymer was removed from the water bath and cooled to room temperature (10). In the DSC experiment, the water crystallized at ca -40°C, which was due to the extremely small size of the clusters, ca. 2 μm.

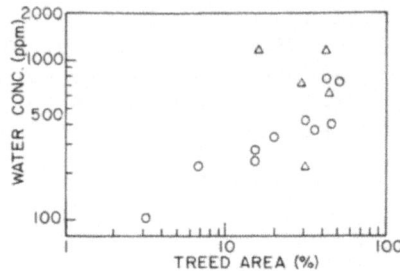

Fig. 4. Water concentration versus treed area for XLPE: (O)
         MEA data, (Δ) DSC data.

        In the present treeing study of XLPE, it was thought that
water in the trees should be free water and, therefore, should
exhibit the crystallization and melting transition of ice. Typical DSC cooling and heating thermograms of a treed XLPE sample are
shown in Fig. 5. As evidenced by the low temperature exotherm
upon cooling and the endotherm near 0°C upon heating, this sample
clearly contains clustered water. The amount of clustered water
can be determined by measuring the enthalpy of the transition and
comparing it to the heat of fusion of ice. This has been done
for all the treed specimens, and the results are given in Fig. 6.
The amount of free water in the treed samples increases with increasing tree damage, and it appears that within experimental
error all of the excess water, i.e., the water in excess of that
absorbed by the control XLPE, is free water. This result strongly
supports the presumption that the water fills the trees, and the
measurement of free water offers an attractive alternative to
microscopy for the quantification of treeing damage in cable
materials. Since the freezing temperature of water is related to
the droplet size (13), the amount of supercooling required to
effect the crystallization of water in the treed specimens may
be used to gauge the tree size. From the DSC curves obtained in
this study the water cluster diameter is estimated to be less than
10 μm. This conclusion is consistent with the recent microscopic

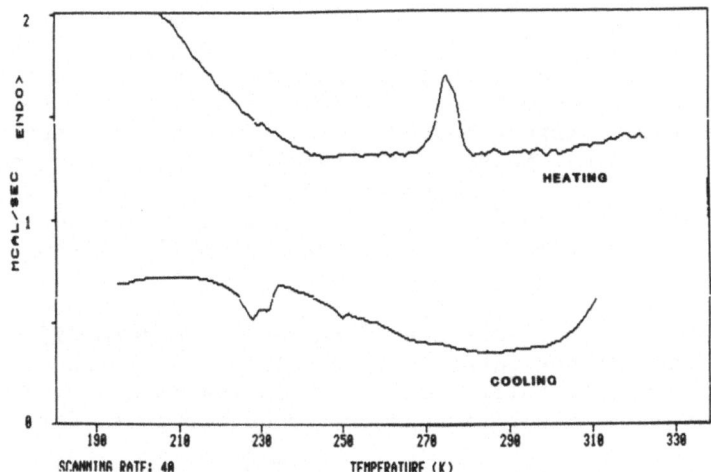

Fig. 5.   Typical DSC thermograms of treed XLPE showing crystalli-
          zation and melting of free water.

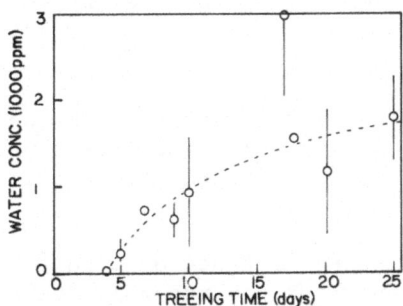

Fig. 6.   Free water concentration versus treeing time for XLPE.

evidence of Bamji et al (5), who found voids in XLPE cable insulation on the order of 1-10 μm.

Diffusion Studies

In a previous study (14), water diffusivities in treed polyethylene were measured at 70°C using desorption data from the MEA.
Relative to undamaged polymers, treed polyethylene was found to
have much higher moisture content, but lower diffusivity. These
values were obtained, however, at extremely low moisture contents,
e.g., <400 ppm, and it could be argued either that any channels
in the polymer had closed up or that the limiting process had
become one of desorption of water from hydrophilic sites in the
aged sample.

An objective of the present work was to measure diffusivities
in XLPE at much higher water concentrations where tree channels,
if present, would be fully swollen. This was done by measuring
water desorption and sorption isotherms on the treed XLPE films.

Raw water absorption and desorption data were analyzed using
the integrated form of Fick's equation for slabs:

$$X = \frac{8}{\pi^2} \sum_{n=1}^{\infty} \exp\left[-\ (2n-1)^2 Dt/H^2\right]/(2n-1)^2] \tag{2}$$

where X is the unaccomplished relative moisture change, D is the
diffusivity (m^2/s); t is time (s); and H is the slab thickness
(m). This equation assumes an initially uniform moisture level in
the slab, a diffusivity which is independent of time or position,
and equal transport through both faces of the slab. The unaccomplished moisture change is given by the expression

$$X = \frac{w(t)-w(\infty)}{w(o)-w(\infty)} \tag{3}$$

In this expression w(t) is the weight of the sample at time t,
including time = 0 and time = ∞. The latter weight is not
experimentally accessible, in principle, and it was decided to
leave it as a parameter.

To handle the fitting of Equation 2 to the weight data, it
is convenient to look at 1-X, or the relative accomplished
moisture change:

$$1-X = \frac{w(o)-w(t)}{w(o)-w(\infty)} \tag{4}$$

In this expression, the numerator is simply the weight change,
while the demonimator -- the infinite time weight change -- can

be left as a multiplicative parameter. The objective function, F,
with parameters a_i is then

$$F = \sum_i \{\ln(\Delta w_i) - \ln a_1 - \ln [1 - (8/\pi^2)$$
$$\sum_n \exp [-(2n-1)^2 a_2 t_i / H^2]/(2n-1)^2]\}^2 \qquad (5)$$

where Δw_i is the absolute value of the weight change after time
t_i, a_1 is the infinite-time weight change, and a_2 is the diffusivity. In practice, the summation over n was truncated after 3
terms.   Iteration was continued until F decreased by no more than
1%.

Table 1 lists the diffusivities and moisture contents for
slabs treed for 5 and 15 days. On comparing the desorption
results for the 5- and 15-day samples we find the expected change
in the water content: the more trees the greater the moisture
content ($t = 7.4$; d.f. = 13; $p < 0.001$) by about a factor of two.
The diffusivities may be different ($t = 1.4$; d.f. = 13; $0.2 < p <$
0.1), but a normalization for the amount of water must first be
made.

For the absorption sequence a lower diffusivity is found along
with lower water contents, the latter being much closer to those
found with the MEA. The pattern of lower diffusivity with lower
water content was also found in previous studies using the MEA
(14). Nevertheless, in spite of the higher water content and more
extensive damage of the 15-day samples relative to the 5-day samples, the diffusivities are not significantly higher when the

Table 1.   Diffusivities and Moisture Contents of
         Treed Polyethylene

Treeing Time, days	Mode	T°C	$(D, m^2/s) \times 10^{13}$	Total Water Content, %
5	Desorption	40	9.2 (±1.2, 7)[a]	0.18 (±0.03,7)
	Absorption	40	1.5 (±0.6, 5)	0.011(±0.005,5)
15	Desorption	40	10.0 (±1.0, 8)	0.34 (±0.05,8)
	Absorption	40	2.4 (±0.8, 6)	0.048(±0.01,6)

[a] Mean (± standard deviation, number of independent
    observations)

comparison is restricted to either the absorption or desorption mode. In fact, if the diffusivities are compared at equal moisture contents, an argument can be forwarded for a decrease in diffusivity with treeing time; this is illustrated in Figure 7. The diffusivities listed in Table 1 are somewhat lower (0.2 to 0.7 decades) than those reported by Yoshimitsu et al (15) for untreed XLPE; we interpret this as further substantiation of the lack of effect of treeing on moisture diffusion.

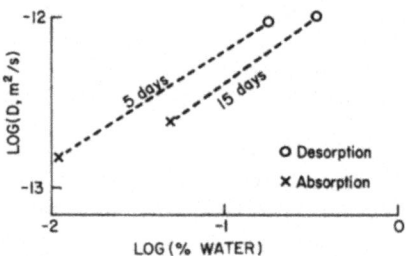

Fig. 7. Diffusion constant versus water concentration for XLPE treed for 5 and 15 days.

CONCLUSIONS

The relationship between water and treeing of XLPE has been studied. It was found that as the treeing damage to XLPE increased the amount of water absorbed by the sample also increased. This excess water is crystallizable, free water, and it is suggested that the trees are filled with water. Since the amount of free water can be measured calorimetrically, DSC analysis of insulation is proposed as a facile method for detecting and possibly sizing water trees. Diffusivities of water calculated from water sorption and desorption isotherms indicate that the diffusion constant may decrease with increasing damage in treed XLPE. This result is in agreement of Bamji et al that trees are unconnected voids.

REFERENCES

1. J. H. Lawson and W. Vahlstrom, IEEE Trans., PAS-92, 824 (1973).
2. G. Bahder, G. S. Eager, and R. G. Lukac, IEEE Ann. Rep., Conf.
    on Elec. Insul. and Dielectric Phenomena, 1974, p. 289.
3. W. Kalkner, U. Müller, E. Peschke, H. J. Henkel, and R. von
    Olshausen, CIGRE 21-07 (1982).
4. S. L. Nunes and M. T. Shaw, IEEE Trans., EI-15, 432 (1980).
5. S. Bamji, A. Bulinski, J. Densley, and A. Garton, IEEE Trans.,
    EI-18, 32 (1983).
6. R. M. Eichhorn, IEEE Trans., EI-12, 2 (1976).
7. S. L. Nunes, M. T. Shaw, S. H. Shaw, and R. A. Weiss, IEEE
    Ann. Rep., Conf. on Elec. Insul. and Dielectric Phenomena,
    1982.
8. S. Yasufuku, IEEE Trans., EI-17, 464 (1982).
9. Handbook of Chemistry and Physics, 59th Edition (1978-79),
    CRC Press, B-272.
10. H. J. Henkel, W. Kalkner, and N. Müller, Siemens Forsch-U.
    Entwickl.-Ber. Bd. 10 Nr. 4, Springer-Verlag, 1981, p. 205.
11. H. E. Bair and G. E. Johnson, in "Analytical Calorimetry",
    Vol. 4, R. S. Porter and J. F. Johnson, ed., Plenum Press,
    New York, 1977, p. 219.
12. H. E. Bair in "Thermal Characterization of Polymeric Materials",
    E. Turi, ed., Academic Press, 1981, p. 845.
13. N. H. Fletcher, "The Chemical Physics of Ice", Cambridge
    University Press, London, 1970, p. 87.
14. M. T. Shaw and S. H. Shaw, Proc. of the 1982 IUPAC Symp. on
    "Interrelations Between Processing, Structure, and Properties
    of Materials", Athens, Greece, 1982.
15. T. Yoshimitsu, H. Mitsui, S. Kenjo, and T. Nakakita, IEEE
    Trans., EI-18, 23 (1983).

APPLICATION OF COMBINED THERMOGRAVIMETRY ATMOSPHERIC PRESSURE

CHEMICAL IONIZATION MASS SPECTROMETRY TO COPOLYMER ANALYSIS

Susan M. Dyszel

U.S. Customs Service
1301 Constitution Ave., N.W.
Washington, D.C. 20229

INTRODUCTION

The U.S. Customs laboratories encounter a wide variety of analytical problems. For example, when a sample of imported rubber arrives in one of the U.S. Customs laboratories, a series of determinations must be made: 1) is it a rubber by definition of the tariff, 2) is it entirely natural rubber, or a synthetic blend, 3) what is its composition by weight. The answers to these questions determine the tariff classification and the applicable rate of duty.

This study combines a thermal analysis technique – thermogravimetry – with atmospheric pressure chemical ionization mass spectrometry and applies the combined technique to the third question. The literature contains references relating to the analysis of styrene butadiene copolymers using thermal analysis techniques (1-5). Pyrolysis – mass spectrometry (5) and vacuum thermogravimetry – mass spectrometry (7) have also been used to investigate polymers such as polystrene and styrene butadiene rubber.

Thermogravimetry – atmospheric pressure chemical ionization mass spectrometry (TG-APCIMS) was introduced for the investigation of polysaccharides (8). This technique relies on the unique capability of the APCIMS to accept samples without a preliminary vacuum or micrometering interface. Thus, the APCIMS is ideal for accepting the effluent from the thermogravmetric balance.

Experimental

The instrumentation used in this study included the Perkin
Elmer TGS-2, temperature controlled by the PerkinElmer System 4
microprocessor controller. The balance output was recorded by
the Bascon Turner 8120 recorder. The APCIMS was the Sciex TAGA
3000. The operating parameters were controlled by a Digital
PDP-11 computer, Digital RX02 dual disk drive, a Tektronix 4025
Terminal and Tektronix 4631 Hard Copy unit. The interface
between the TGS-2 and the inlet of the TAGA consisted of teflon
tubing into a glass tee swept by zero air. This interface was
discussed previously (8). The operating parameters are listed in
Table 1.

Samples of styrene butadiene rubber having known styrene
content and identified as block copolymers were obtained from
Aldrich Chemical Co. Samples of Cariflex SBR were obtained from
the Los Angeles Customs laboratory. No specific information is
known of additives for the samples except S-1502 which has an
aromatic oil additive. Samples were cut into small cubes to fit
into the balance pan and weighed between 3 and 12 milligrams.
The samples run in this work are listed in Table 2.

Table 1   Experimental Conditions

TGS-2

Temperature programmed 60 to 800°C at 40°/minute
N$_2$ purge at 100ml/min. for pyrolysis

Air at 100ml/min. for final combustion after 15 minutes had
elapsed (660°C)

INTERFACE

Zero Air Sweep gas at 0.9 1/min

TAGA

Masses scanned 15-200 amu
1 scan each 30 seconds
Positive ionization mode

Table 2   SBR Samples

100% Styrene – Crystalline Polystyrene
85% Styrene – Styrene Butadiene Copolymer
45% Styrene – Styrene Butadiene Copolymer
30% Styrene – Styrene Butadiene Copolymer
23% Styrene – Styrene Butadiene Copolymer
5% Styrene – Styrene Butadiene Copolymer
Cis Polybutadiene
Cis-Trans Polybutadiene

Cariflex S-1011 Styrene Butadiene
Cariflex S-1013 Styrene Butadiene
Cariflex S-1502 Styrene Butadiene
Cariflex S-1712 Styrene Butadiene
Cariflex S-1009 Styrene Butadiene Divinyl Benzene

Results

The molecular ions for sytrene ($M/e^+ = 104$) and butadiene ($M/e^+ = 54$) were selected as the basis for relating the stryene content to the observed results. Figures 1, 2, and 3 are examples of a single mass scan for polystyrene, 45% styrene butadiene copolymer and cis-trans polybutadiene respectively. For each of the sample mass scans, the intensity of ions 54 and 104 was tabulated. It was observed that if the system background was measured by the intensity of 54 and 104 in the first scan of the run (effective sample temperature 60 to 80°C), those intensities were sometimes greater than the intensity of the same ions at higher temperatures. This phenomenon was also noted for system blank runs. The simple substraction of the background intensity thus was precluded.

Therefore, the following relationships were established and computed:

$$A_n = \frac{\text{ion intensity of } M/e^+ = 104 \text{ for scan n}}{\text{ion intensity of } M/e^+ = 104 \text{ for scan 1}}$$

$$B_n = \frac{\text{ion intensity of } M/e^+ = 54 \text{ for scan n}}{\text{ion intensity of } M/e^+ = 54 \text{ for scan 1}}$$

$$C_n = A_n + B_n$$

where n equals the TAGA scan number.

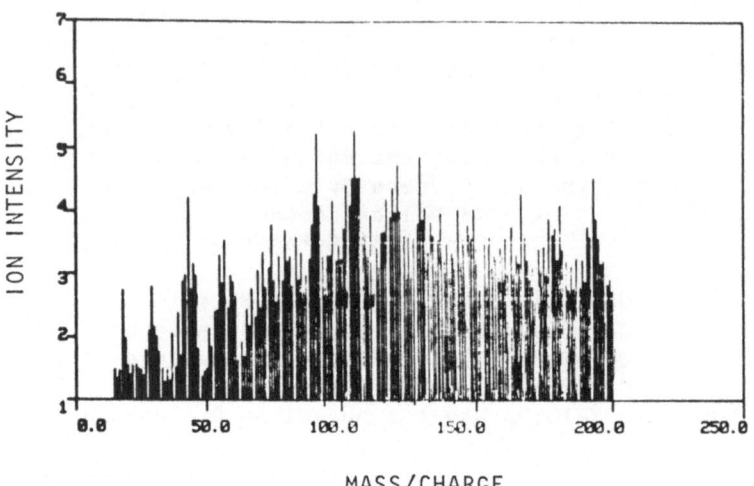

Fig. 1 Mass Scan for Polystyrene between the limits of 15 and
        200 amu.

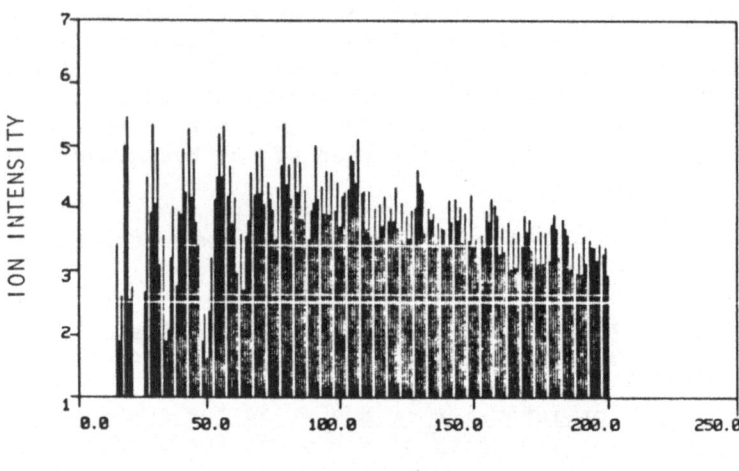

Fig. 2 Mass scan for 45% Styrene in Styrene Butadiene copolymer
        between the limits of 15 and 200 amu.

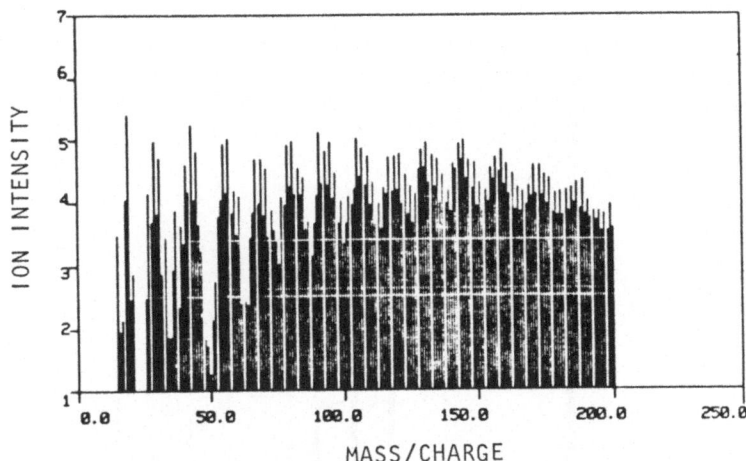

Fig. 3 Mass scan for cis, trans Polybutadiene between the limits
of 15 and 200 amu.

Figures 4, 5 and 6 illustrate graphically what happen to the
ion intensity ratio when the ion intensities seen in Figures 1, 2
and 3 are divided by the initial mass scan in each sequence. The
computed A_n, B_n and C_n are tabulated and then plotted for
each sample as of function of the scan number. Examples of these
plots are shown in Figures 7-11. Figure 12 plots the same ratios
determined for a system background run. Note the general absence
of peaks for any of the 3 plotted quantities. These figures
indicate that for increasing amounts of styrene in the copolymer
sample the intensity ratio also increases. The peak scan number
can be converted to a temperature scale following the temperature
program by the following equation:

$$T_n^O = [Scan \ n-1 \ x \ 20^O/scan] + 60^O.$$

This equation assumes linearity in the temperature program.
Table 3 lists the ion intensity peaks and peak temperatures for
the known sample copolymer tested.

The themogramivetric curves for the preceeding data are shown
as Figures 13 and 14. Figure 13 shows the family of TG pyrolysis
curves produced under the stated TG conditions for varying amounts
of sytrene in the SBR blend. It is observed that the decompositon
temperatures are similar as might be expected providing the
pyrolysis conditions are kept constant.

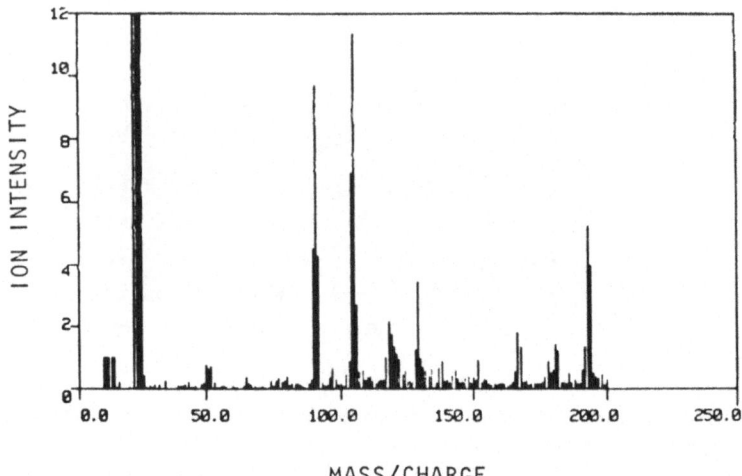

Fig. 4   Ratio of peak mass scan divided by first mass scan for
        Polystyrene.

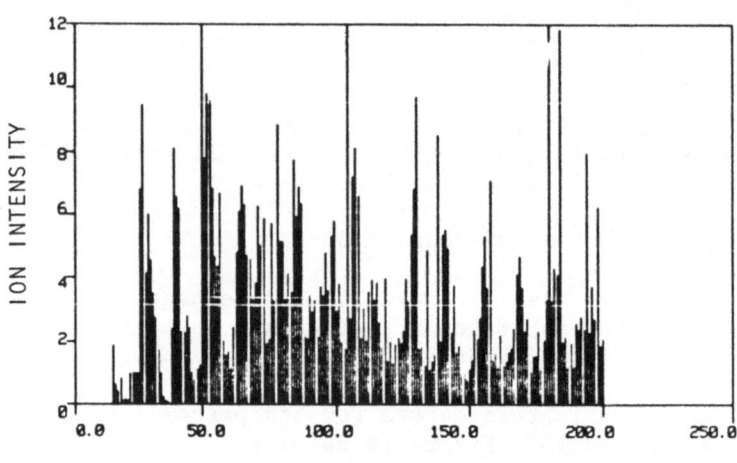

Fig. 5   Ratio of peak mass scan divided by first mass scan for
        45% styrene in Styrene Butadiene copolymer.

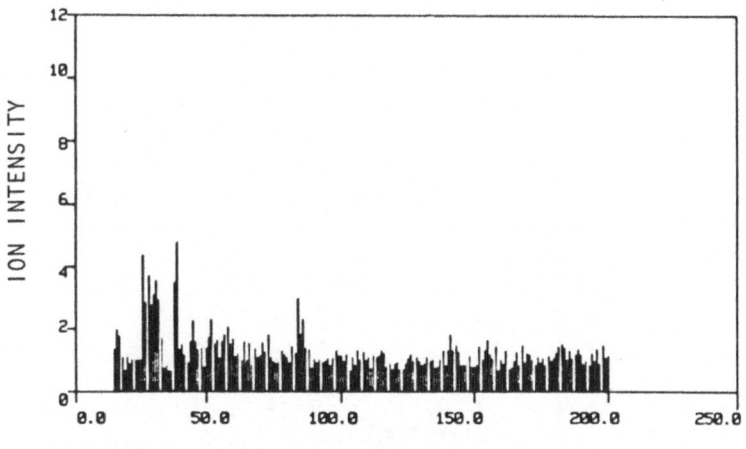

Fig. 6   Ratio of peak mass scan divided by first mass scan for
         cis, trans polybutadiene.

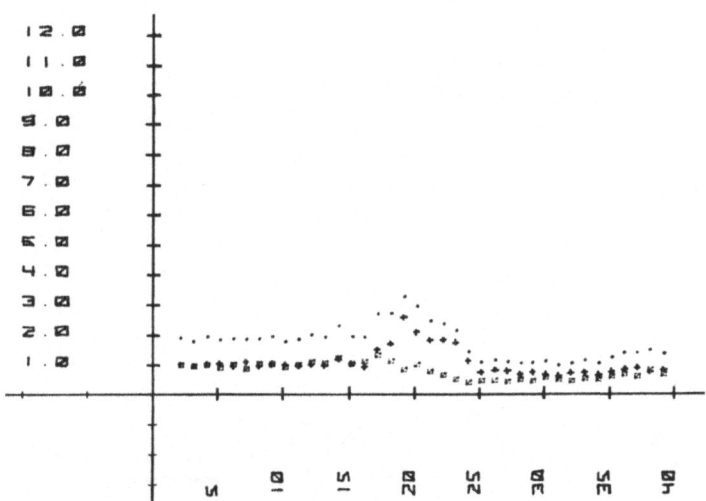

Fig. 7   Intensity ratio plotted as a function of scan number ▨A,
         +B, · C for cis, trans Polybutadiene.

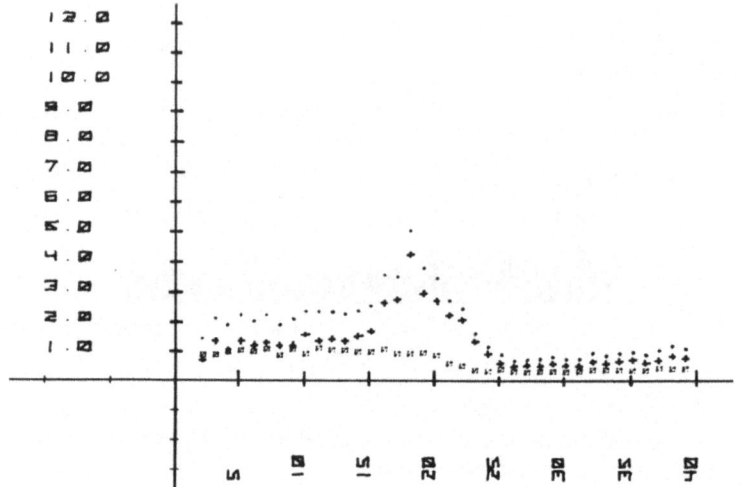

Fig. 8   Intensity ratio plotted as a function of scan number ▨ A,
         +B, · C for 5% Styrene in Styrene Butadiene copolymer.

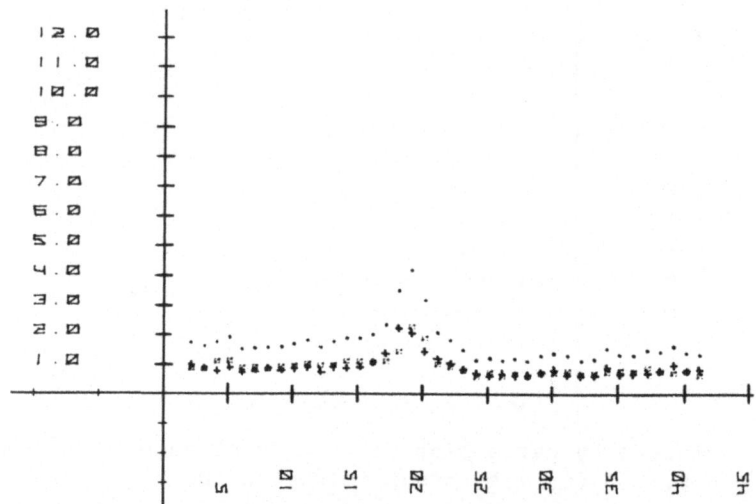

Fig. 9   Intensity ratio plotted as a function of scan number ▨ A,
         +B, · C for 23% Styrene in Styrene Butadiene copolymer.

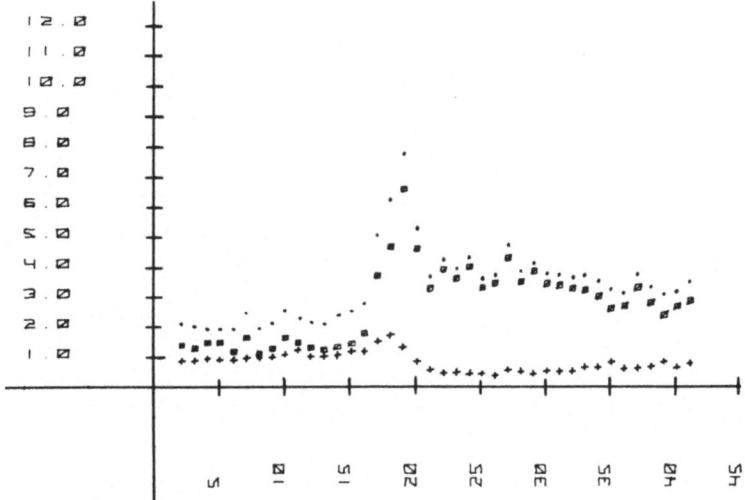

Fig. 10   Intensity ratio plotted as a function of scan number ⊠ A,
          +B, · C for 85% Styrene in Styrene Butadiene copolymer.

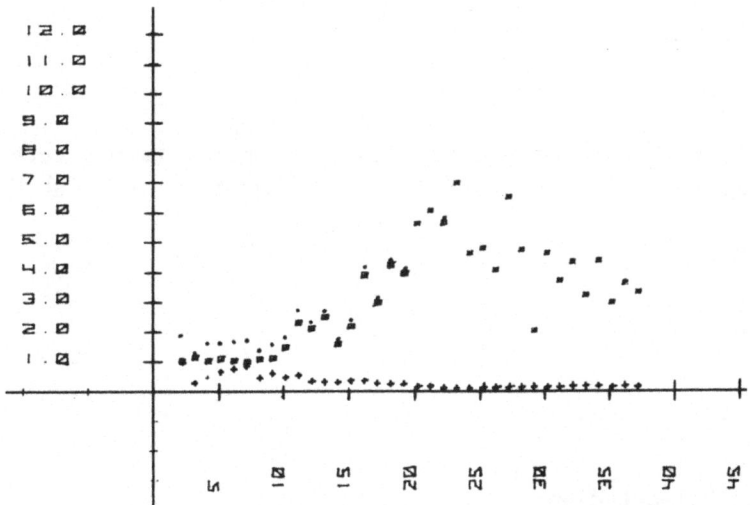

Fig. 11   Intensity ratio plotted as a function of scan number ⊠ A,
          +B, · C for Polystyrene.

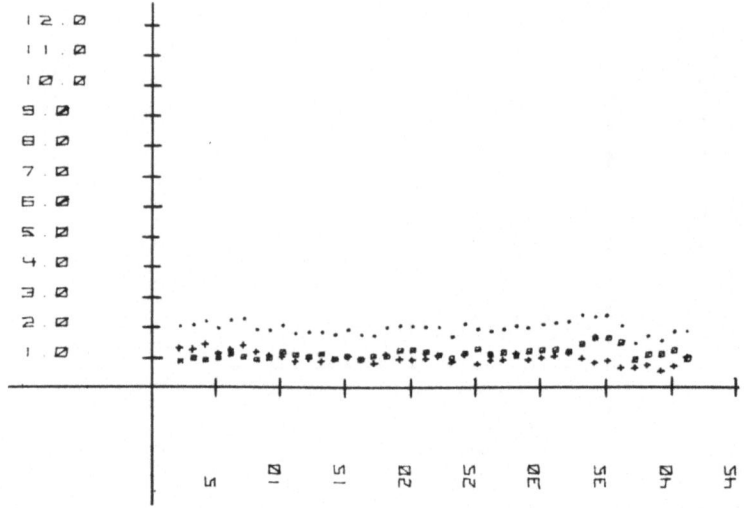

Fig. 12 Intensity ratio plotted as a function of scan number ⊠ A,
         +B, · C for system background.

Table 3

	Scan number at peak A	Temperature °C at Peak A	Scan number at peak B	Temperature °C at Peak B
5% S SBR	16	360	18	400
23% S SBR	19	420	18	400
30% S SBR	19	420	17	380
45% S SBR	18	400	18	400
85% S SBR	19	420	18	400
100% S	23	500	7	200
CIS Polybutadiene				
	6	160	19	420
CIS-trans Polybutadiene				
	17	380	19	420

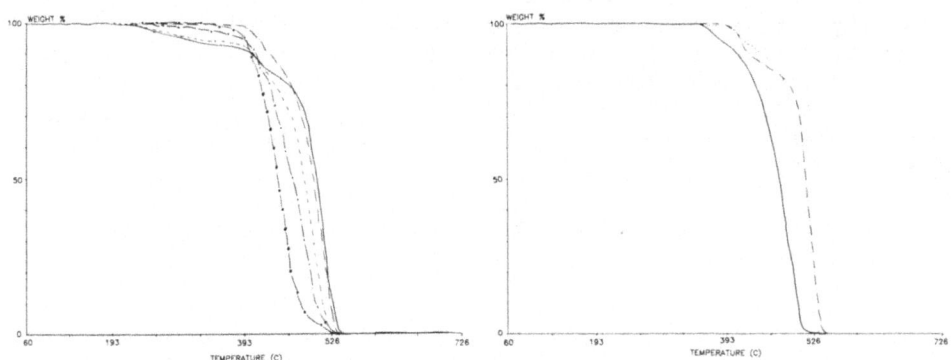

Fig. 13 Thermogravimetric curves   Fig. 14 Thermogravimetric Curves
     for SBR copolymers

————————— 5% Styrene   ————————— Polystyrene

——— ——— 23% Styrene   · · · · · Cis, trans

——·—— 30% Styrene   polybutadiene

——+—— 45% Styrene   ————————— Polybutadiene

——o—— 85% Styrene

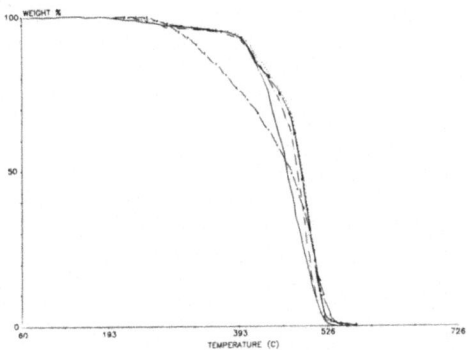

Fig. 15 Thermogravimetric Curves for SBR Copolymers

————————— S-1013

—— —— —— S-1011

——·—— S-1502

——+—— S-1712

——o—— S-1009

To demonstrate the application of this method to unknown
blends five blends were selected. Figure 15 shows the pyrolysis
TG curves for these SBR blends and the plotted A, B and C
intensity ratios are shown as Figure 16-20. By comparison with
the A, B, C plots of the known blends, these blends rank in the
following order of increasing styrene content S-1502   S-1009
S-1712   S-1011   S-1013. The low ranking of S-1502 is further
confirmed by the presence of an aromatic oil modifier in the
blend.

The intensity peak temperature correlation was made and is
tabulated in Table 4.

Table 4

Sample ID	Scan number at peak A	Temperature °C at Peak A	Scan number at peak B	Temperature °C at Peak B
S-1009	18	400	17	380
S-1712	19	420	17	380
S-1502	19	420	19	420
S-1013	18	400	17	380
S-1011	19	420	18	400

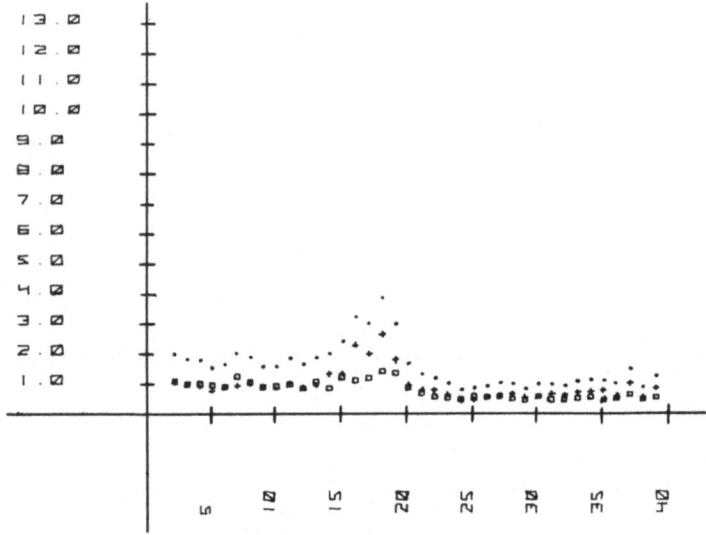

Fig. 16   Intensity ratio plotted as function of scan number   A,
          +B, ·C for SBR copolymer S-1502.

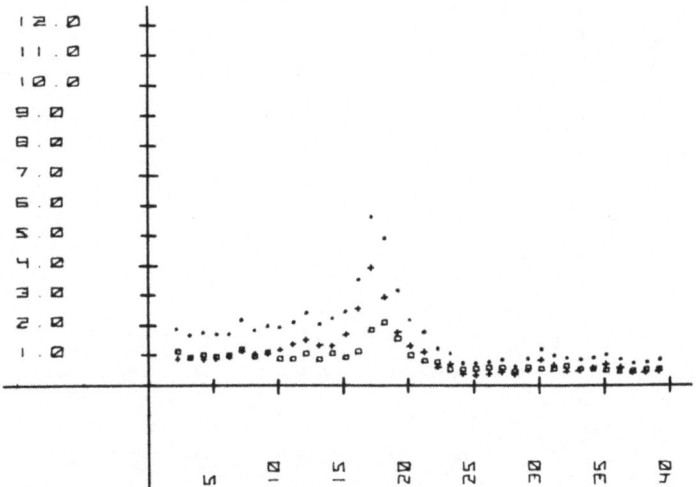

Fig. 17   Intensity ratio plotted as function of scan number ▧ A,
+B, ·C for SBR copolymer S-1009.

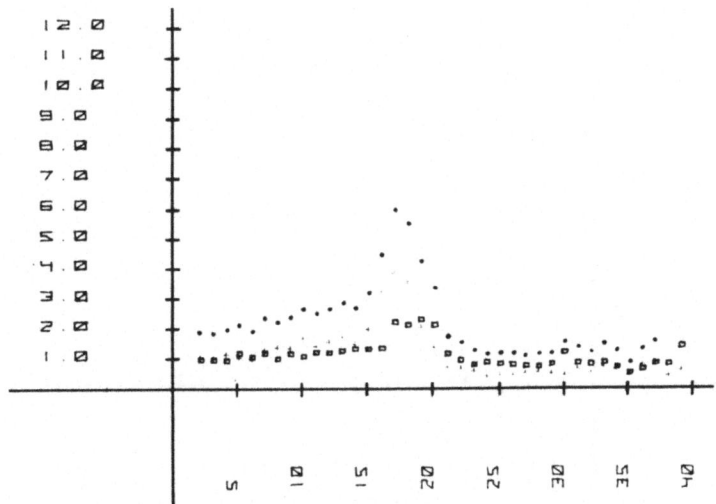

Fig. 18   Intensity ratio plotted as function of scan number ▧ A,
+B, ·C for SBR copolymer S-1712.

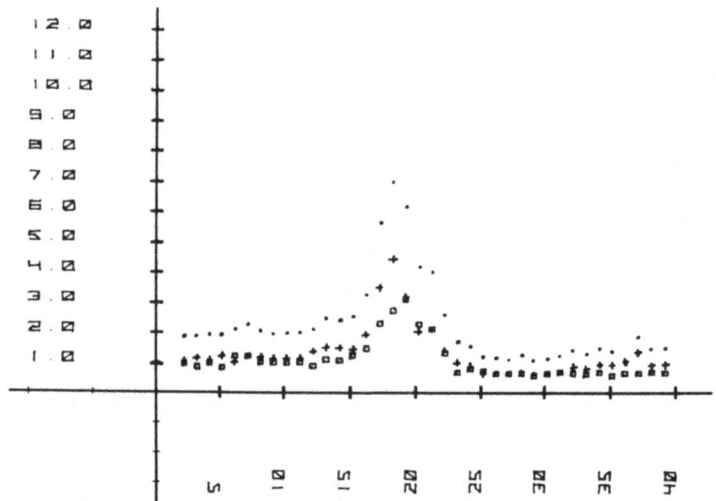

Fig. 19   Intensity ratio plotted as function of scan number ⊠ A,
          +B, ·C for SBR copolymer S-1011.

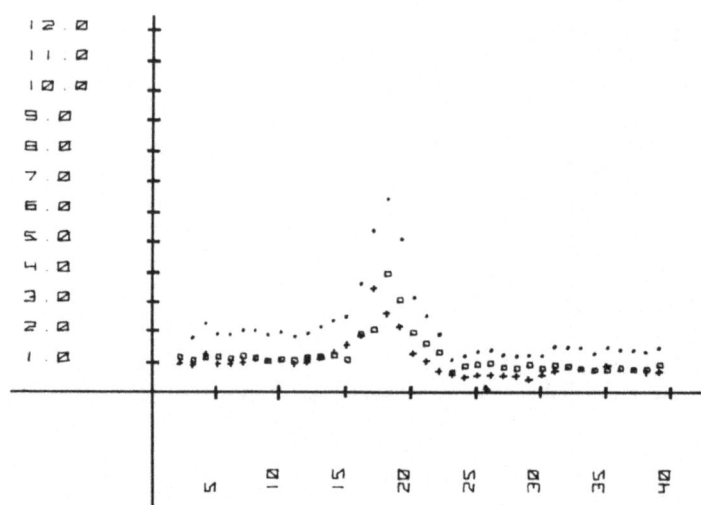

Fig. 20   Intensity ratio plotted as function of scan number ⊠ A,
          +B, ·C for SBR copolymer S-1013.

Conclusion

The molecular ions of 54 for butadiene and 104 for styrene are shown to increase in intensity as the pyrolysis of the styrene butadiene copolymer takes place. The value of the ease of interface construction and design simplicity is modified to some extent by the high observed background. This background proved to be consistent in behavior and a method for dealing with it has been shown.

The ions selected for study are non-exclusive in nature as shown by the presence of the 104 ion in the polybutadiene samples. The system is sensitive to conformational changes in the polymer system as demonstrated by the observed differences between the mass intensity ratio behavior of cis and cis-trans polybutadiene.

We have shown some examples of the type of results currently obtained by this combined thermogravimetric atmospheric pressure chemical ionization mass spectrometric system. This study leads the way to further developmental work in the identification and quantitation of polymer pyrolysis reactions.

References

1) Wyden, H., Thermogravimetric Determination of the Main Component in Technological Elastomer Systems, Kunstst. - Plast. 29: 9 (1982).

2) Sicar, A.K., Identification of NR and IR in Blends of BR and SBR by Thermal Analysis, Thermochim. Acta 27: 367 (1978).

3) McCreedy, K. and Keskkula, H., Application of Thermogravimetric Analysis to the Thermal Decomposition of Polybutadiene, J. Appl. Polym. Sci. 22: 999 (1978).

4) Brazier, D.W., Thermal Analysis in the Rubber Industry, in: "The State of the Art of Thermal Analysis", O. Menis, H.L. Rook and P.D. Garn, eds., U.S. Government Printing Office, Washington, D.C. (1980).

5) Maurer, J.J., Advances in Thermogravimetric Analyses of
       Elastomer Systems, J. Macromol. Sci., Chem. 8: 73 (1974).

6) G. Czybulka, H. Dunker, H. J. Dussel, H. Logemann and
       D.O.Hummel, Pyro-feldionenmassenspektrometrische
       Untersuchungen an unvulkanisierten und vulkanisierten
       Kautschuken mit komponenten geringer Konzentration
       (Comonomere, Beschleumiger, Metalloxide,
       Antioxidantien), Die Angewandte Makromolekulare Chemie
       100: 1 (1981).

7) Wilson, D.E., and Hamaker, F.M., Thermal Degradation Analysis
       of Polymeric Materials By Combined Mass Spectrometric and
       Thermogravimetric Analysis, in: "Thermal Analysis", R.F.
       Schwenker, Jr. and P.D. Garn, eds., Academic Press, New
       York (1969).

8) Dyszel, S.M., Coupling a PE-TGS-2 to the Sciex TAGA 3000 for
       Evolved Gas Analysis, in: "Thermal Analysis", B. Miller,
       ed., Wiley Heyden Ltd., Chichester (1982)

SEMI-AUTOMATED MEASUREMENT OF

COMPOSITES FLEXURAL STRESS RELAXATION

P. S. Gill and J. D. Lear

Clinical and Instrument Systems Division
Photo Products Department
E. I. du Pont de Nemours & Co.
Wilmington, DE

INTRODUCTION

Composite materials offer the opportunity to formulate combinations of mechanical properties unobtainable in single component materials. High stiffness fiber-reinforced thermoplastic and thermosetting resins are replacing metals and ceramics in many applications where anisotropic stiffness and light weight are advantageous. Viscoelasticity becomes an important concern in non-metalic composites because they usually show significant time-dependence of their mechanical properties at lower temperatures than metals. The problem of predicting composite viscoelasticity from that of the components has, of course, been addressed theoretically.[1,2] There is a need, however, for sensitive and convenient experimental methods to investigate the many different combinations of fibers and resins being generated in materials research. Small strain, temperature scanned dynamic mechanical analysis (DMA) is one such method found useful in characterizing composites by the intensity and position of mechanical damping peaks.[3,4] Traditional time-domain measurements such as creep and stress relaxation are also needed since these directly probe the load-bearing stability so important to designers.

The temperature dependence of time-domain measurements finds important application in estimating material viscoelasticity under conditions of complex loading and temperature cycles which may be inconvenient or impossible to test directly. This can be done, with cautions discussed below and elsewhere,[5] using "master curves," obtained by plotting a measured viscoelastic property against log time at a series of temperatures, then adding or subtracting log time increments to each separate curve to produce the best possible super-position of data into a single continuous curve at a chosen reference temperature. The combination of measured shift factors and master curve of one particular property can also be analyzed, to varying degrees of depth, to infer other properties using the well-developed methods of linear viscoelasticity theory.[6,7] Obtaining and analyzing the data needed to obtain shift factors and master curves can be greatly facilitated with modern computerized instruments.[8,9] In this work, a commercially available frequency domain instrument, the Du Pont 982 DMA, together with a microprocessor-based temperature control/data acquisition unit, the Du Pont 1090 Thermal Analyzer, was applied to the semi-automated measurement of small strain flexural stress relaxation master curves, for aramid and graphite fiber-reinforced epoxy resin composites. Data for a commercial polycarbonate thermoplastic resin is included for comparison, and application of time-temperature shift factors to frequency normalization of resonance DMA data is demonstrated.

MATERIALS AND METHODS

Samples, chosen for ready availability and to illustrate a large range of properties, are listed in Table 1. Dimensions were chosen to keep instrument compliance corrections less than 10%. Removal of moisture and residual cure potential of the composites required repetitive DMA scans under nitrogen to achieve $\pm 1\,°C$ damping peak reproducibility in successive scans. 982 DMA temperature scans (shown in Figure 4) were also made after the stress relaxation experiments, to check property stability. Stress relaxations were all done below the maximum temperature attained in the DMA pre-scans.

TABLE 1: SAMPLE SUMMARY

CODE	POLYCARBONATE	AR/SP306/0	AR/SP306/90	T300/3501-6	T300/EPX
FIBER	None	ARAMID	ARAMID	GRAPHITE	GRAPHITE
RESIN	"TUFFAK"(a)	SP306 EPOXY(b)	SP306 EPOXY	3501-6 EPOXY(c)	EXP. EPOXY(d)
LAY-UP	----	10 PLY UNI, 0°	10 PLY UNI, 90°	8 PLY UNI, 0°	QUASI-ISOTROPIC(e)
LENGTH (mm)	64.90	66.29	33.38	56.45	53.60
WIDTH (mm)	13.48	10.98	12.70	10.68	10.02
THICKNESS (mm)	4.62	1.79	1.79	1.07	1.32
ARM GAP	67.93	69.73	41.82	59.89	69.55
MAX. STRAIN (%)	0.0005	0.0003	0.0007	0.00016	0.0004
$E/2G - 1$(f)	0.5	15.0	0.5	40.0	10.0

(a) Trademark of Rohm and Haas Company.

(b) Product of 3M Company.

(c) Product of Hercules Company.

(d) Experimental composition modified for impact resistance.

(e) Lay-up 90/45/0/-45/90/45//sym. 0° = "L" direction

(f) Used for small shear correction to flex modulus calculation.

Experiments were automated by using the 1090's linked
methods to program the necessary changes in temperature. Sample strain was programmed between "off" (no
applied stress) and "on" (stress applied to bring the 982
DMA arm position to electrical zero) by using the 1090's gas
switching accessory contact closure to activate a relay
wired into the 982 DMA "mode" switch. This allowed program
control of the DMA's "align" (unstrained) and "cal"
(constant strain) modes. Arm displacement (signal B) and
autozero current (signal A) were measured throughout the
experiments at a 5 second sampling interval and the data
stored in disc memory for subsequent analysis. Raw data
(e.g., Figure 1) were examined prior to analysis to insure
acceptable equilibration of arm position prior to the
initial stress application at each temperature.

Acceptable data sets were transferred via the 1090's
RS-232 port to an external computer (DEC PDP-10) and
compressed by point deletion to accommodate a logarithmic
time scale. Zero time was defined as that just prior to the
first observed increase in stress signal caused by mode

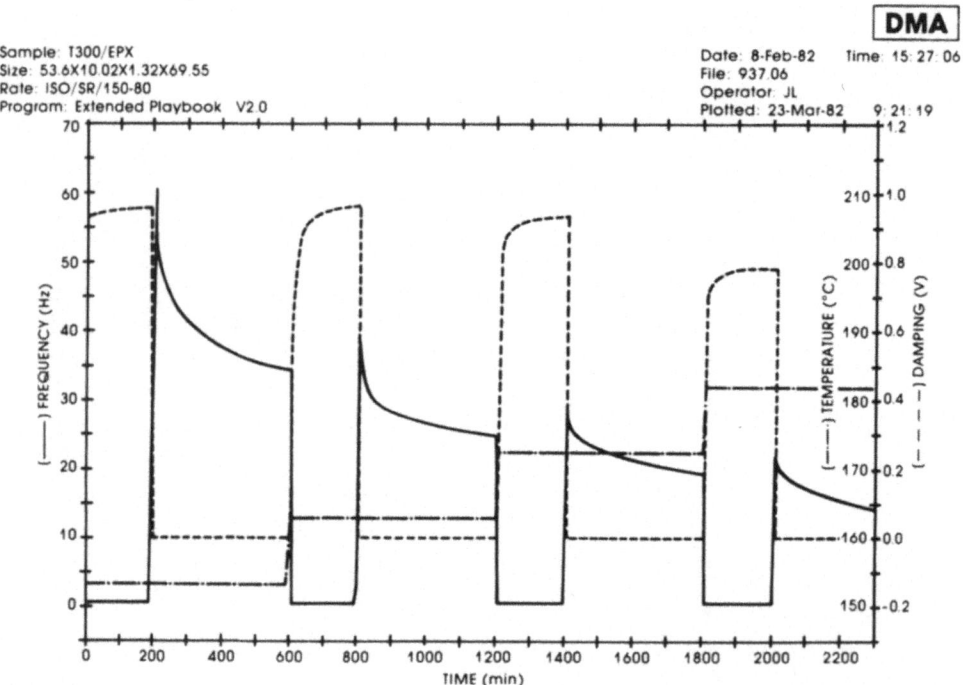

Figure 1. Stress relaxation of graphite epoxy, expanded scale.

switching. This was an adequate definition since only data beyond 10 seconds were considered, and the uncertainty in log time was negligible in shift factor determinations. Sample end displacement was measured from the difference in time zero offset position signal and that maintained during stress application (a very small baseline correction). Initial offsets were pre-set by LVDT adjustment to between

$$E' = 2(1 + \sigma) \frac{8\pi^2 \, Jf^2 \text{-} K}{B^2} \frac{(L + \Delta L)}{A} \left[\alpha + \frac{(L + \Delta L)^2}{24k^2 \, (1 + \sigma)} \right] \beta(f)$$

and $E' = 2G' (1 + \sigma)$

Where: σ = Poisson's Ratio
           A = Sample cross-sectional area
           k = Sample cross-sectional radius of
               gyration

($k = T/\sqrt{12}$ for flat samples, $k = R/2$ for cylindrical samples)

           B = Arm Spacing (Distance Between
               Pivot Centers)
        $\beta(f)$ = Instrument compliance correction
           α = Shear distortion factor
           J = Moment of inertia
           K = Pivot spring constant
           L = Sample length
        ΔL = Sample length correction
           T = Sample thickness
           E' = Elastic Modulus
           G' = Shear Modulus
           f = Resonant Frequency

Figure 2. Modulus equation for DuPont 982 DMA.

0.1 and 0.5 mm sample end displacement which, for the long span samples used in this work, gave calculated surface tensile strains less than 0.001% (see Table 1). The ratio of force to displacement signals was used, together with the sample dimensions, instrument compliance, and end corrections (determined under ambient conditions using procedures recommended in the Du Pont 982 Operator's Manual) to compute flexural modulus (E') using the equations in Figure 2 with the substitution:

$$f^2 = 3.743 \times 10^4 \; C' \; (V-V_o)/(A-A_o)$$

where   f   =   Resonant frequency (Hz) used in modulus equations

          C'   =   Du Pont DMA tan delta constant (Hz^2-mm/mV)

          V   =   Du Pont DMA Signal "A" (Hz units)

          A   =   Du Pont DMA Signal "B" (mV units)

and the "0" subscripts denote the baseline values of signal A in "align" and signal B in "cal" modes. The numerical factor is calculated from factory calibrated electronics constants which relate arm movement and motor current to the electrical signals (see Appendix). Using these equations, modulus vs log time curves were calculated with the DEC-10 and displayed on a graphics terminal. Shift factors were measured to achieve a visually judged good superpositioning of the modulus curves, then plotted against temperature to determine a 150°C value. This was subtracted from the experimental values at each temperature to provide a common reference (zero shift) temperature for comparison of the different samples.

RESULTS AND DISCUSSION

This work set out to demonstrate that time domain visco-elastic measurements could be made conveniently with a commercially available, dynamic mechanical analysis instrument. The microprocessor control of temperature, strain program, and data acquisition made possible with the Du Pont 1090 Controller allowed many of these measurements to be made unattended during evenings and weekends, and the data analysis was facilitated by the 1090's ability to transfer data to a more general purpose computer. The stress relaxation master curves, corresponding DMA scans and shift factors shown in Figures 3 to 8, illustrate what can be done with the Du Pont 982 DMA instrument. The time domain curves (Figure 3) resemble the DMA temperature scans (Figure 4) in showing earlier and more complete loss of stiffness for the unreinforced thermoplastic polycarbonate, and increasing modulus with longitudinal fiber reinforcement of the cured epoxy composites. The AR/SP306/90 and T300/3501-6 data shown in expanded scale in Figures 6 and 7 show evidence for

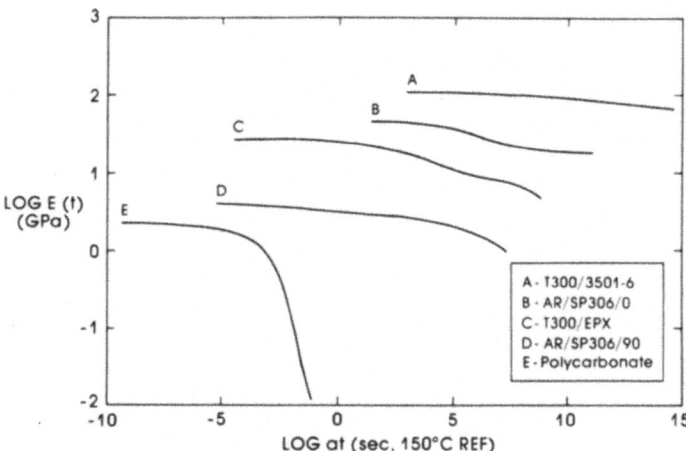

Figure 3. Master curve summary for a variety of composites (using 150° C reference temperature).

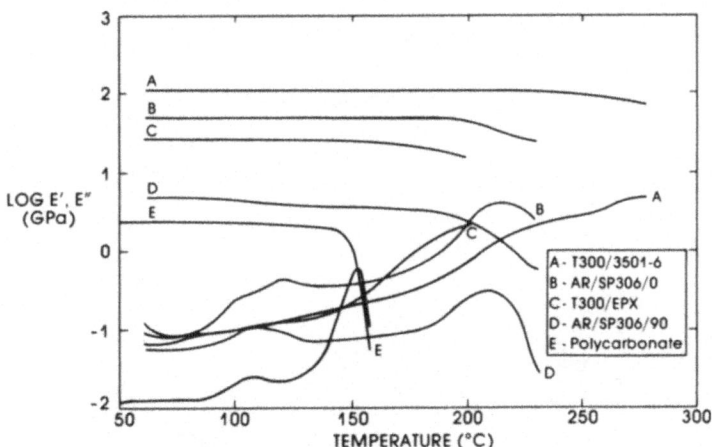

Figure 4. 982 DMA temperature scan summary for a variety of composites. 0.5°C/min.

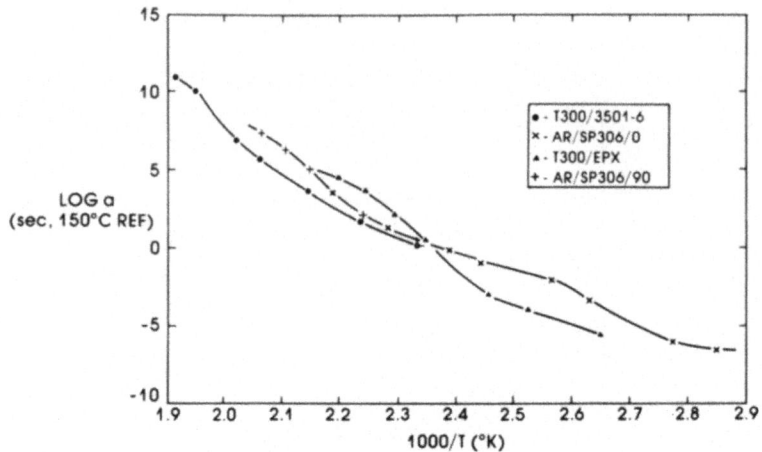

Figure 5. Calculated shift factors for various composites.

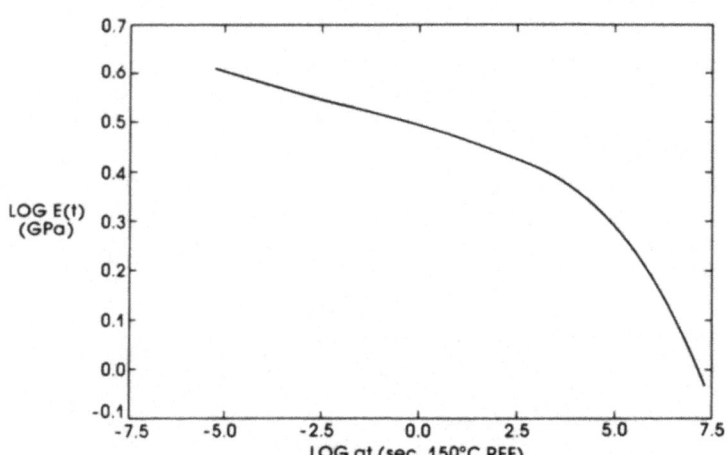

Figure 6. Aramid/epoxy, expanded scale.

the secondary relaxations appearing in the DMA damping
measurements. The shift factors for the composites are
summarized at a common reference temperature (150°C) in the
Arrhenius plots of Figure 6 and show the significant
non-linearity expected for measurements made below or close
to the glass transition. Shift factors for the 0 and 90
AR/SP306 composites appear to define a single curve (see
also Figure 8) as might be expected for a resin-dominated
property.

An attractive aspect of master curve representations is
their invitation to extrapolate mechanical properties beyond
experimentally accessible time or frequency scales. Such
extrapolations require cautious consideration of error
propagation and recognition that the fundamental assumption
of time-temperature superposition ("thermorheological simplicity," Ref 1) often holds operationally valid over very
limited time-temperature ranges. The use of master curves
for engineering design is accordingly a risky proposition
where, "if the predictions are reliable, then the superposition approach is justified".[5] In practice, master curves
are probably best used to help design direct test protocols,
by mathematical model calculations of viscoelastic response
to strain and temperature cycles, likely to be encountered
in specific applications. An example of this is the use of
shift factors to estimate the influence of frequency on the
temperature dependence of dynamic mechanical data. This has
value when instrument limitations prevent direct measurements in a frequency range of interest, or, as with
resonance-type instruments such as the torsion pendulum or
the Du Pont DMA, frequency changes during the measurement.[8,9]

In general, to transform a measurement made at frequency
f and temperature T_1 to one at a frequency f_o, the
temperature is shifted to T_2 to satisfy the relationship:

$$Log\ (f/f_o) = F(T_2) - F(T_1)$$

where the F's are temperature-dependent shift factors such
as those measured in this work. A common practice is to
assume either a WLF or Arrhenius form for the shift factors:

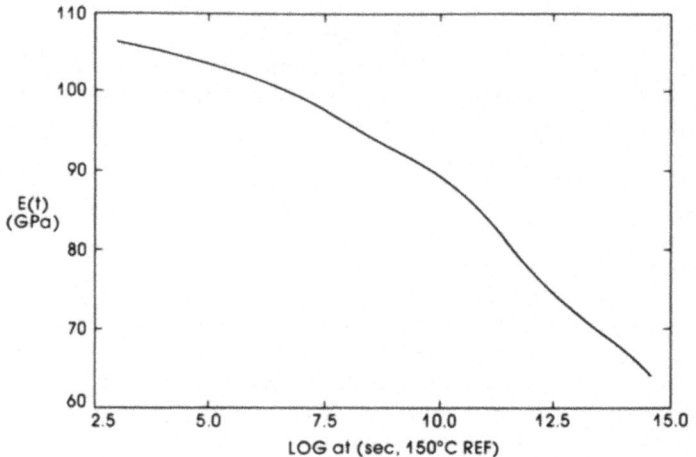

Figure 7. Graphite/epoxy, raw data.

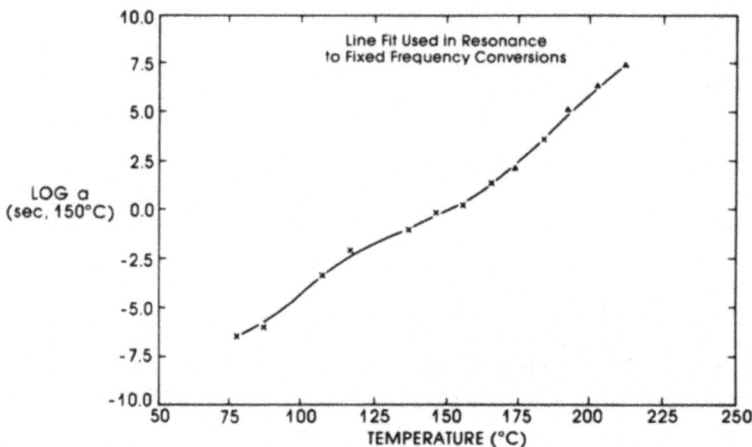

Figure 8. Aramid/epoxy shift factor curve.

WLF Form

$$F(T) = -C_1(T - T_o) \, [C_2 + (T - T_o)]$$

Arrhenius Form

$$F(T) = C(T_o - T)/T_o T$$

This approach is limited, however, to data obtained within the temperature range where the chosen type of function is applicable. A wider range is possible using empirical shift factors as done in this work. Figure 9 depicts the algorithm employed to transform resonant (varying frequency) loss modulus vs temperature curves into calculated constant

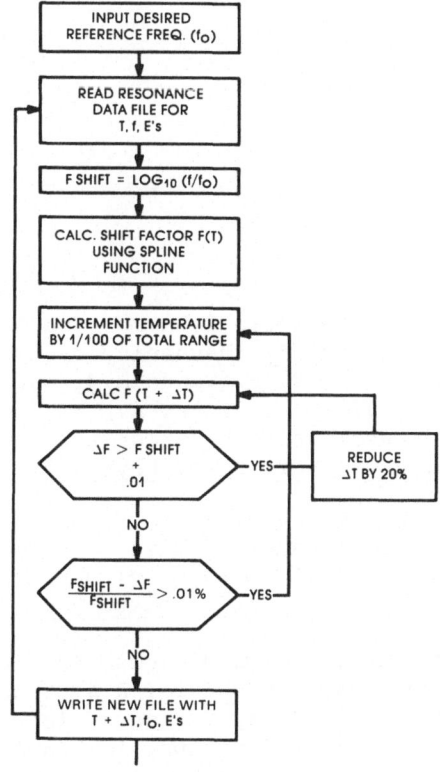

Figure 9.

frequency curves, as exemplified in Figure 10 for the
AR/SP306/90 sample, using the shift factor curve shown in
Figure 8 with a cubic spline function representation of the
smoothed data. The two relaxation peaks are predicted to
shift at different rates; 6 deg/decade of frequency change
for the primary and 10 deg/decade for the secondary relaxation, which also can be seen to broaden with increasing
frequency. Note that the temperature shifting has been con-
strained to stay within the bounds of the measured shift
factors, so the highest frequency primary, and the lowest
frequency secondary relaxation peaks do not appear in this
calculation. Extrapolation of shift factors could, of
course, be done by assuming some appropriate analytic
continuation of the measured data, a procedure which might
well be justified if the form of the shift factor dependence
were known from data accumulated for similar materials (e.g.
Ref 8). In any case, as with time-shifting master curves,
the resulting predictions should be recognized as estimates,
and tested experimentally wherever possible before being
relied on for engineering design.

ACKNOWLEDGEMENT

    Wilburn Smith (Du Pont) supplied samples used in this
work. His cooperation and encouragement are greatly
appreciated.

REFERENCES

1. R. A. Schapery, Viscoelastic Behavior and Analysis of
   Composite Materials, in "Mechanics of Composite
   Materials," Vol. 2, G. P. Sendeckyj, ed., Academic
   Press, N.Y. (1974).
2. Y. T. Chen, "A Constitutive Equation for Composite
   Systems," J. Poly. Sci. Poly. Phys., Ed., 11, 2013
   (1973).
3. J. D. Keenan, J. C. Seferis, and J. T. Quinlivan,
   Effects of Moisture and Stoichiometry on the
   Dynamic Mechanical Properties of a High Performance
   Structural Epoxy, J. Appl. Polym. Sci. 24 2375
   (1979).
4. P. S. Gill, Characterization of Composites by Dynamic
   Mechanical Analysis, Industrial R & D, Apl. (1982).
5. E. Passaglia and J. R. Knox, Viscoelastic Behavior and
   Time-Temperature Relationships, in "Engineering
   Design for Plastics," E. Baer, ed., Van Nostrand
   Reinhold (1964).

6. J. D. Ferry, "Viscoelastic Properties of Polymers," 2nd ed., John Wiley, NY (1971).
7. R. M. Christenson, "Theory of Viscoelasticity," Academic Press, NY (1971).
8. R. P. Chartoff and J. L. Graham, "Computerized Viscoelastic Master Plots for Vibration Damping Applications" ACS Symposia Series, No. 197, 1982.
9. D. J. Townend, "A Computerized Technique for Producing WLF Shifted Data from a Single Frequency Temperature Scan," in "Proceedings of the 7th International Conference on Thermal Analysis," B. Miller, ed., P. 313 (1982).
10. J. D. Lear and P. S. Gill, "Theory of Operation of the Du Pont 982 DMA" Du Pont Thermal Analyzers reprint No. E-42400, 1982.

APPENDIX

The Du Pont DMA equations of motion are easily solved for static deformation by equating first and second derivative terms to zero.[10] The resulting equation, solved for modulus (E), has the same form as the equation derived for the resonance mode with the substitution of M/ϕ for $8 \quad {}^{2}Jf^{2}$ where:

M = moment of force applied to sample end

ϕ = angular displacement of arm around flexure pivot

J = single arm inertial movement

f = resonance frequency in dynamic operation

M is linearly proportional to the DC motor current signal "A" in "cal" mode, and ϕ to signal "B" in "align" mode. Consequently, since J is constant, the ratio of signal A (cal) to signal B (align) is proportional to f^2. The proportionality constant can be measured directly using a non-viscoelastic sample such as spring steel. This is, in fact, done in the procedure for calibrating the DMA's tan delta constant (C') which relates the moment amplitude in the sine wave drive, to the rms motor current signal "B" in the DMA "quant" mode, and to the change in LVDT amplitude signal on the arm displacement. The factory-calibrated relationship is:

$$C' = \frac{\text{Sig B (align) } f^2 \text{ (1.17 x } 10^{-3} \text{ mm/mV)}}{\text{Sig A (cal) (20 mV/Hz) (2.19)}}$$

where the units of signal A are mv and B and f are in Hz. The 2.19 factor relates the driver D.C. current to the rms value measured in "quant" mode.

SOME USES OF DIFFERENTIAL SCANNING

CALORIMETRY IN BIOMEMBRANE RESEARCH

Donald L. Melchior

Department of Biochemistry
University of Massachusetts Medical School
Worcester, Massachusetts   01605

The application of Differential Scanning Calorimetry (DSC) to biological problems is relatively recent. DSC is proving to be a powerful approach not only for investigating structural problems, but for studies on physiological processes as well. DSC has had its most notable biological successes in studies of lipid containing systems (1,2). The intent of this chapter is to briefly describe some calorimetric approaches our laboratory has used to obtain information on these systems.

Those biological molecules classified as lipids are primarily either amphipathic or nonpolar. In aqueous suspension (the normal biological state) they can exist in a variety of aggregated forms displaying a full range of thermal behavior. This is illustrated in Figure 1 which shows thermograms for several types of lipids. Figure 1A is a scan of the fatty acid, stearic acid. It undergoes a first order transition from a crystalline to a fluid state. A thermogram of a cholesterol ester, cholesterol oleate, is seen in Figure 1B. The thermal behavior of cholesterol esters is quite dependent on thermal history (3). In the example shown here, the sample was cooled at about 1 C°/sec from a temperature above the endotherm, and then scanned up in temperature at 10 C°/minute. At the lowest temperatures it exists as a supercooled liquid, then with increasing temperature it crystallizes giving rise to an exotherm followed by an endothermic melting of the crystal resulting in an isotropic liquid. The thermal behavior of cholesterol esters can be quite complex with various liquid crystalline transitions occuring depending on impurities and molecular packing constraints as well as thermal history.

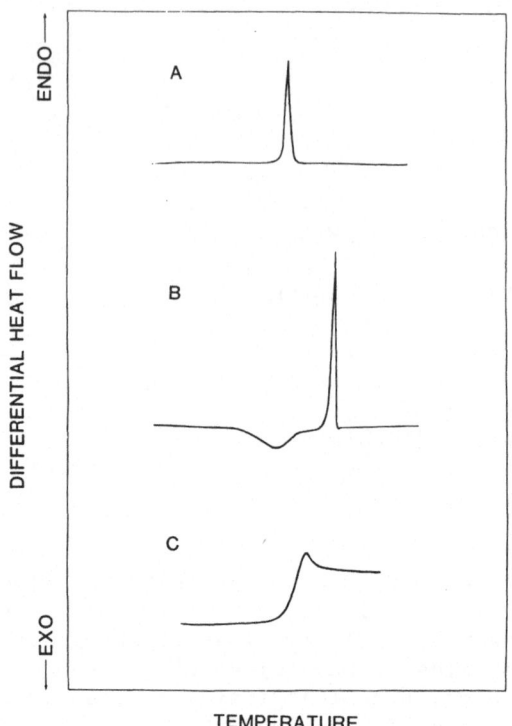

Fig. 1.   Calorimeter heating scans of various lipids.   (A)
Stearic acid.   (B) Cholesterol oleate.   (C) A calcium
precipitate of cardiolipin.

Figure 1C shows a portion of a thermogram of a calcium complex
of the phospholipid cardiolipin.   Here a glass transition is
evident (4).

    Figure 2 illustrates the phase behavior of an isolated
class of naturally occuring phospholipids, the
phosphatidylethanolamines, in excess water.   At low temperatures
these molecules are organized in a crystalline bilayer with
their polar head groups oriented toward the bulk water phase and
their hydrocarbon tails within the bilayer interior.   With
increasing temperature, the bilayer undergoes a reversible
endothermic transition to a fluid bilayer.   With further
increase in temperature, individual molecules due to thermal
motion become progressively more cone-like in shape until
eventually the bilayer breaks down and the lipids assume an
hexagonal conformation.

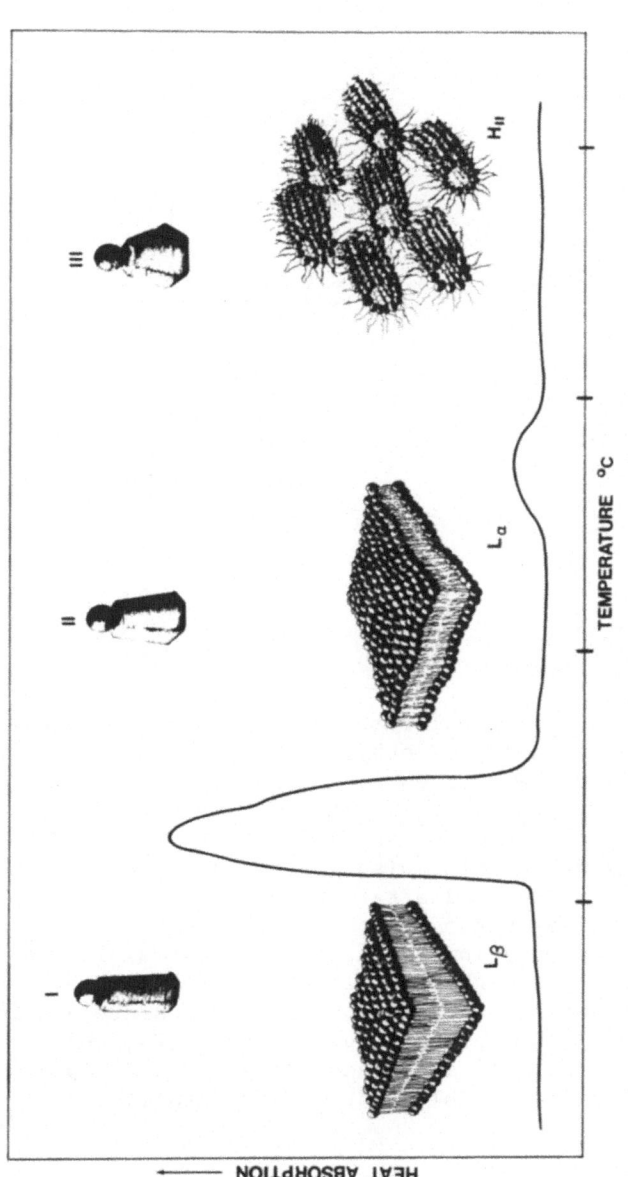

Fig. 2. The phase behavior of mixed-chain phosphatidylethanolamine in excess water. At low temperature this lipid exists as crystalline bilayers (the gel or L_α state). With increasing temperature the bilayers undergo a reversible endothermic transition commonly called the bilayer phase transition. The resulting state (the liquid crystalline or L_β state) is composed of fluid bilayers. The lipid hydrocarbon chains are more disordered in fluid bilayers than in crystalline bilayers. This results in more cone-shaped molecules as illustrated in the top of the figure. With further increase in temperature the lipid molecules become still more cone-shaped and their preferred conformation is the inverted hexagonal H_{11} state.

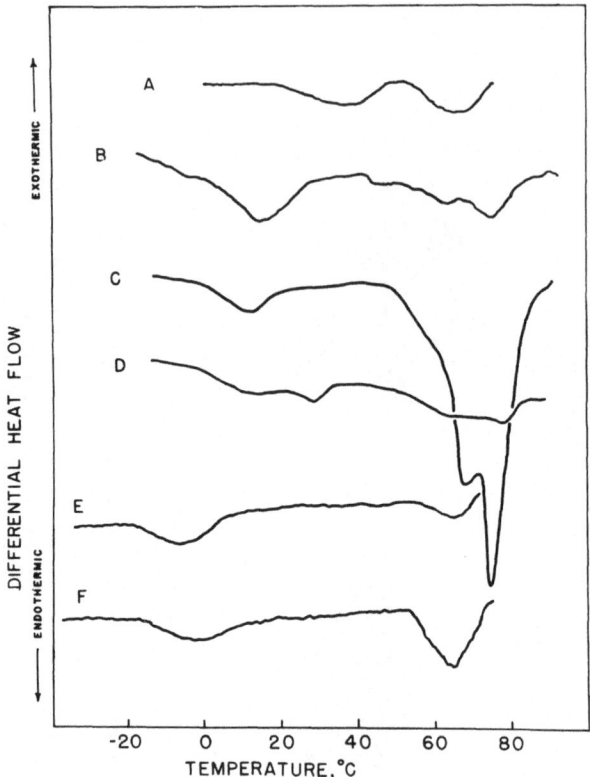

Fig. 3. Differential scanning calorimetry scans of biomembrane
transitions, all obtained with 50% ethylene glycol/water as
solvent. (A) A. laidlawii membranes from cells grown in
tryptose medium at 37°C; (B) M. lysodeikticus membranes from
cells grown in brain heart infusion at 37°C; (C) E. coli K12W945
whole cells grown in minimal salts with glucose at 20°C; (D) the
same cells as in (C), but scanned after thermal protein
denaturation; (E) rat liver microsomes; (F) rat liver
mitochondria. In all cases a lower temperature reversible lipid
transition is followed by a higher temperature irreversible
protein peak. The protein denaturation peaks are featureless in
(A), (E), and (F), but show fine structure in (B), (C), and (D).
Unlike other organisms, E. coli after heating shows two lipid
transitions and residual reversible protein denaturation, as
seen in (D).

Biomembranes serve a multifunctional role in life
processes. They provide charged barriers between the insides
and outsides of cells and compartments within cells. Together
with associated proteins, membranes function as selective

barriers for regulating the transport of various molecules and ions. Biomembranes are also attachment sites for enzymes, antigens, and hormone receptors and provide a medium for reactions involving hydrophobic reactants.

The basic structure of biological membranes is the bilayer. While certain isolated classes of normally occuring membrane lipids such as the phosphatidylethanolamines can assume non-bilayer conformations, in nature the mixtures of lipids in biomembranes exist in a bilayer conformation. Under normal physiological conditions membrane bilayers are in a fluid state. When cell membranes become crystalline due to changes in membrane composition or environmental temperature, normal physiological processes are inhibited.

Figure 3 shows a series of thermograms for various native biomembranes. The bilayer transitions are seen to be broad, extending over a range of as much as 40C°. The shape and temperature range of these transitions is determined by the melting points and relative proportions of the many different species of lipids present in the membrane. As the membrane melts or freezes it behaves much as any other multicomponent system. During melting, fluid domains of lipid progressively enriched in higher melting point lipids grow at the expense of crystalline membrane domains.

As might be expected, many membrane activities reflect the physical state of the membrane bilayer. An example of this is seen in Figure 4. Here the kinetics of protein mediated sugar transport through a lipid bilayer is seen to follow the melting of the bilayer. For these studies the human red blood cell hexose transport protein was removed from its native membrane and reconstituted into artificial membranes with a predetermined thermal profile. Sugar transport was monitored by a turbidometric method to obtain the parameters V_{max} (maximum rate of sugar transport) and K_m (that concentration of sugar at which the rate of transport is half maximum) (5). V_{max} is seen to nicely follow the bilayer transition (sugar flux below the transition was too low for the K_m to be determined over the entire temperature range). In native membranes the kinetics of sugar transport also follow the bilayer transition (6). The ability to reconstitute membrane enzymes into bilayers of predetermined thermal profiles provides an approach for investigating the role of the bilayer in modulating enzymatic activities. Correlated DSC and enzyme kinetics should answer such questions as whether enzymes associate with specific lipids in the membrane and if so, how their activity reflects this association.

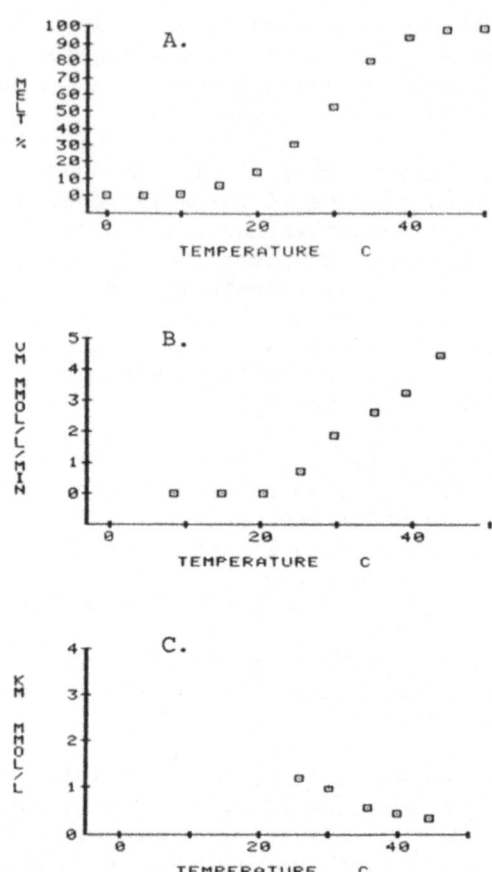

Fig. 4. Correlation of sugar transport with membrane bilayer physical state. For this study the human red blood cell sugar transport system was inserted in artificial membranes with a preselected thermal profile. The panels show as a function of temperature (A) extent of the bilayer transition (B) V_{max} for sugar transport and (C) the K_m for transport.

A fluid or at least partially fluid lipid bilayer is essential for cellular function. Abnormally high transition temperatures reflect abnormally crystalline membranes, and are associated with cell leakage, changes in active transport and the activities of some membrane-associated enzymes, prolonged generation times, eventual loss of viability, and even cell

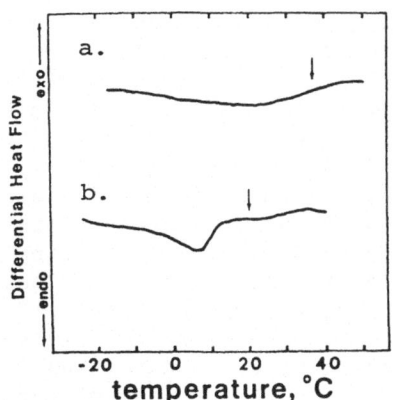

Fig. 5. The effect of growth temperature upon the position of
the cytoplasmic membrane transition in E. coli W945 grown in
minimal medium. Calorimeter scans were run in 50% ethylene
glycol-water. In (a), from cells grown at 37°C, the transition
extends from about − 10 to 40°C. In (b), from cells grown at
20°C, it extends from about − 5 to 15°C. In both cases the
membranes are fully fluid at or just below growth temperature.

lysis. Since transition temperatures depend upon the fatty acid
composition of membrane lipids, low transition temperatures are
assured by the biosynthesis of appropriate fatty acids or their
selection from exogenous sources. To maintain a sufficiently
low transition temperature, membrane lipids must contain fatty
acids possessing low melting points.

Furthermore, the composition must be responsive to
temperature in such a way that the membrane transition always
remains below or mostly below the temperature of growth. This
necessity implies a control mechanism, one that senses
temperature or the physical state of the membrane and directs
the incorporation of proportionally more unsaturated or other
low-melting fatty acids into membrane lipids as the temperature
decreases. This phenomenon is seen in higher organisms, but is
particularly important in bacteria and other microorganisms
where the membrane transition is often in the neighborhood of
growth temperature (7). Compare, for example the membrane
transition of E. coli W945 grown in minimal medium at 37°C
(Figure 5a) with that of the same organism grown at 20°C (Figure
5b). In both cases the transition is almost entirely below
growth temperature.

Crucial to understanding the temperature-dependent fatty acid selection process is the realization that fatty acids seem to be selected on the basis of melting point, a thermodynamic property that only indirectly reflects molecular structure. Although unsaturation is the usual route to low melting point, the same goal is attained in many organisms by employing structural alternatives, such as branched chains in many gram-positive bacteria or cyclopropane-containing chains in many gram-negative bacteria (8). A more convincing argument that the principal consideration is thermodynamic rather than structural is based on the fact that a given organism, if forced to do so, will choose any low-melting point fatty acid to accomplish its goal of lowering transition temperatures and controlling fluidity. The best illustration is the microorganism Acholeplasma laidlawii. This organism is capable of incorporating large amounts of exogenously supplied fatty acids into its membrane lipids, fatty acids of progressively lower melting points being required as growth temperature is decreased. Cis-unsaturates serve the purpose admirably even at the lowest temperatures, but growth is also normal if cis-unsaturates are removed from the growth medium and replaced by branched-chain or cyclopropane acids or by elaidate, an unnatural trans-unsaturated compound. Unsaturated fatty acid auxotrophs of E. coli show similar behavior and will accept elaidate or even bromostearate. If in fact the temperature-sensing selection mechanism is thermodynamically determined, and depends upon an ensemble of molecules rather than upon the structure of individual molecules, it is difficult to imagine it to be based upon enzyme specificity. The binding of substrates to enzymes reflects the molecular structure of the ligand and interaction occurs on a one-to-one basis. In such interactions the thermodynamic properties of the membrane bilayer have no meaning.

Correlations of fatty acid incorporation and binding by A. laidlawii and model systems suggest that the sensor might be the bilayer itself (9). Figure 6a is a thermogram of membranes from cells grown in tryptose medium at 37°C. As usual for a biomembrane, the melt is broad, extending from about 20°C to about 45°C. The ratio of palmitate to oleate incorporated into the membrane lipids is shown in Figure 6b. To obtain the plot, cells were grown in tryptose at 37°C, then aliquots were incubated for five minutes at various temperatures in the same medium containing ^{3}H-labeled oleate and ^{14}C-labeled palmitate. The lipids were extracted, separated from unesterified fatty acids by thin-layer chromatography, and assayed in a scintillation counter for the amounts of esterified palmitate and oleate. The plot shows the expected behavior, the palmitate/oleate ratio increasing with increasing temperature.

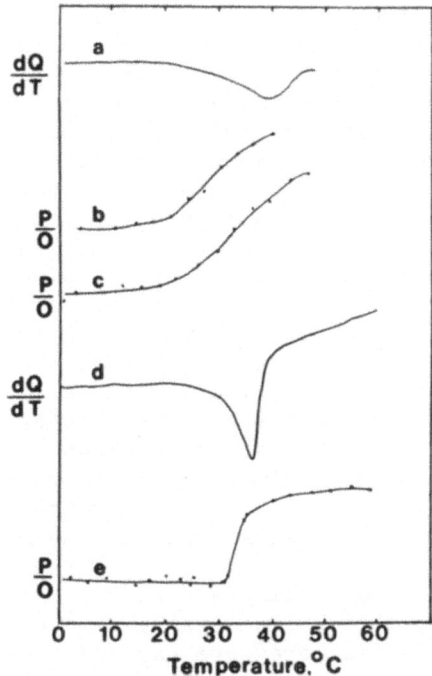

Fig. 6. The bilayer as a temperature-sensing selector of fatty
acids. a is a DSC scan of membranes from A. laidlawii grown at
37°C in tryptose. The ratio of palmitate to oleate (P/O) taken
up from the medium and esterified into membrane lipids after
short-time incubation at various temperatures is shown in b.
Physical binding of palmitate and oleate to protein-free lipids
is shown in c. A mixture of dipalmitoyl phosphatidylcholine and
egg phosphatidylcholine (25%) produced the DSC scan in d and the
fatty acid binding curve in e.

Interestingly the shape of the curve follows the membrane
transition, which is a direct indication of the physical state
of the bilayer. If many such experiments are carried out using
cells with different transition temperatures, the results are
always the same: the ratio parallels the membrane transition.
In the extreme case, when the transition is at -20°C for cells
grown in an oleate-enriched medium, the ratio above 0°C is a
straight line with a gentle positive slope but no inflections.
The total fatty acid analysis of A. laidlawii grown in oleate
similarly shows almost no temperature dependence. To verify
that the response is not necessarily enzyme dependent, the

physical binding of free palmitate and oleate to lipids
extracted from the cell membrane was examined as a function of
temperature. Small pieces of filter paper were impregnated with
the lipids dissolved in chloroform, dried, incubated in buffer
containing labelled oleate and palmitate, then finally blotted
to remove excess liquid and counted. Calorimetry verified that
bilayers were formed within the paper. Again the ratio, shown
in Figure 6c, follows the lipid transition. The protein-free
lipid bilayer always mimics the behavior of the living cells and
like the cells, the lipids select a progressively higher
proportion of oleate as the temperature drops. Furthermore, the
ability to act as a temperature sensor and selector may be a
general property of any phospholipid bilayer. Figure 6d is a
thermogram of bilayers formed from a mixture of 25% egg lecithin
in synthetic dipalmitoyl lecithin. The oleate/palmitate binding
ratio of the same lipid preparation is shown in Figure 6e. In
this case the slope of the palmitate/oleate ratio below the
transition is negative; its sign can be controlled by varying
the content of unsaturated fatty acids in the bilayers.

These and related experiments suggest a mechanism
for the control of membrane lipid fatty acid composition.
The experiments described here were carried
out with exogenous fatty acids and A. laidlawii and the
postulated mechanism may apply only to this case, but the same
principle could regulate lipids, including triglycerides,
biosynthesized from endogenous sources. The enzyme or enzymes
necessary for the transfer of fatty acyl groups from acyl
coenzyme A to phospholipids are imbedded in the membrane bilayer
but they cannot sense the physical state of the membrane and,
unaided, cannot strongly distinguish between high-melting and
low-melting fatty acids. They accept and use essentially any
fatty acid presented to them. The temperature-programmed
selectivity, and hence the control of the membrane transition
temperature and the fluidity of the bilayer, reside in the
bilayer itself. From another point of view the enzyme is not
simply a protein but a lipoprotein, a protein imbedded in a
bilayer. The catalytic function is assumed by the protein and
the selectivity directly by the lipid.

The preceeding examples used DSC to investigate sugar
transport and lipid metabolism. Another application of DSC to
biomembranes is to study associations between various membrane
lipids. Perhaps the most important of these associations is
that of cholesterol with membrane phospholipids. It has been
shown that cholesterol suppresses bilayer transitions putting

the bilayer into a state intermediate between crystalline and fully fluid (10).

This plasticizing effect of cholesterol on membrane bilayers is reflected in dynamic processes occurring in the bilayer. For example, in cholesterol-rich bilayers, simple processes such as the transbilayer diffusion of water (11) or more complex processes such as protein-mediated sugar transport (12), are enhanced relative to crystalline bilayers but decreased relative to fully fluid bilayers.

Whereas on a bulk level cholesterol is considered to put bilayers into a state of intermediate fluidity, on a molecular level the details of cholesterol-phospholipid interaction are not well understood. This has been investigated in membranes by dilatometry as well as DSC. Figure 7, taken from a dilatometric study of dipalmitoylphosphatidylcholine (DPPC) bilayers (13), demonstrates the action of cholesterol as a plasticizer. At temperatures below the phase transition, increasing cholesterol content causes the bilayer volume to increase while at temperatures above the phase transition increasing cholesterol content causes the bilayer volume to decrease. Increasing cholesterol content progressively eliminates the bilayer transition. These investigations allowed construction of a three-dimensional surface with dimensions of mole fraction cholesterol, temperature, and apparent partial specific volume. Much of the phenomenology reported for DPPC and DPPC/cholesterol bilayers appears and can be integrated on this surface. In addition to the thermotropic events associated with the system, two cholesterol-induced events at 17.5-20 and 29 mole % cholesterol are particularly in evidence. Dilatometry together with DSC suggests that DPPC-cholesterol mixtures belong to either of two quite distinct systems. The first system extends from $0 < X_{chol} < 20$ and consists of domains of nearly pure DPPC melting abruptly at about 40°C and domains of a 4:1 DPPC-cholesterol complex melting with a broad transition centered at about 40°C. The second system extending from $20 < X_{chol}$ is characterized by a broad symmetrical transition which for 20 mole percent cholesterol has a maximum at about 41.5°C. With increasing cholesterol content this transition broadens, shifts up in temperature, and decreases in size, disappearing at $X_{chol} = 50$. In the 20 to 50% region an event occurs at $X_{chol} = 29$. Below the thermotropic transition it appears as a decrease in rate of change of the apparent partial molar volume of the bilayer with increasing cholesterol content.

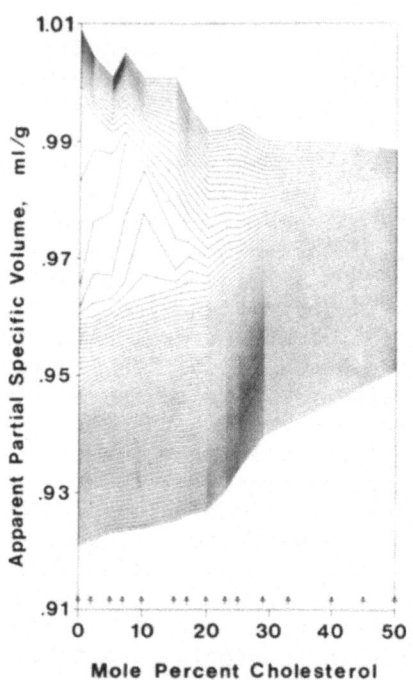

Fig. 7. The apparent partial specific volume ($\overline{V}_a$) of pure and
cholesterol-containing DPPC bilayers over the temperature range
of 0 to 50°C. $\overline{V}_a$ is plotted against mole % cholesterol (X_{chol})
at half-degree temperature intervals extending from 0°C (bottom)
to 50°C (top). The concentration of bilayer cholesterol in the
dilatometer runs used to construct the plots are indicated by
the 15 arrows along the X_{chol} axis at X_{chol} = 0,2,5,7,10,15,17,
20,23,25,29,33,40,45, and 50. Experimental points at these
concentrations are connected by straight lines. Since vertical
cuts at the arrows reproduce the experimental volume-temperature
curves, the vertical spacings between the lines are a measure of
the coefficient of expansion at various cholesterol
concentrations.

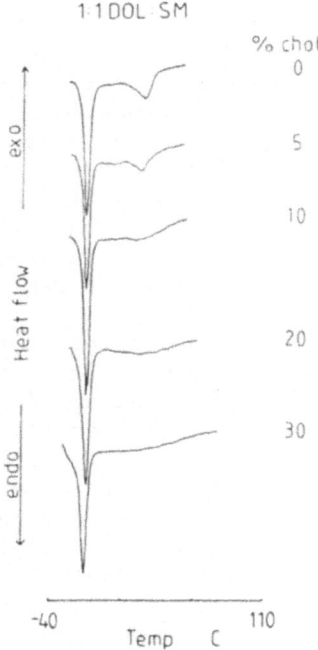

Fig. 8. Thermograms of bilayers containing dioleoyl lecithin
(DOL) and sphingomyelin (SM). There is clear phase separation
between DOL (low melting point) and SM (high melting point). As
the cholesterol content of the bilayers is increased step-wise
to a 0.3 molar ratio, the sphingomyelin transition is suppressed
in preference to the DOL transition.

        The interaction of cholesterol with membrane phospholipids
is further complicated by findings that cholesterol is not
necessarily uniformly dispersed in mixed phospholipid bilayers.
In DSC studies on bilayers composed of two lipids of the same
class, but with melting points sufficiently different to give
two peaks in the calorimeter, cholesterol associates with the
lower melting point lipid. That is, it appears to be frozen out
of solid crystalline regions of the bilayer. In addition,
however, cholesterol has a preference for specific lipid classes
that overides its preference for lower melting point lipids

within a specific class (Figure 8). The order of affinity of
cholesterol for five major classes of lipids was determined by
DSC as sphingomyelin >>phosphatidylserine = phosphatidyl
glycerol>phosphatidylcholine>> phosphatidylethanolamine (14).
Because of a tendency for preferential association with various
lipid classes, the presence of cholesterol in a bilayer may
promote bilayer heterogeneity by producing cholesterol-rich
regions containing specific lipids. Crystallization would not
occur in such patches, and lipid-protein associations might be
different from cholesterol poor bilayer regions.

The use of DSC may give insight into the mode of action of
some hydrophobic drugs. An example of this is a study of the
psychoactive drug Δ^9-tetrahydrocannabinol (THC) (15). It was
found by DSC studies on cholesterol free bilayers that above a
certain concentration, THC complexes stoichiometrically with
various phospholipids, Fig. 9. Further, complex formation
differs for lipids with either saturated or unsaturated
hydrocarbon chains. When bilayer phospholipids are sufficiently
dissimilar for phase separation to occur, THC preferentially
associates with the lower melting lipid. When cholesterol is
added to lipid bilayers below 20 mole % cholesterol,
THC/phospholipid complex formation is enhanced. Above 20 mole %
cholesterol there is no indication of THC/phospholipid complex
formation. In addition to supporting the idea that at 20 mole %
cholesterol there is a bilayer phase rearrangement, these
calorimetric findings give some insight into possible mechanisms
for the action of hydrophobic drugs such as THC.

The mode of action of THC is unknown. Since THC has a high
membrane-water partition coefficient, an association between THC
and membrane lipids could be biologically significant. A
complex between THC and a specific membrane lipid could serve as
a substrate for enzyme action initiating a cascade of events.
Alternatively, the association of THC with bilayer lipids may
somehow alter membrane organization. While model bilayer
studies are not necessarily germane to native membrane systems,
the preceeding study suggests specific possibilities for the in
vivo mode of action of THC. If THC/phospholipid complex
formation is important for THC action, high membrane cholesterol
concentrations could block or modulate THC activity while low
cholesterol concentrations would enhance THC activity.
Alternatively, if THC/phospholipid complex formation inhibits
THC action, high concentrations of cholesterol would enhance its
activity by allowing it to act as a free molecule. The
concentrations of THC used in the DSC study described may or may

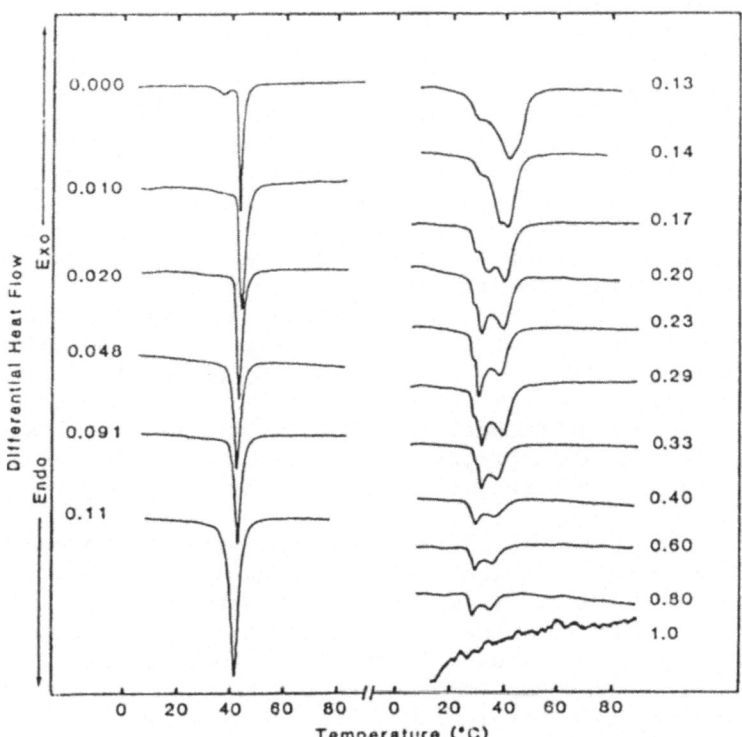

Fig. 9. Increasing THC in DPPC Bilayers. Thermograms are of
DPPC bilayers with increasing concentrations of THC. The
scaling of the thermograms is not precise, but represents
approximately equal amounts of DPPC. The mole fraction of THC
is indicated beside each thermogram. The sensitivity of the
thermogram of THC (X_{THC} = 1.00) is about 20 times greater than
the other thermograms.

not be relevant to in vivo action since physiological
concentrations within specific cells are unknown. Further, THC
could preferentially segregate into specific membrane domains to
produce high local concentrations.

   As is evident from the work described up to this point,
membranes may very well possess regions or domains differing in
lipid composition and physical properties. The existence of
heterogeneity in biomembranes could have great biological

significance. Since the spectra of melting points of
biomembrane lipids are continuous and broad, individual
endotherms corresponding to the melting of membrane regions
greatly enriched in a particular lipid species do not occur.
One means of studying bilayer heterogeneity is to use various
probe molecules. These probes are usually lipid analogues
designed to give differing signals depending on their
environment. Such probes use a variety of reporter moieties,
e.g. fluorescent, ESR, NMR. A present limitation in the use of
probe molecules is an understanding as to how they partition
among various membrane regions. In this laboratory using DSC we
have obtained the concentration phase diagrams for a series of
fluorescent carbocyanine probes in various phosphatidylcholine
bilayers (16). A correlation of the phase preference of probes
in model systems, as deduced from DSC, with their fluorescence
spectra allows an interpretation of the spectra obtained when
these probes are inserted into actual cell membranes.

     DSC has also been used to investigate membrane associated
cholesterol esters. Cholesterol esters have a very low
solubility in lipid bilayers and are not normally found
associated with biomembranes. Cholesterol esters, however, are
found in substantial amounts in atherosclerotic lesions.
Because of its occurence in diseased tissue, the mode of
association of cholesterol esters with biomembranes is of
interest. Possible modes of association could be droplets
within the hydrophobic core of the membrane bilayer, binding to
membrane protein or as part of membrane attached serum
lipoproteins. A potentially useful model system for
investigating this association is the membrane of the
microorganism Mycoplasma capricolum. The Mycoplasma due to
their simplicity have served as model membrane systems in many
studies. As mentioned previously, cholesterol esters show
complex behavior that is a function of thermal history,
impurities and physical packing constraints. Using DSC on
native membranes and extracted membrane material, it was
possible to demonstrate that the majority of cholesterol esters
associated with the membranes of M. capricolum exist as
relatively large and pure liquid droplets (17).

     All the preceeding studies were carried out on material
cooled at relatively slow rates before being run in the DSC. A
new approach being taken by this laboratory is to quickly cool
samples from a desired temperature to 85°K (4). The sample is
then loaded into the DSC under conditions in which sample
temperature does not rise above 90°K. The cooling rate is
sufficiently fast (approx. 1,000-10,000°K/sec) that the
near-equilibrium conditions obtained at normal cooling rates

does not occur. Under these conditions it is found that the resulting thermograms differ from those for conventionally slow cooled samples. This is most likely a consequence of rapid cooling not allowing the phase separations of lipids and proteins that occur at slow cooling rates (1). The application of quick cooling techniques with the approaches described in this chapter should make DSC an even more powerful technique for investigating problems of biomembrane structure and function.

References

1. D. L. Melchior and J. M. Steim, Lipid-associated thermal events in biomembranes, Prog. Surf. Membr. Sci. 13:211-296 (1979).

2. R. N. McElhaney, The use of differential scanning calorimetry and differential thermal analysis in studies of model and biological membranes, Chem. Phys. Lipids 30:229-259 (1982).

3. D. M. Small, The physical state of lipids of biological importance: Cholesterol ester, cholesterol, triglycerids, Adv. Exp. Med. Biol. 7:55-83 (1970).

4. D. L. Melchior, E. P. Bruggemann and J. M. Steim, The physical state of quick-frozen membranes and lipids, Biochim. Biophys. Acta 690:81-88 (1982).

5. A. Carruthers and D. L. Melchior, Cytosolic modulation of human erythrocyte hexose transfer, Biochim. Biophys. Acta 728:254-266 (1983).

6. L. Thilo, H. Trauble, and P. Overath, Mechanistic interpretation of the influence of lipid phase transitions on transport functions, Biochemistry 16:1283-1290 (1977).

7. D. L. Melchior and J. M. Steim, Thermotropic transitions in biomembranes, Annu. Rev. Biophys. Bioeng. 5:205-238 (1976).

8. D. L. Melchior, Lipid phase transitions and regulation of membrane fluidity in prokaryotes, in: "Current Topics in Membranes and Transport, Vol. 17" S. Razin and S. Rottem, eds. Academic Press, New York, (1982).

9. D. L. Melchior and J. M. Steim, Control of fatty acid
    composition Acholeplasma laidlawii membranes, Biochim.
    Biophys. Acta 466:148-159 (1977).

10. R. A. Demel and B. de Kruyff, The function of sterols in
    membranes, Biochim. Biophys. Acta 457: 109-132 (1976).

11. R. Bittman and L. Blau, The phospholipid-cholesterol
    interaction. Kinetics of water permeability in
    liposomes, Biochemistry 11:4831-4839 (1972).

12. D. L. Melchior and M. P. Czech, Sensitivity of the
    adipocyte transport system to membrane fluidity in
    reconstituted vesicles, J. Biol. Chem. 254:8744-8747
    (1979).

13. D. L. Melchior, F. J. Scavitto, and J. M. Steim,
    Dilatometry of dipalmitoyllecithin-cholesterol bilayers,
    Biochemistry 19:4828-4834.

14. R. A. Demel, J. W. C. M. Jansen, P. W. M. van Dijck, and L.
    L. M. van Deenen, The preferential interaction of
    cholesterol with different classes of phospholipids,
    Biochim. Biophys. Acta 465:1-10 (1977).

15. E. P. Bruggemann and D. L. Melchior, Alterations in the
    organization of phosphotidylcholine/cholesterol bilayers
    by THC, J. Biol. Chem.: in press.

16. M. F. Either, D. E. Wolf and D. L. Melchior, A Calorimetric
    investigation of the phase partitioning of the
    fluorescent carboxyanine probes in phosphatidylcholine
    bilayers, Biochemistry 22:1172-1178 (1983).

17. D. L. Melchior and S. Rottem, Organization of cholesterol
    esters in the membranes of Mycoplasma capricolum, Eur.
    J. Biochem. 117:147-153 (1981).

THE EFFECT OF SAMPLE TEMPERATURE GRADIENTS

ON DSC THERMOGRAMS AT THE GLASS TRANSITION TEMPERATURE

Edward Donoghue[*], Thomas S. Ellis and Frank E. Karasz

Department of Polymer Science and Engineering
University of Massachusetts at Amherst
Amherst, MA 01003

INTRODUCTION

Differential scanning calorimetry is currently being applied
with increasing frequency to relatively heavy, and thus thick,
samples of a variety of thermally insulating materials.[1]
Rationale for such applications are, for example, to reduce noise
in measurements of small specific heat changes at the glass transition or to reduce, during measurement, the effects of ongoing
processes such as the evaporation of plasticizer. As is shown
below, however, surprisingly large temperature gradients typically
occur in samples of this kind and, although these gradients do not
usually affect the determination of specific heat by differential
scanning calorimetry, at the glass transition these large gradients do, in fact, mask the true specific heat variation with
temperature.

We present experimental data to establish the magnitude of the
sample gradients and the variation of these with heating rate and
sample thickness, and we present computer simulations of DSC thermograms for idealized sample materials with sigmoidal "glass
transitions" of various widths in order to demonstrate the discrepancy between DSC scans and the corresponding specific heat dependence on temperature at the glass transition. We present in
addition intuitively plausible physical explanations for the beha-

[*]Department of Mathematics, Amherst College, Amherst, MA 01002

vior of the sample gradients and for the form of the DSC thermograms in terms of the temperature drop across the sample and the
absorption of heat by the sample. We defer to a later paper[2],
however, the lengthy derivations of the mathematical models and of
the solutions of these upon which the intuitive picture is based.

The large temperature drop across a thick sample of material
with low thermal conductivity is due in part to a correspondingly
large temperature drop across the sample cup itself. This latter
temperature drop, which we presume is the same for both sample and
reference cups, results from two distinct temperature gradients:
one due to absorption of heat by the platinum sides and lid of the
cup, and the other caused by the temperature difference between
the heater beneath the cup and the purge gas flowing around it.
The temperature distribution in the cup produced by the flow of
heat from heater to purge gas is the same for non-zero heating
rates, after the passage of an initial transient time, at a given
temperature as it is in the steady state at the same given temperature. This pseudosteady state distribution is modified at
non-zero heating rates by a second, time-independent temperature
distribution in order to accomodate the absorption of heat by the
platinum cup as the temperature of this increases. Because the
platinum components are very thin, however, the amount of heat
absorbed per unit length of heat flow by these is small and, as a
result, this second, dynamic temperature gradient is much smaller
than the pseudosteady state gradient. In the next section these
two gradients are characterized quantitatively by an equation
derived from our data that gives the true temperature on the
calorimeter cup lid in terms of the programmed temperature. In
this formula the total temperature drop across the cup is seen to
be linearly dependent on the heating rate and on the temperature
difference between heater and purge gas. Over the range of
programmed temperatures from 300K to 500K, this total temperature
drop increases from 6K to 16K.

The effect on the sample of this substantial temperature drop
across the cup can be determined with greater generality with the
sample placed directly on the bottom of the cup rather than in a
sample pan. With this arrangement it is clear that the temperature
drop across the cup induces a flow of heat from the bottom of the
cup through the enclosed sample and the gas that surrounds it to
the sides and lid of the cup. As in the case of the cup itself,
the external temperature drop produces within the sample a pseudosteady state temperature distribution that depends on the temperature of the bottom and lid of the cup in exactly the same way
in the true steady state as at non-zero heating rates (after the
passage of a transient period). This distribution in the sample,
as with that in the calorimeter cups, is modified at non-zero

heating rates by a time-independent dynamic distribution due to
heat absorption within the sample. In marked contrast with the
calorimeter cups, the heat absorption by the sample per unit
length of heat flow is large; sufficiently so, in fact, that the
dynamic temperature drop across the sample is generally larger
than the pseudosteady state one. In the third section the dependencies on heating rate and on sample thickness of the temperature
range of DSC operation the dynamic gradient is large enough that
the temperature of the top of the sample is less than the temperature of the sample cup lid.

The effect of the substantial sample temperature gradients on
the measurement of specific heats by differential scanning calorimetry is slight if, as is usually the case, the specific heat
varies only slowly with temperature. Then even though the temperature drop across a sample is large, the change in specific
heat across it is small and, as a result, the heat absorbed per
unit temperature increase is approximately the same for all infinitesimal slices of the sample taken parallel to the cup bottom.
The total heat absorbed per unit time at the bottom of the sample
is, in this case, very nearly what it would be if the temperature
throughout the sample were constant, and the DSC thermogram can
validly be interpreted as being essentially the variation in specific heat with temperature.

In the neighborhood of the glass transition temperature, T_g,
however, the specific heat increases abruptly and by a substantial
amount. The effects of this transition on the rate of heat absorption by the sample are shown in the fourth section to be more
easily discerned in the limiting case where the variation of specific heat with the temperature at T_g is a step function, rather
than an extended transition as actually observed.[3] Then with temperatures at top and bottom of the sample bracketing T_g, the
variation of the specific heat with height through the sample is
abrupt in a neighborhood of the level at which the temperature is
T_g. The rubber below this level in the sample has a substantially
higher specific heat, and thus absorbs more heat per unit temperature increase, than does the glass above it. The instantaneous rate at which heat is absorbed by the sample is then some
average of two values characteristic of rubber and of glass. As
the temperature through the sample increases at the programmed
rate, so that the glass/rubber boundary sweeps upward through the
sample, the total heat absorbed shifts upwards from an average
value close to that for the glass to one typical of the rubber.
The DSC thermogram in this region is shown to increase almost
linearly, reflecting the movement of the glass/rubber boundary
through the sample rather than following some underlying variation
in specific heat with temperature. In addition, the movement of

the glass/rubber boundary through the sample is shown to leave the
temperature distribution within the sample far from the limiting
temperature distribution that is valid over long times and
corresponds to the new specific heat.   Thus there follows a transient period during which the temperatures within the sample decay
exponentially to the limiting distribution; during this transient
the rate of heat absorption by the sample, and thus the DSC scan,
increases exponentially to its limiting value for the rubber.
Thus it is not until the end of the transient regime that the DSC
thermogram once again manifests the underlying variation of the
specific heat with temperature.   The comparisons made in the
fourth section between simulated DSC thermograms and the specific
heat variation with temperature from which these were derived
demonstrate that the indicated temperature of the midpoint of the
transition in the DSC scan, for 2 mm thick samples of a material
such as polystyrene, is, in fact, about 10K above T_g (defined as
the midpoint of the "true" specific heat vs. temperature plot)
for a heating rate of 20K/min.   In addition, the width of the DSC
scan at T_g is shown to be largely independent of the corresponding
width of the underlying specific heat variation, even for
realistically wide transitions with half widths of 5K to 10K.

THE TEMPERATURE DROP ACROSS THE CALORIMETER CUPS

    The thermal analyses discussed in this report were made in a
Perkin-Elmer DSC-2 cooled with a dry ice/acetone mixture so as to
maintain the temperature of the DSC head T_H at a sufficiently
low value that the two calorimeter cups can be cooled rapidly to a
temperature about 100K below the range of experimental interest.

    With the temperature of the cups maintained under program
control, the effect of the purge gas cooling on the temperature of
the bottom of the cups is negligible.   In fact, the true temperature at the bottom of the cups, $T_B(t)$, differs from the indicated temperature, which we take to be the programmed temperature,
$T_p(t)$, by an amount that depends linearly on the heating rate, T,
but which is independent of the purge gas temperature and of the
thermal properties of any enclosed sample.   The Perkin-Elmer DSC-2
manual expresses this in the form

$$T_p(t) - T_B(t) = C\dot{T} - D \tag{1}$$

where C and D are constants.   The temperature shift CT is presumably due to heat absorption within the base of the calorimeter
cup, and thus the constant C is determined by the construction of
the cup.   From an analysis of our data as described below we have
obtained for C a value of 0.058 min.   The value of D is determined
entirely by the temperature calibration of the DSC.   By conven-

tion, the calibration is set so that $T_B(t) \equiv T_p(t)$ at a heating rate T =10K/min; for this calibration, D = 0.58.

We have obtained the relation, analogous to Eq.(1), that gives the true temperature of the calorimeter cup lids, $T_L(t)$, as a function of the programmed temperature $T_p(t)$ and heating rate T. The relation was derived from measurements made at various heating rates of differences in the programmed temperatures for melting of metallic standards placed on the bottom and lid of the sample cup.*

Firstly, we operated the DSC-2 in a quasi-static mode: incrementing the indicated temperature by 0.1K at T = 5K/min and then waiting for the elapse of the transient period for the steady state to be re-established. In this way we obtained the indicated temperatures for melting on lid and bottom for four metals: mercury, indium, tin and lead. We found that the difference between these observed melting points is a <u>linear</u> function of T_M, the true melting temperature of each standard, and that this difference vanishes when $T_M = T_H$, the purge gas temperature. That is, we found that the linear least squares fits to the observed melting points for the standards on the lid and on the bottom intersected at a temperature of 197K, in good agreement with the sublimation temperature of dry ice.

We next operated the DSC dynamically and obtained the indicated temperatures for melting on lid and bottom for the four metals. In each case the difference between the observed melting points for lid and bottom was found to be a linear function of T. The slopes of these fits for the four metals were the same, to within experimental accuracy. The least squares fits to the data for the bottom melting for each standard yielded the values for the constants in Eq.(1). The extrapolations of the least squares fits to the data for each metal to T = 0 agreed very well with the quasi-static values obtained as described above.

All the experimental data, as shown in Fig.1, fit an equation of the form

$$T_L^M - T_B^M = A(T_M - T_H) + C\dot{T} \tag{2}$$

where T_L^M and T_B^M are, respectively, the <u>indicated</u> temperatures for melting on the lid and bottom of a standard with <u>true</u> melting temperature of T_M. From Eq.(1) we similarly have that

---

*This method was originally suggested to us by J.H. Flynn, Polymer Science and Standards Division, National Bureau of Standards, Washington, D.C. 20234

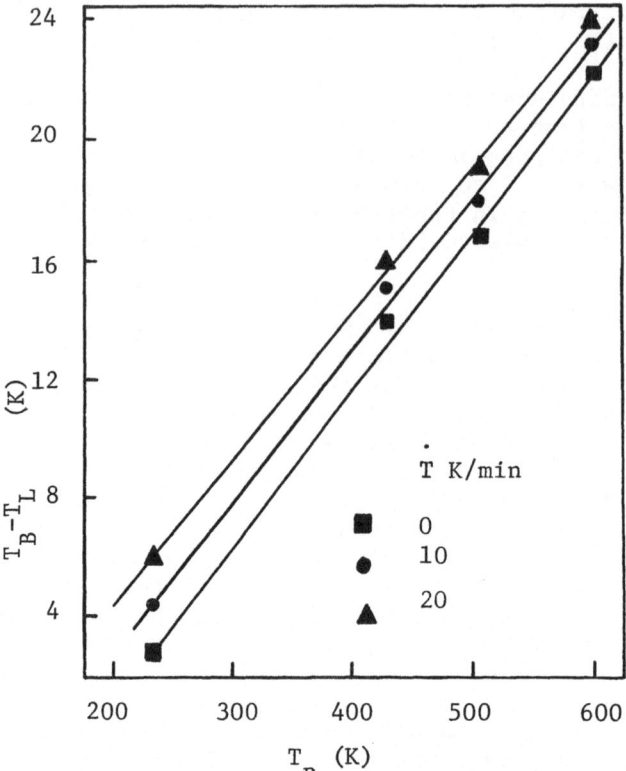

Fig. 1. The temperature drop across a calorimeter cup plotted
as a function of indicated temperature for three
heating rates. The drop is a linear function of
temperature, and is surprisingly large, between 10K
and 20K, at the higher temperatures.

$$T_B^M = T_M + C\dot{T} - D \qquad\qquad\qquad (3)$$

assuming that no time lag occurs between the melting of the standard and the sensing of the absorption of latent heat. Equation
(3) specifies T_B entirely in terms of known quantities and thus
can be used to eliminate T_B from Eq.(2). Then with the identification in Eq.(2) of T_L as the programmed temperature $T_P(t)$ at
which $T_L(t)$, the true temperature of the calorimeter cup lid,
equals T_M, we get

$$T_L(t) = T_H + F\ [T_B(t) - T_H - C'\dot{T}] \qquad (4)$$

where F=1/(1+A) and where we have made use of Eq.(1) a second
time. Note that the validity of Eq.(4) also depends on there
being no lag between the time at which melting occurs on the <u>lid</u>
of the calorimeter cup and the time at which the absorption of
latent heat is sensed. From our least squares fits to the data we
find that $C' = 0.052$ min and that $F = 0.95$. The near agreement
between C and C' indicates that the absorption of heat by the side
and lid of the calorimeter cup is nearly the same as the absorption of heat in the base of the cup. Just as the time rate of
change of $T_B(t)$ is $\dot{T}$, the programmed heating rate, the time rate
of change of $T_L(t)$ is $F\dot{T}$, as can be derived from Eq.(4). With $F =$
0.95, the rate at which $T_L(t)$ increases is 19K/min if $\dot{T} = 20$K/min.
Thus $T_L(t)$ lags further and further behind $T_B(t)$ as the two of
these increase with time, with $T_B(t) - T_L(t)$ ranging from about
zero for $T_p(t) \simeq T_H \simeq 200$K to about 6K at $T_B(t) \simeq 300$K and to
about 16K at $T_p(t) \simeq 500$K.

A comparison of Eqs.(1) and (4) reveals the different
dependencies of $T_B(t)$ and $T_L(t)$ on the programmed temperature.
Whereas Eq.(1) shows that $T_B(t)$ and $T_p(t)$ differ by an amount that
is <u>independent</u> of the programmed temperature Eq.(4), when put in
the form

$$T_B(t) - T_L(t) = (1 - F)\ [T_B(t) - T_H] + FC'\dot{T} \qquad (5)$$

shows that the temperature drop across the calorimeter cup between
bottom and lid is a linear function of the temperature of the cup,
as well as a linear function of the heating rate. The difference,
of course, is due to the effective cooling of the calorimeter cup
lid by the purge gas, which has no cooling effect on the calorimeter cup bottom.

Implicit in Eqs.(1) and (5) is the independence of these temperatures on the thermal properties of any sample enclosed in the
cup. This independence has been tested by us with experiment as
described in the next section. Because of this independence,
Eqs.(1) and (4) or (5) specify the thermal environment to which
all DSC samples are subjected; these equations thus specify the
boundary conditions in the heat equation solved numerically, as
reported in the last section, to determine the distribution of
temperature within DSC samples through the glass transition
region.

TEMPERATURE DISTRIBUTIONS WITHIN THE SAMPLE

While we know that temperature differentials of order 10K do occur across the calorimeter cup under ordinary operating conditions, we cannot immediately conclude from this that similarly large drops occur across samples in the cup. Indeed if the ratio of the thermal conductivities of sample and of surrounding gas is large, so that the gas in effect provides thermal insulation, then the steady state temperature drop across the sample will indeed be small. Polystyrene and other good insulators, however, have thermal conductivities which differ from that of a gas only by a factor of about four, and samples of such material will be shown below to have in fact steady state gradients the same order of magnitude as those across the calorimeter cups.

The specific heat of polystyrene is in addition sufficiently large that the dynamic gradients across samples of this material are equal to or larger than the pseudosteady state gradients across these at standard heating rates. The total temperature differentials across polystyrene samples are thus in the same range, and commonly greater than, the total temperature drop across the calorimeter cup.

The formula derived in the last section for the temperature drop across the cups is extremely simple, being linear in heating rate and programmed temperature with coefficients that are constants. The "constants" are, however, dependent on the material and dimensions of the cups and are fixed only so long as the cups themselves remain unchanged. The corresponding formula for the temperature drop across a sample is similarly linear in heating rate and in the temperature drop across the sample cup. The coefficients of these are similarly dependent on the dimensions and thermal properties of the sample; since the latter vary widely among samples, the coefficients can no longer be taken as constants, however. We have obtained the functional dependence of these coefficients on the dimensions and thermodynamic properties of the sample, but we defer to a later article the lengthy derivation of these formulas and the detailed comparison of these with our data. For the present we restrict our analysis of this data to verifying certain quantitative features of these large sample gradients in thick samples. The low thermal conductivity of polystyrene, nearly matching that of the surrounding gas, and its sizeable specific heat, which were identified earlier as jointly producing the large sample gradients, are seen below to produce in addition a wide variation in the total sample gradients with heating rate and sample thickness. For this reason, polystyrene is an especially suitable subject material for the study of these variations.

In order to simplify both experimental procedures and the
sample gradients themselves, and in order to obtain a wide
variation in gradients, the samples used in our experiments were
placed directly in the sample cup (i.e. they were not sealed in
sample pans). Small pieces of indium standard (weighing a few
micrograms) on small pieces of aluminum foil were placed on the
bottom of the cup and top of the lid to indicate the programmed
temperatures at which the true temperatures at these locations
reached the melting point of the standard. For the same reason, a
small piece of indium was also placed directly on top of the
sample. The differences between the programmed temperatures for
melting were recorded for samples of varying weight and thickness
heated at the standard heating rates. The results of these
measurements are summarized in Fig.2.

The indicated temperature corresponding to the melting point
of the standard on the bottom of the cup shows only a small
variation, within experimental accuracy, as the sample weight
varies from zero (i.e. no sample) up to about 60 mg. This verifies that $T_B(t)$ as given by Eq.(1) is indeed independent of sample
properties and thus can serve as a boundary condition in a mathematical model of the heat flow through sample and surrounding gas.
The indicated temperature corresponding to the melting point of
the standard on the lid varies more widely, by a degree or so, but
does not vary systematically with sample weight. (The temperature
of the lid is very sensitive to its seating position, and so quantitative agreement with formulas derived with use of these as
boundary conditions can be expected only for data gathered when
the lid is seated properly. The lowest value of the indicated
temperature for melting on the lid when repetitively obtained was
taken as the optimum lid seating position, as a poorly seated lid
is cooler since heat flows into it at a slower rate.) Thus
Eq.(5), much as Eq.(1), gives $T_L(t)$ validly for all samples, and
these expressions together define the boundary conditions seen by
all samples.

Because the properties of the sample have no effect on the
temperature at the bottom or on the lid of the cup, the additional
assumption of one dimensional heat flow vertically through sample
and gas to the cup lid implies that the temperature on the top of
the cylindrical sample is independent of the diameter of the
cylinder. This is so because with this assumption the temperature
distribution along any vertical line through the sample and gas is
determined by the temperatures on this line at sample cup bottom
and lid, which are the same for all vertical lines. The steady
state and dynamic temperature drops across the sample are thus
assumed hereafter to be independent of sample weight, although
certainly dependent on sample thickness.

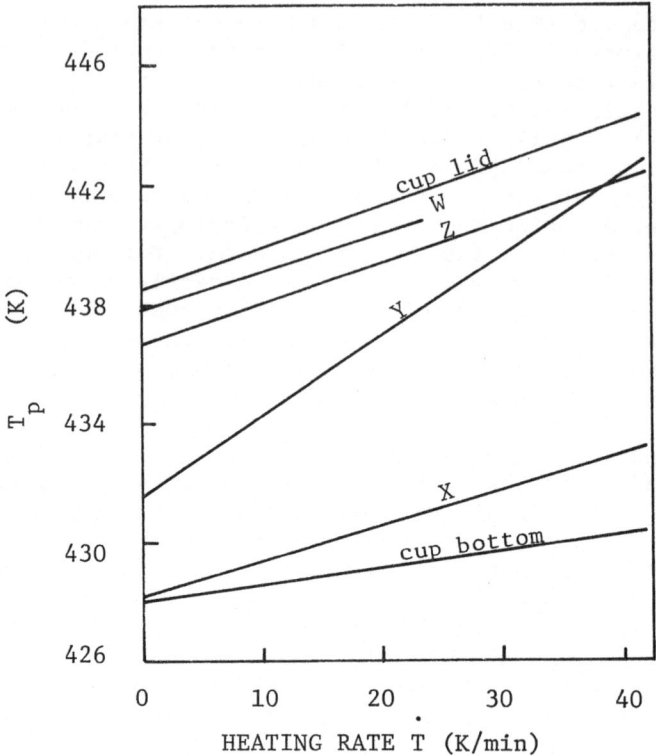

Fig. 2.   The temperature drops across four cylindrical samples
          of polystyrene plotted as a function of heating rate.
          The respective weights (mg) and thicknesses (mm) of
          the samples are: X = (11.0, 0.57), Y = (10.4, 1.95),
          Z = (7.3, 1.5) and W = (52.9, 2.75). Note that the
          variation in temperature drop across a sample is
          greatest at intermediate thickness (i.e. Y).

     The strong dependencies of the sample gradient on sample
thickness and on the heating rate are evident in Fig.2, in which
are plotted the differences in indicated temperatures for melting
of indium on the cup lid and bottom and on the top of four
polystyrene samples of varying thickness. The sample gradients
for the steady state (T = 0) are seen to vary from nearly zero for
sample X to almost the total drop across the sample cup itself in
the case of sample W. In our mathematical model these steady
state gradients are independent of the sample specific heat but
dependent on the ratio of thermal conductivities of sample and

surrounding gas. The variation with sample thickness is linear only in the case where this latter ratio is unity. For the more realistic case where this ratio is greater than one, the gradient across the sample remains close to zero over a wider range of sample thickness than for the linear variation.

The variation of the temperature drop across the sample with heating rate is evidently linear over the entire range of the heating rates of practical interest. In our mathematical model the slopes of these lines are proportional to the sample specific heat and are inversely proportional to the sample thermal conductivity. The variation of slope with sample thickness is more complex and interesting: the slope attains a maximum at a certain sample thickness, about half the depth of the sample cup. That such a maximum does occur is evident in Fig. 2, where the dynamic gradient for sample Y increases more rapidly with T than for either the thicker samples W and Z or the thinner sample X. This means that the sample temperature drop is largest for samples of intermediate thickness. For sample thicknesses close to but above this stationary value, the total sample temperature gradient is larger than that across the cup itself over a broad range of heating rates below 40K/min (in the figure, the gradient across sample Y reaches this crossover point only above 50K/min). The physical explanation of the local extremum in the slope is intuitively clear: as sample thickness increases, the total heat absorption along the path of heat flow increases, but the effect on the sample of the temperature of the lid becomes more important as the insulating air gap thickness decreases and the total heat flow through the sample to the lid increases.

DSC THERMOGRAMS AT THE GLASS TRANSITION

Under the assumption that the flow of heat through the sample is largely the one-dimensional, vertical flow from the calorimeter cup bottom through the sample and surrounding gas to the lid, the temperature distribution within the sample and gas is determined by the one-dimensional heat equation,

$$\frac{\partial}{\partial x}\left(k(t)\frac{\partial T}{\partial x}(x,t)\right)=\rho c_p(t) \frac{\partial T}{\partial t} \tag{6}$$

as applied separately to sample and gas, together with four boundary conditions. Two of these are the requirements for continuity of temperature and of heat flow at the boundary between sample and gas. The requirement that the temperature at the bottom of the sample (at x=0) equal the temperature of the calorimeter cup bottom gives the third boundary condition,

$$T(0,t) \equiv T_B(t) = T_P(t) - C\dot{T} + D \tag{7}$$

while the corresponding requirement at the cup lid (at x=L) gives $T(L,t) \equiv T_L(t)$, where $T_L(t)$ is as given by Eq.(5).

Because the heat capacity of a gas is only one-thousandth that of typical solids, heat absorption by the gas can be neglected. Then the right-hand side of the heat equation for the gas is zero and so by the left-hand side the heat flow through the gas is independent of position. From this there follows immediately the relation

$$k_G(T) \frac{\partial T}{\partial x}(1,t) = k_G(T) \frac{T(L,t)-T(1,t)}{L-1} = \theta_G[T_L(t) - T(1,t)] \tag{8}$$

where $\theta_G = k_G/(L-1)$, and where $T(1,t)$ is the temperature at the top of the sample. With this relation used as the boundary condition for the top of the sample the numerical solution of the heat equation is greatly simplified, since the computation of the distribution in the gas and the matching of this solution with that of the sample at x=1 is dispensed with.

The Crank-Nicolson[4] method for the numerical solution of the boundary value problem defined by Eqs.(6)-(8) is valid for the case of discontinuous coefficients, $k(T)$ and $\rho C_p(T)$, as obtains, or nearly obtains, at T_g. This method provides the temperature distribution at time $t+\delta t$, given the distribution at time t, as the solution of a certain non-linear system of algebraic equations, which are solved with the use of the Newton-Raphson method[5] and the Thomas Algorithm[6]. The Crank-Nicolson method is more easily (and more generally) applied to the heat equation with boundary conditions of the kind given by Eq.(8), rather than by Eq.(7). For the numerical solutions by the Crank-Nicolson method discussed below, then, the boundary condition,

$$-k_S(T)\frac{\partial T}{\partial x}(0,t) = \theta[T(0,t)-T_B(t)], \tag{9}$$

was used in place of Eq.(7); the value of θ used was sufficiently large, however, that $|T(0,t)-T_B(t)|$ remained less than 0.01K at all times.

In order to simulate the temperature distribution and DSC thermogram for a sample material at its glass transition we

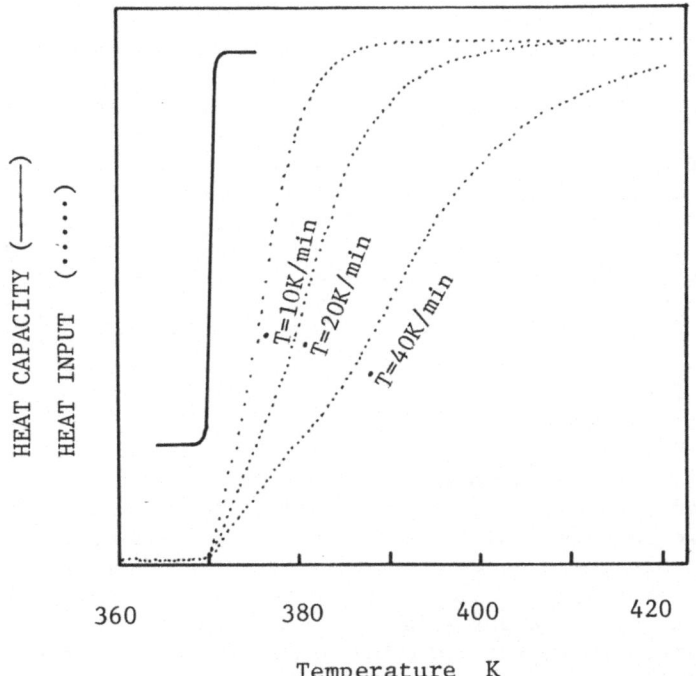

Fig. 3.   Simulated thermograms (dotted) for three heating rates
          for a sample thickness 2 mm with a heat capacity (solid)
          that increases by 15% over a 2K temperature range. The
          large sample temperature gradients mask this underlying
          variation of the heat capacity while producing scans
          composed of two distinct regimes (as is especially
          evident at $\dot{T}$ = 40K/min).

adopted thermodynamic properties with typical numerical values and
idealized analytic properties: we took the thermal conductivity
as temperature-independent, and the heat capacity per unit volume
to increase sigmoidally, as a hyperbolic tangent, from one
constant (glassy) value to a second (rubbery),

$$k = 3.0 \times 10^{-4} \text{ cal/cm}^2\text{secK} \tag{10}$$

$$\rho Cp(T) = \tfrac{1}{2}[0.441+0.398] + \tfrac{1}{2}[0.441-0.398]\tanh[A(T-T_g)/2] \text{ cal/cm}^3 \text{ K}$$

where we set T_g = 370K. The program was run starting at an initial programmed temperature sufficiently far below T_g that $\rho C_p(T)$,
as given above, was approximately equal to its lower asymptote.

We used as the initial condition the solution of Eqs.(6), (8) and (9) that is valid over the longest time scale: this is the post-transient solution consisting of the pseudosteady state and dynamic distributions discussed in the previous two sections.

After minimal experimentation with the numerical solutions at T_g for idealized "sample materials", as defined by Eq.(10), it is evident that the structure of the DSC thermogram is most clearly revealed with "glass transitions" in which the variation of C_p with T is substantially more abrupt than for real glass transitions. With the parameter A in Eq.(10) set to 3.15, the specific heat varies from its lower to upper asymptote over a range of only 1K on either side of T_g. The simulated thermogram for such material for a heating rate of 20K/min and for a sample of thickness 2 mm is shown in Fig.3.

The temperature drop across this sample is substantial at all programmed temperatures, increasing from 8.8K at 360K to 10.2K at 420K. Two regimes, which are shown by the computations to result from this drop, are evident in the plot: a linear portion over the range from 370K to 380K, and an exponential decay over the range from 380K to about 410K. As explained in the Introduction, and as reference to the detailed temperature distributions verifies, the linear portion of this scan is produced by the movement of the glass/rubber boundary through the sample, and the following monotonic increase is merely transient exponential decay to the limiting solution, valid over long times, that is imposed on the sample by the programmed increase in temperature. The two regimes evident in the simulated scan thus both result from the effects of large thermal gradients in the sample, and so the shape of the DSC scan in the region of the glass transition is determined by these thermal gradients rather than by the underlying variation in C_p with temperature. This latter is plotted as well in Fig.3 where it is evident that the shape of the scans reveal little of the actual shape of the C_p vs. T plot.

The thermogram is, in fact, determined by two <u>times</u>: the time for the glass/rubber boundary to traverse the sample, and the longest transient time for thermal equilibration. Since these times are independent, or are only weakly dependent, on the heating rate, the widths of the two regimes are roughly proportional to $\dot{T}$. Thus, as seen in Fig.3, the scan for $\dot{T} = 10K/min$ is about half as wide as that for a heating rate of 20K/min, while the scan for $\dot{T} = 40K/min$ is about twice this latter width.

Estimates of T_g obtained by the midpoint method for the three heating rates of Fig.3 are, respectively, 7K, 11K and 21K above the midpoint of the C_p vs. T plot. Although it is not shown here,

the width of the simulated scans depends as well on the thickness
of the sample and on its thermodynamic properties. Thus, while
the "midpoint method" is of use in comparing the change in
T_g between samples of approximately the same thickness and thermodynamic properties, it is not a trustworthy method of obtaining
T_g absolutely.

The thermograms in Fig.3 are, of course, all derived from an
underlying variation in C_p with T that is unrealistically narrow;
the possibility remains that the midpoints of broader, and more
realistic, model "glass transitions" will be better approximated
by the midpoints of the scans. That this is not the case is evident in Fig.4 where are plotted the three scans for "glass
transitions" of widths 2K (as in Fig.3, and corresponding in
Eq.(10) to A=3.14), 10K (A=0.58) and 20K (A=0.29) the latter two
bracketing the range of realistic glass transition widths. The
midpoints of the three scans are seen to coincide with one another
at about 381K, eleven degrees above the common midpoint of the
three "glass transitions". The scans are thus insensitive to the
width of the underlying glass tansition to a surprising extent,
and so we see the gradients across the sample mask this feature of
the underlying C_p vs. T dependence as well.

The scans for the "glass transitions" of more realistic widths
do differ significantly from the scans shown in Fig.3 for the
narrow transition in that the two regimes evident in the latter
are not obvious for the wider transitions. Reference to the
detailed temperature distributions verifies what one would certainly suppose: that the two regimes nevertheless occur as
before. Despite this, the thermograms are smooth, sigmoidal
increases from one constant value to a second, and thus resemble
qualitatively, without being quite as symmetrical, the underlying
C_p vs. T plots from which they are derived.

The two regimes that have been shown to compose the scan at T_g
are the same as the two that are well-known to compose the peak in
the scan of a first-order phase transition: a linear portion
corresponding to the interconversion of one phase into the other,
followed by an exponential transient. In the case of T_g, however,
the exponential transient is in the same direction as the linear
portion, rather than opposite it as in the case of melting, so
that the scan at the glass transition increases monotonically,
rather than having a sharp maximum. Thus, the only indication of
the two regimes at T_g that can be evident in DSC thermograms is
the subtle discontinuity in slope between the two regimes. This
discontinuity is visible only for unrealistically narrow model
glass transitions. In realistically broad model transitions the
discontinuity is smoothed out, and no evidence of the two regimes
is then apparent.

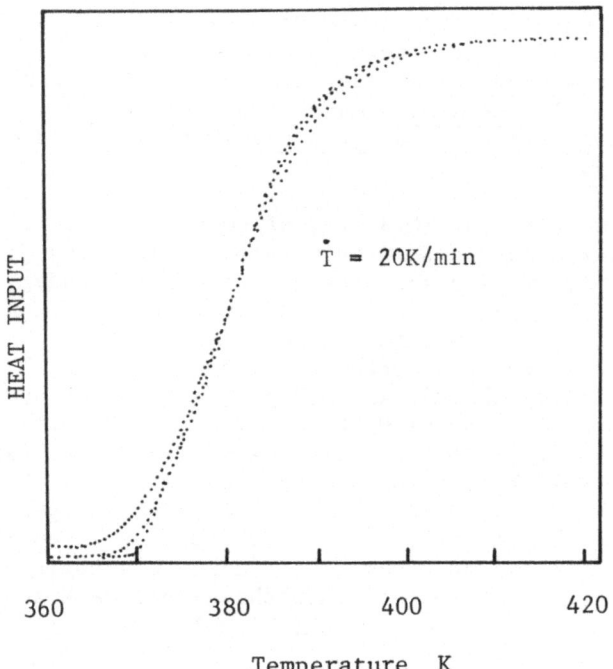

Fig. 4.   Simulated thermograms for samples 2mm thick with three
          different temperature variations at the glass
          transition: one unrealistically narrow, 2K wide (as in
          Fig. 3), and two others, one 10K and one 20K wide,
          which bracket the range of realistic glass transition
          widths. The sample temperature gradients completely
          obscure these underlying variations in the specific
          heat.

     DSC thermograms in the region of the glass transition are
thus doubly deceptive: the two regimes constituting them join
smoothly in the scan, giving no evidence of their separate
occurrence, and, in addition, the smooth scan that results
resembles, and thus is easily confused with, the underlying specific heat variation with temperature, which is known from the
results of adiabatic calorimetry.

ACKNOWLEDGEMENT

     The authors thank AFOSR 82-1101 and the Center for University of
Massachusetts-Industry Research on Polymers (CUMIRP) for financial
support.

REFERENCES

1. S. Kraus, M. Isleandar and M. Iqbal, "Properties of Low Molecular Weight Block Copolymers. 1. Differential Scanning Calorimetry of Styrene-Dimethylsiloxane Diblock Copolymers." <u>Macromolecules</u> 15: 105 (1982). T.S. Ellis, F.E. Karasz and G. ten Brinke, "The Influence of Thermal Properties on the Glass Transition Temperature in Styrene/Divinylbenzene Network-Diluent Systems." J. Appl. Polym. Sci. 28: 23 (1983).

2. To be published.

3. F.E. Karasz, H.E. Bair, and J.M. O'Reilly,"Thermal Properties of Atactic and Isotactic Polystyrene." J. Phys. Chem. 69: 2657 (1965).

4. J. Crank, "The Mathematics of Diffusion." Oxford University Press, Oxford (1956).

5. G. Dahlquist and A. Bjorck, "Numerical Methods." (translated by N. Anderson), Prentice-Hall, Englewood Cliffs (1974), Section 6.9.2.

6. Reference 5, Section 5.3.1.

CHARACTERIZATION OF COAL LIQUEFACTION RESIDUES

BY THERMAL METHODS OF ANALYSIS

Charles M. Earnest
The Perkin-Elmer Corporation
Main Avenue - M/S 131
Norwalk, CT 06856

INTRODUCTION

Recent years have seen vast experimentation with many different process designs for the liquefaction of coals. The degree of coal conversion and composition of the product oil vary with both the coal rank, maceral composition, mineral matter content, and conversion process. Whereas much attention has been focused on the separation and characterization of the product oil by chromatographic and spectroscopic means, less work has been done on the unconverted or process altered residues from liquefaction processes. Although many of the processes do incorporate some sort of "bottoms processing", other possible uses of these residues include road materials, carbon electrodes, coal gasification feedstocks, and as direct combustion fuels. Recently, coal conversion by-products have been used as raw materials in the synthesis of thermosetting polyesters[1].

FACTORS WHICH AFFECT LIQUEFACTION YIELDS

A listing of the factors which affect the degree of conversion in coal liquefaction processes is given in Table 1. At an ACS Symposium held in 1974, it was pointed out by Davis, Spackman, and Given[2] that the highest liquefaction yields are obtained from coals of the high volatile bituminous rank. Coals of either higher or lower rank than this give lower yields. Furthermore, from the results of their work involving petrographic analysis (microscopic techniques) and a set of batch autoclave techniques (Penn State and Gulf Procedures), it was clearly shown that the level of the vitrinite maceral component, or those derived from woody tissue or

343

bark of trees, is directly related to the degree of conversion. Also, that the pseudovitrinite maceral, which possesses varying degrees of inertness in coking processes, can show at least some reactivity in a liquefaction process. The waxy and resinous exinite group of macerals also proved to be advantageous components for coal liquefaction feedstock. The inertite group of macerals (fusinite, semifusinite, etc.) make little contribution to the liquefaction process.

Table 1.   Factors Which Affect Degree of Conversion
in Coal Liquefaction Processes

---

1. Coal Rank

2. Maceral Composition

3. Mineral Matter

    a. Composition

    b. Physical State

4. Nature of Conversion Process

In contrast to all other ranks of coal, there is no evidence of correlation of maceral content with liquefaction yields obtained from lignite samples. It has been observed[3] that lignites with the highest total mineral and pyrite levels give the highest liquefaction yields. Mineral matter apparently exhibits catalytic effects on the reaction rate. Tarrer, et al.[4] have reported that the rate of liquefaction increases directly with the concentration of mineral matter. This work also showed that the physical state, as well as the chemical composition, of the mineral matter affect hydrogenation and hydrodesulfurization activity during coal liquefaction. Given, et al.[3] reported a catalytic effect of sodium ion as an inorganic constituent of lignite. The sodium ion did not increase the liquefaction yield but gave a product oil of much lower viscosity.

The severity of the liquefaction process with respect to temperature and pressure, length or time of processing, and whether or not the initial state of the process is catalytic, solvent extraction, etc., will also determine the relative degree of conversion and subsequent composition of both the product oil and residues. Although the examples which could be cited here are many, Davis, et al.[2] found that in varying the temperature in the batch autoclave experiments (mentioned above) from 316°C to a maximum of 427°C, the

solid residues showed varying degrees of conversion of vitrinite as
the temperature of the process was increased. At the lower temperatures unaltered exinite was observed. At about 399°C, part of the
vitrinite was transformed to a pitch-like substance while the rest
was converted to a carbonized form. At the maximum temperature of
the experiment, the residue contained large amounts of fusinite and
semifusinite.

RESIDUES USED IN THIS STUDY

With all of this in mind, we embarked on a project to determine
what information may be derived from thermal methods of analysis in
characterizing such coal liquefaction residues. For the purposes of
our studies, solid residues were obtained from two different pilot
plants representing two completely different routes to coal liquefaction. One involved solvent extraction while the other employed
catalytic hydrotreatment as the initial stage of the process. In
all subsequent work, the residues will be referred to as "Process #1"
(solvent extraction) and "Process #2" residues, respectively. In
the solvent extraction process, the initial liquefaction stage is
carried out at about 455°C and ca. 2000 psi using no catalyst. The
resulting liquid and gaseous products are separated by vacuum and
atmospheric distillation leaving the residue as a "bottoms slurry".
The catalytic process (Process #2) on the other hand employs a
$Co-Mo/SiO_2-Al_2O_3$ catalyst in the reactor. The gases are separated
in a single step from the liquid product and solid residue. The
solids are then removed by either centrifuging or by filtration.
The liquefaction yields in terms of product oil output is reported[5,6]
to be ca. 3 barrels per ton of coal feedstock in both Process #1 and
#2.

CHARACTERIZATION OF THE RESIDUES

The two residues were physically very different when observed
at room temperature. The residue from Process #2 was soft and tar-
like and could be compressed with a push of the finger. The Process
#1 residue was rigid and resembled the original coal to some extent.
This fact alone was the first suggestion that the thermal properties
should be quite different.

ELEMENTAL ANALYSIS

The two solid residues were first subjected to elemental analysis using the Perkin-Elmer Model 240C Elemental Analyzer and Model
240DS Data Station. The analyzer was employed using both the normal
CH&N mode as well as the sulfur mode of operation. The ash content
of the residues were established by weighing the sample before and
after analysis in the 240C Elemental Analyzer. The results of these
studies are tabulated in Table 2. As is seen in the table, the most
outstanding difference in the results is the ash and sulfur values

for the two residues. The difference in ash content causes the other values for the Process #2 residue to appear much lower as can be seen by comparing the CH&N results on an ash free basis. The higher sulfur residue in Process #2 is not unexpected since the feedstock was known in this case to be a Western Kentucky bituminous coal blend of relatively high sulfur content. Furthermore, it has been shown that the residual sulfur is essentially all organic and is found mostly in the benzene insoluble portion of the residue[6]. The C/H ratios and H/C ratios are listed in the table. The Process #2 residue is seen to have the higher H/C ratio. These elemental analysis results will be used later in this paper for calculating the calorific value of these residues.

Table 2.   Results of Elemental Analysis of Coal
           Liquefaction Residues.

			Ash Free Basis	
	Process 1	Process 2	Process 1	Process 2
%C	69.39	35.64	86.53	73.49
%H	4.28	2.38	5.34	4.91
%N	1.42	0.67	1.77	1.38
%S	2.8	5.94	---	---
%O*	2.3	3.87	---	---
% Ash	19.8	51.0	---	---
C/H Ratio	----	----	1.35	1.25
or H/C Ratio	----	----	0.74	0.80

*By difference

THERMOGRAVIMETRIC STUDIES

As with coals, thermogravimetry offers an excellent means of observing volatilization profiles of coal conversion residues when the analysis is performed in a dynamic, inert atmosphere. The volatilization profiles of the two residues under study were obtained

using the Perkin-Elmer TGS-2 Thermogravimetric Analyzer used in conjunction with the System 4 Microcomputer Controller and Perkin-Elmer
XY_1Y_2 Recorder. The thermal curve shown in Figure 1 shows the volatilization profile for the residue from Process #1 obtained using a
nitrogen purge of 40 cc/min and a heating rate of 20°C/min. As can
be seen in the Figure, the total volatile content of this residue is
29.7% and the extrapolated onset for the volatilization is 387°C.
The TG thermal curve for the volatilization of the residue from
Process #2, obtained using the same conditions as that used for the
Process #1 residue, is shown in Figure 2. A total volatile content
of 31.4% is observed for this residue with 29.9% lost in the initial
state (100°C-750°C) and a second decomposition (750-950°C), believed
to be associated with the mineral matter, contributes an additional
2.4% to the total weight loss. Although the two residues were also
characterized by TG using an air (oxidizing) purge, Figure 3 shows
how one may switch the purge gas at the end of the volatile characterization run and obtain both the fixed carbon and ash values for
the two residues. Hence, the proximate or compositional analysis may
be obtained in one single unattended thermogravimetric run. The tabulated results for the TG studies are presented in Table 3. The lower temperature of volatilization onset (172°C) for the Process #2
residue is not surprising since solubility studies reported[6] on this
type of residue have shown an oil content (hexane soluble fraction)
of 21-29% depending upon the severity of the pressure used in the
process. It should be mentioned that when proximate analysis is the
only objective, recent publications by Earnest and Fyans[7],[8] detail a
microcomputer controlled thermogravimetric procedure in which the
average time of analysis of such coal derived materials is only
twenty minutes.

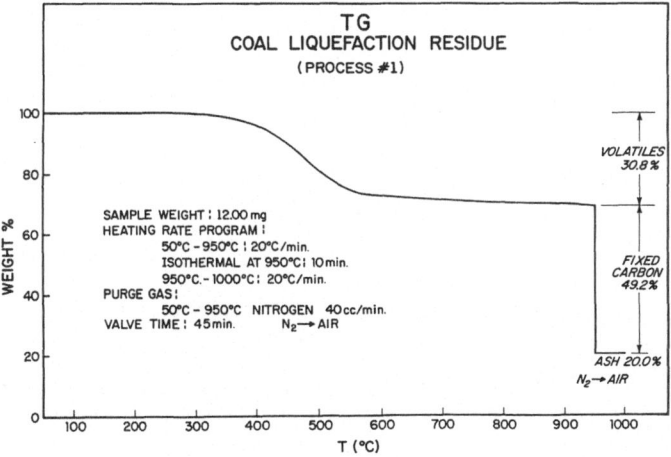

Fig. 1.   Volatilization Profiles for Process #1 Residue As
          Observed by Thermogravimetry and Derivative Thermogravimetry

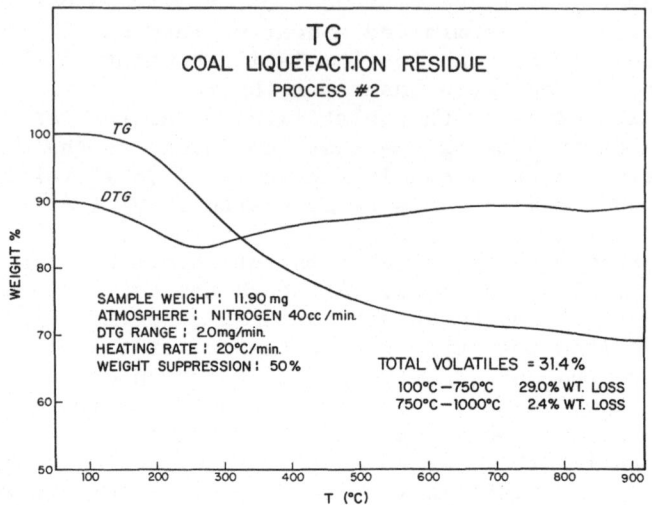

Fig. 2.   Volatilization Profiles for Process #2 Residue as Observed by Thermogravimetry and Derivative Thermogravimetry

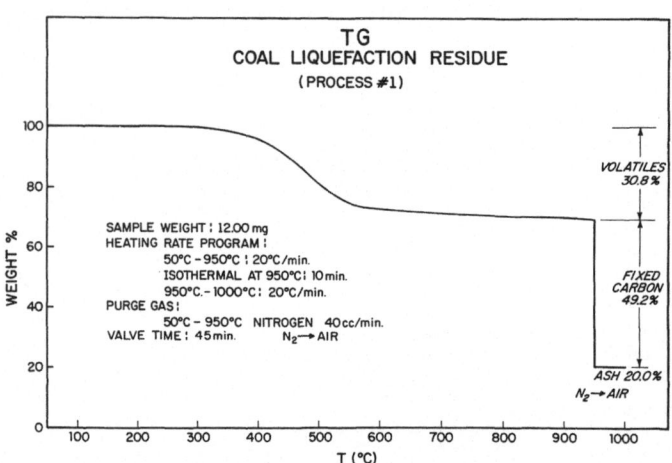

Fig. 3.   TG Volatilization Followed by Automatic Purge
Gas Switching to Ash the Residue Specimen

Table 3. Summary of TG Compositional Analysis Results and Volatilization Onset Temperatures

	Process #1	Process #2
% Volatiles	29.7%	31.4%
% Fixed Carbon	50.3%	17.1%
% Ash	20.0%	51.5%
Volatilization Onset	387°C	172°C

DIFFERENTIAL SCANNING CALORIMETRY (DSC)

Both liquefaction residues were studied by DSC in both inert (dynamic N_2) and oxidizing (dynamic air) atmospheres. A computerized DSC system (Perkin-Elmer DSC-2C/TADS) was used in these studies. A flow-thru cover was used with the DSC sample holder assembly in all of these studies. Standard gold sample pans were employed for the oxidative profiles obtained in dynamic air atmosphere. For the lower temperature studies conducted in dynamic N_2 atmosphere, experiments showed that the results obtained were the same regardless of whether standard aluminum or standard gold DSC pans were employed. All figures given here are hard copy printouts from the Perkin-Elmer Thermal Analysis Data Station.

Figure 4 shows the computer optimized DSC thermal curves obtained for both a powdered and a pelletized sample of the Process #1 residue obtained in dynamic nitrogen atmosphere. As is described by these thermal curves, this residue exhibits a weak two stage exothermic ordering (thermosetting) behavior in the 90°C to 270°C temperature region. The pelletized sample shows an additional exothermic event which is only weakly observed in the powdered sample. This enhancement is possibly due to the release of stress caused by the severe compaction of the pellet press.

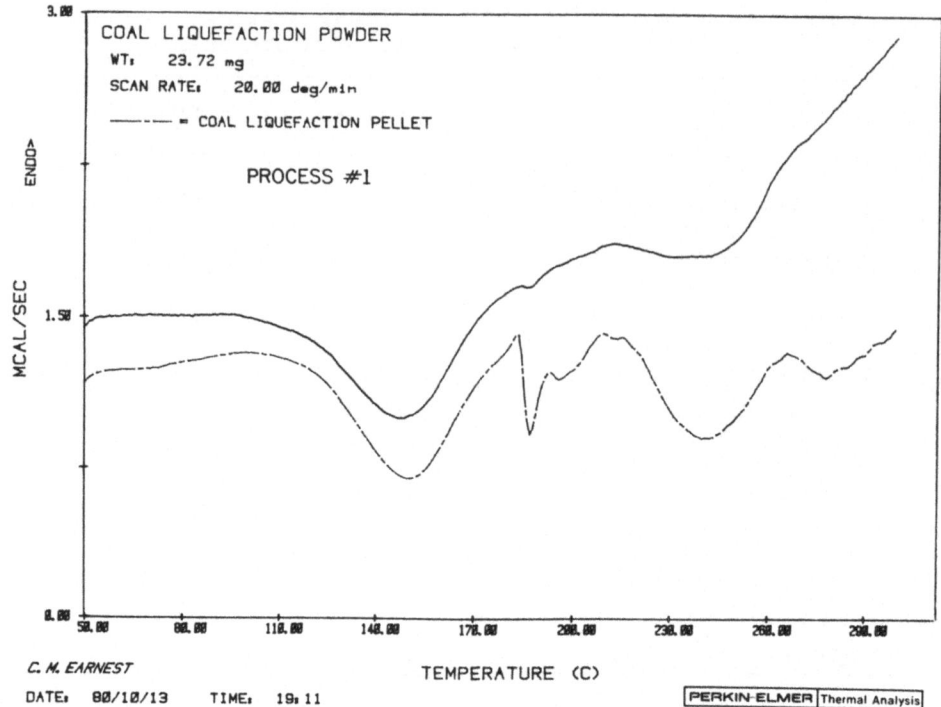

Fig. 4.   Power-compensated DSC Thermal Curves for Both a
           Powdered   and Pelletized Sample of Process #1 Residue

        Peak analysis, as performed by the TADS Computer System, showed
that the first stage of this ordering activity was observed at an
onset temperature of 115°C and an exothermic peak area of -12.17
joules/gram (-2.91 calories/gram). The second exothermic ordering
event in the DSC thermal curve was assigned an onset temperature of
212.8°C and an energy of -5.85 joules/gram (-1.4 calories/gram).
Thus, a total of -18.02 joules of exothermic ordering per gram of
the Process #1 residue was observed on heating to 280°C in inert
atmospheres.

        On second heating, the Process #1 residues exhibited a glass
transition which is observed in the DSC thermal curve, shown in
Figure 5, as a marked baseline shift or change in specific heat
of the residue. Figure 6 shows the TADS computer assignment of
the glass transition temperature (Tonset = 134.3°C and a midpoint
temperature of 145.5°C) and the Δc_p (change in heat capacity
associated with this weak transition) is determined to be 3.63 x
10^{-2} Cal/g deg. This second order ("glass") transition is associated
with the amorphous nature of the Process #1 residue.

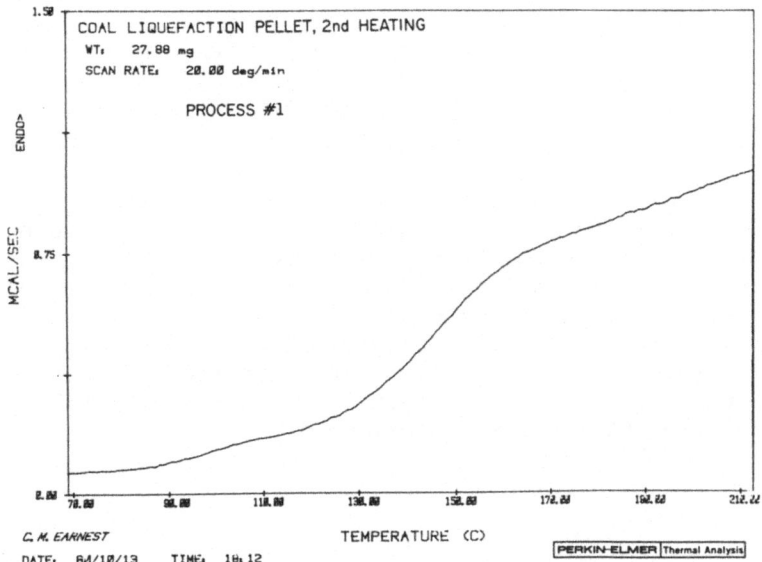

Fig. 5.   DSC Thermal Curve Showing Glass Transition Observed
on Second Heating of the Process #1 Residue

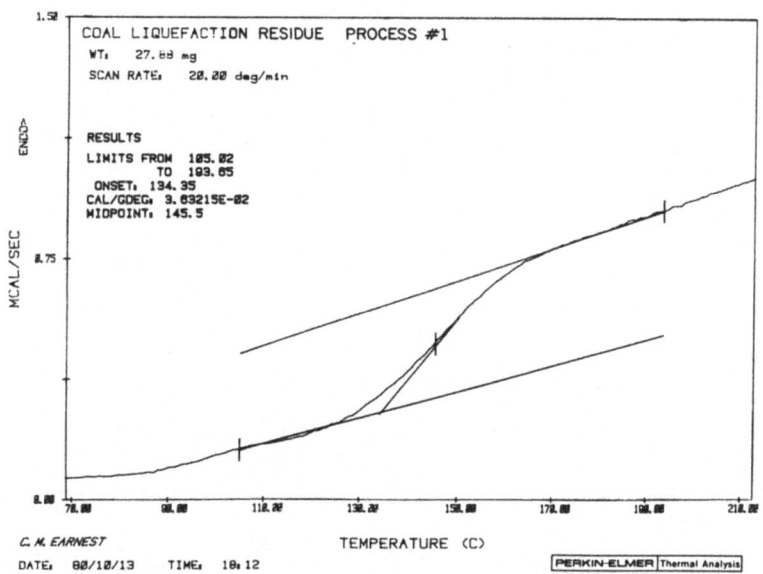

Fig. 6.   Assignment of Glass Transition Temperature and Specific
Heat Change (ΔC_p) Using the TADS/DSC-2C Standard Software

On studying the Process #2 (catalyzed) residue in a similar
fashion, no such exothermic ordering was observed. The Process #1
residue was observed to give a strong endothermic signal in the DSC
thermal curve at temperatures above 110°C due to the volatilization
of the oil (hexane soluble) fraction.

DSC studies of coal liquefaction residues, as well as coal and
coal products in general, conducted in dynamic air atmospheres, give
an oxidative profile of the material under study. In this study, the
DSC oxidative profiles for liquefaction residues were compared with
that of high volatile bituminous feedstock coal. The coal liquefaction residues exhibited a higher oxidative stability than the
original feedstock coal. This is demonstrated by the normalized
DSC thermal curves given in Figure 7 for a high volatile bituminous coal and the Process #1 residue.

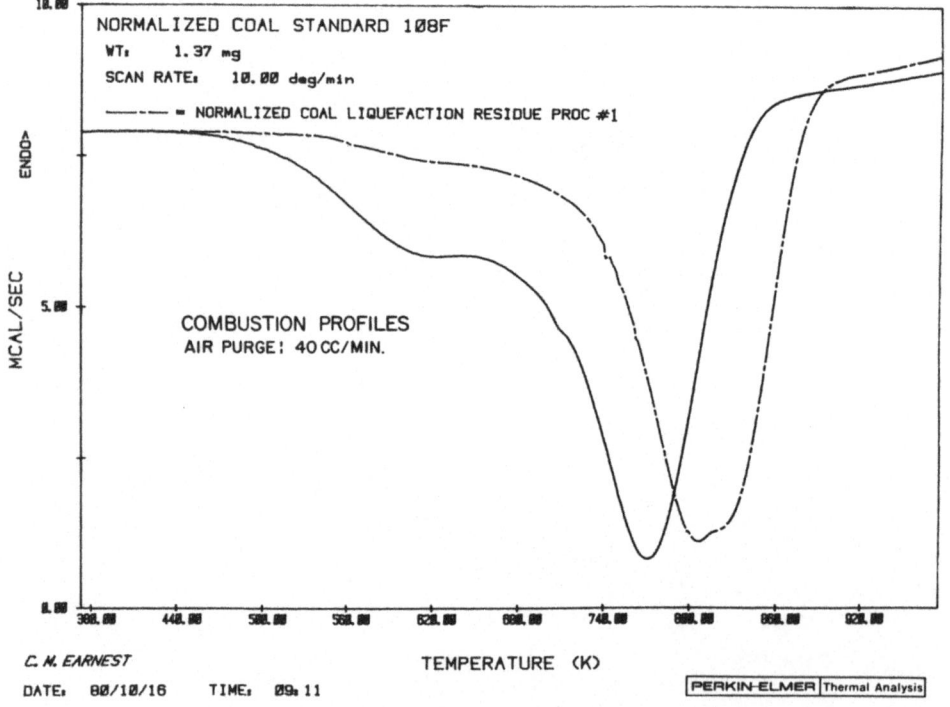

Fig. 7.   Normalized DSC Oxidative Profiles for Process #1
           Residue and a High Volatile Bituminous Coal Standard

Figure 8 shows the normalized oxidative comparison of the two
different process residues. Both are of similar oxidative stability
and exhibit at least a two stage combustion profile on heating at
10°C/min.

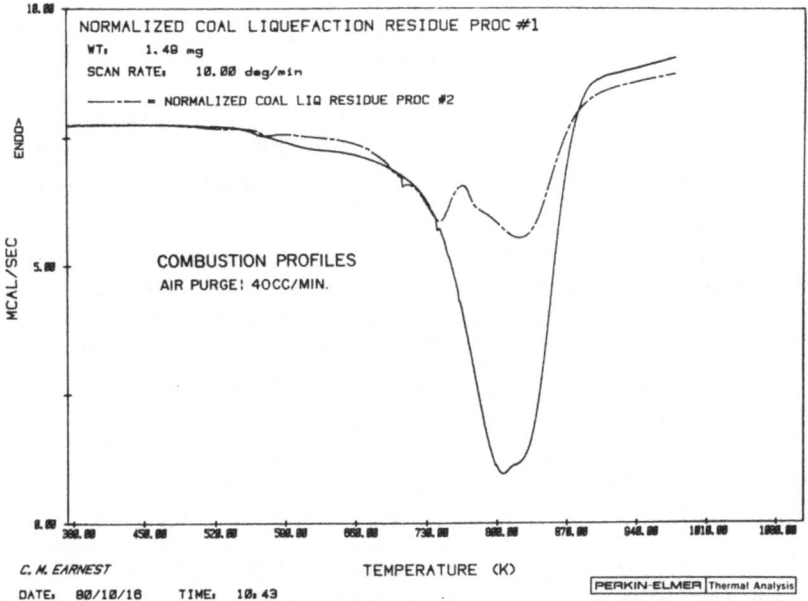

Fig. 8.   Normalized Comparison of DSC Oxidative Profiles
            for Process #1 and Process #2 Residues

The calorific values (heats of combustion) were also assigned
from these oxidative profiles by comparing the total normalized peak
areas, given by the peak analysis routine of the TADS, to that of a
high volatile bituminous coal of known calorific value (Alfa Resources
108F Coal Standard). The calorific values were then calculated by
the relationship

$$Q_u = \frac{Q_S\ A_u}{A_S}$$

where Q_u = calorific value of specimen under study,
       Q_S = known calorific value of Coal Standard 108F,
       A_S = DSC peak area for combustion of Coal Standard 108F,
       A_u = DSC peak area for combustion of specimen under study.

The calorific values which were obtained in this manner are
given in Table 4.

Table 4.   Calorific Values Obtained Using Power
            Compensated Differential Scanning Calorimetry

| | | Calorific Value | |
Residue Specimen		As Received	Ash Free
Process #1		12,256 BTU/lb	15,283 BTU/lb
	(or)	28.48 MJ/Kg	35.51 MJ/Kg
Process #2		6,122 BTU/lb	12,624 BTU/lb
	(or)	14.22 MJ/Kg	29.33 MJ/Kg

As a means of checking these values, the calorific content may be calculated from the elemental analysis data given earlier in this work. Several equations are available for calculating the calorific value. The equation below, published by Mason and Gendhi[9],

$$Q \ (\ BTU/lb \ ) = 198.11C + 620.31H + 80.93S + 44.95 \ Ash - 5,153$$

was employed in this work and has been included as a feature of the Analysis Program of the Perkin-Elmer Model 240DS Software package. In the above equation, C,H,S, and Ash represent the weight fractions of carbon, hydrogen, sulfur, and ash in the residues. When the values given in Table 2. are used with this equation, an "as received" calorific value of 12,365 BTU/lb (28.76 MJ/Kg) is obtained for the Process #1 residue and a value of 6,157 BTU/lb (14.32 MJ/Kg) is obtained for the Process #2 residue. To say the least, excellent agreement is observed for these two different methods of obtaining the calorific content of these coal liquefaction residues. Furthermore, one may observe from these values that the Process #1 residue shows excellent potential as a direct combustion fuel source.

THERMOMECHANICAL ANALYSIS (TMA)

The Perkin-Elmer TMS-2 Thermomechanical Analyzer was used in conjunction with the System 4 Microcomputer Programmer and Perkin-Elmer XY_1Y_2 recorder to study the two liquefaction process residues. The thermomechanical analyzer was employed using both the compression and expansion probes. Figure 9 shows the behavior of the Process #1 residue using the compression probe with a 3.0 gram (net) loading. When heated at $10°C$/min this residue shows a uniform compression profile occurring from $166.5°C$ (T_{Onset}) to $212°C$. The maximum rate of compression, as shown by DTMA, is observed at $191°C$. However, when a sample which was preheated to $290°C$ was run in the same fashion (Figure 10), a multi-stage compression profile is observed by the DTMA thermal curve. This effect was very reproducible.

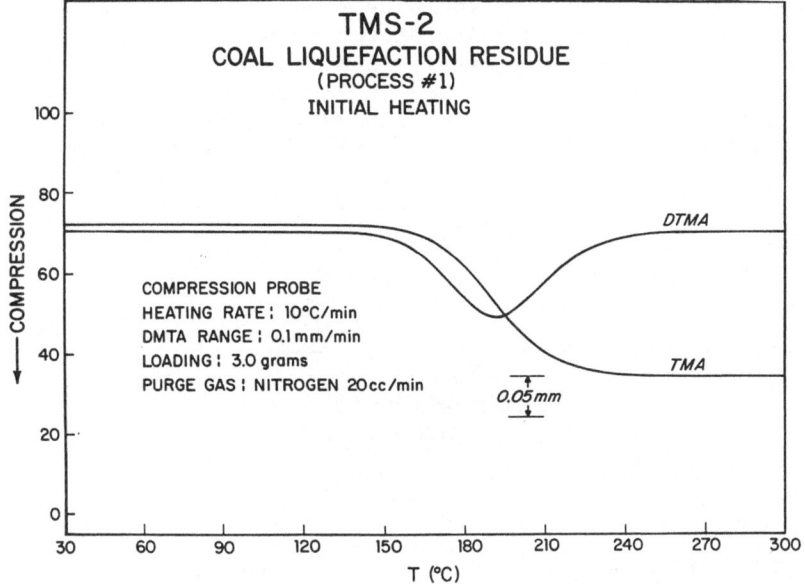

Fig. 9.   TMA and DTMA Thermal Curves Observed on the
Initial Heating with the Process #1 Residue

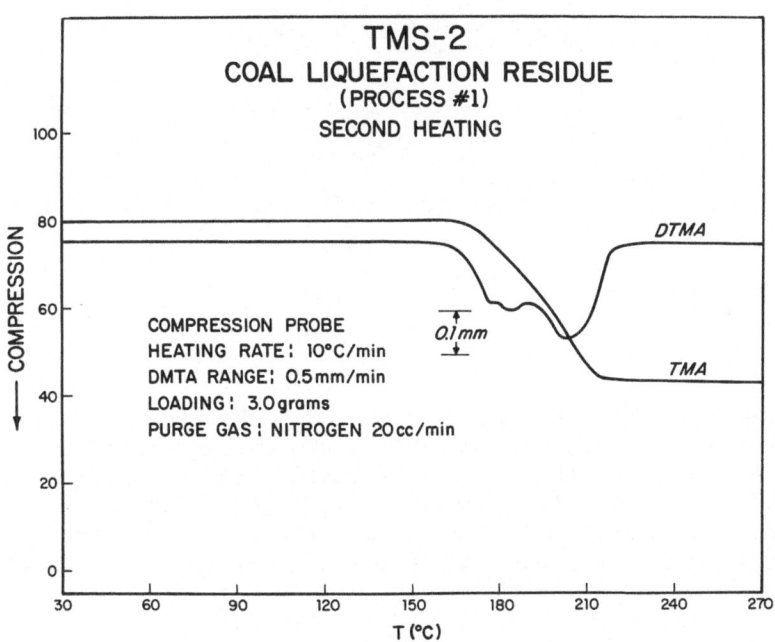

Fig. 10.   TMA and DTMA Thermal Curves Observed on a Specimen
Which Had Been Previously Heated to 290°C

Since the residue from the catalyzed liquefaction process was observed to volatilize at temperatures above 110°C, the compression study on this specimen was carried out from 10°C to 90°C. As is shown in the thermal curves given in Figure 11, the Process #2 residue is observed to undergo softening over a broad temperature range with a maximum rate of compression observed at 34°C in the DTMA curve.

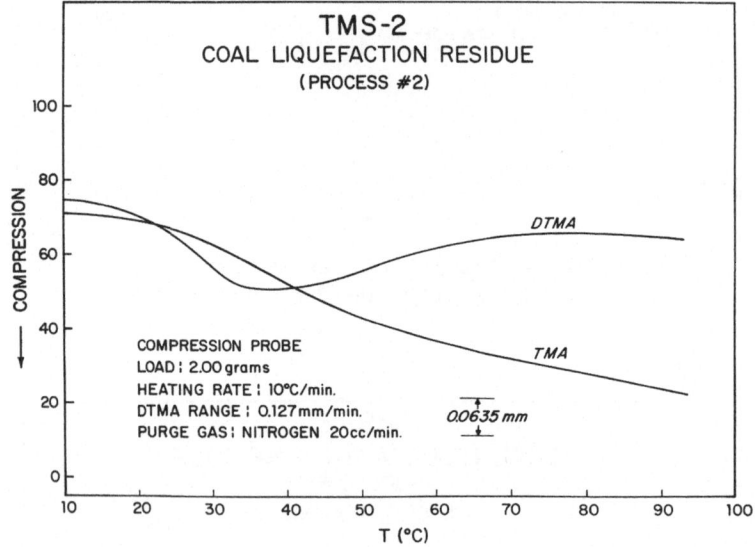

Fig. 11.   TMA-DTMA Compression Profile for Process #2 Residue

Figure 12 shows the use of the TMS-2 Thermomechanical System with an expansion probe to establish the average coefficient of linear expansion of the Process #1 residue in the temperature range from 60°C to 140°C. As is shown in the figure, the expansion coefficient was calculated to be 4.56×10^{-5}. An exploratory run using the expansion probe in the TMS-2 Thermomechanical Analyzer and liquid nitrogen coolant in the TMS-2 furnace dewar revealed a change in expansion coefficient in the Process #2 residue of -33°C, (shown in Figure 13). This is within the temperature region where many asphalts (bitumens) show second order transitions. The ex-

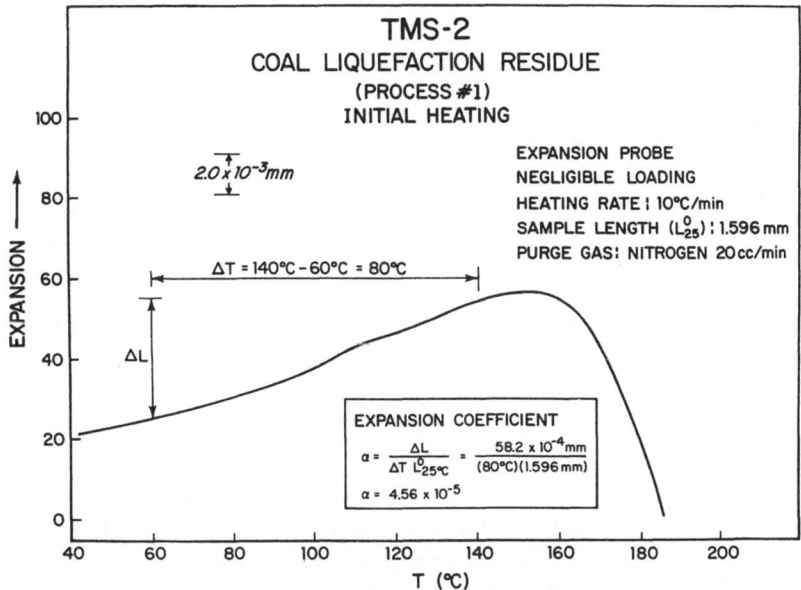

Fig. 12.   Assignment of Linear Expansion Coefficient of Process #1 Residue Using TMA

pansion probe in this case is weighted to a point slightly above "zero loading". Therefore, at temperatures above 12°C, a compression effect is observed which corresponds to the information given previously in Figure 11.

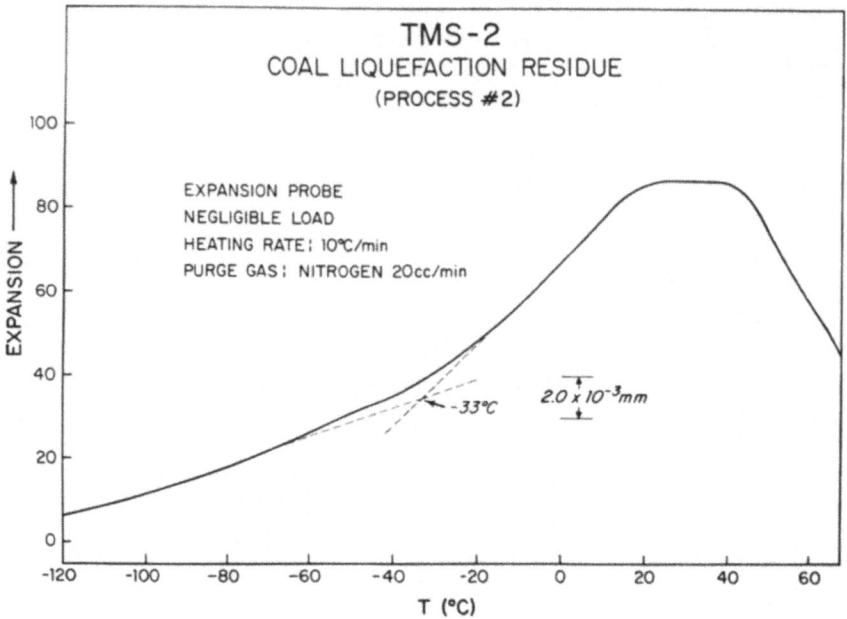

Fig. 13.   TMA Expansion Profile for Process #2 Liquefaction Residue

CONCLUSION

The use of thermal methods of analysis for the characterization of coal liquefaction residues can provide much information regarding the physical nature and composition of such coal conversion by-products. Their thermal and oxidative stabilities are readily assigned from both TG and DSC methods. Heat history effects may be observed from both DSC and TMA techniques. Volatile matter content, fixed carbon values, and ash content are readily obtained from thermogravimetry. Calorific values are accurately determined by dynamic DSC studies in flowing air atmospheres. Furthermore, volatilization profiles by TG and combustion profiles by both DSC and DTG may be used as tools for fingerprinting such residues. The use of computerized DSC offers the additional concept of normalized comparisons and subtraction of these profiles versus temperature. A combination of the information obtained from thermal methods of analysis, elemental analysis, microscopic and spectroscopic techniques, and solubility studies may lead to the total characterization of coal conversion by-products.

REFERENCES

1. C. Giori, W. Eisenberg, T. Yamauchi, S. Shelfo, and I. Solomon, Syntheses of thermosetting polymers from the by-products of coal conversion processes, J. Appl. Polym. Sci. 24:2323(1979).

2. A. Davis, W. Spackman, and P.H. Given, The influence of the properties of coals on their conversion into clean fuels, presented at the American Chemical Society National Meeting, Symposium on the Role of Technology in the Energy Crisis, Sept. 9, 1974, Atlantic City, New Jersey.

3. P.H. Given, D.C. Cronauer, W. Spackman, H.L. Lovell, A. Davis, and B. Biswas, Dependence of coal liquefaction behavior on coal characteristics.I. Vitrinite-rich samples; II. Role of petrographic composition, Fuel 54:34-49 (1975).

4. A.R. Tarrer, J.A. Guin, W.S. Pitts, J.P. Henley, J.W. Prather, and G.A. Styles, Effect of coal minerals on reaction rates during coal liquefaction, in: "Liquid Fuels from Coal", R.T. Ellington, ed., Academic Press, New York.

5. R.D. Williams, Industrial coal conversion processes, American Chemical Society Southeast Texas Section Continuing Education Course on The Chemistry of Synthetic Fuels and Their Impact on the Environment, Lecture #4, April 10, 1979.

6. S. Akhtar, N.J. Mazzocco, M. Weintraub, and P.M. Yavorsky, Synthoil process for converting coal to non-polluting fuel oil. Presented at the 4th Synthetic Fuels from Coal Conference, Oklahoma State University, Stillwater, OK, May 6-7, 1974.

7. C.M. Earnest and R.L. Fyans, Recent advances in microcomputer controlled thermogravimetry of coal and coal products, Thermal Analysis Application Study No. 32, Perkin-Elmer Corp., Norwalk, CT, 1981.

8. C.M. Earnest and R.L. Fyans, A thermogravimetric method for the rapid proximate and calorific analysis of coals and coal products, in: "Thermal Analysis, Vol. II, Proc. of the 7th ICTA", B. Miller, ed., John Wiley, New York, 1982, p. 1260.

9. D.M. Mason and K. Gendhi, Formulas for calculating the heating values of coals and chars, published by the Institute of Gas Technology, Chicago, 1980.

COMPUTER AUTOMATION OF THE DYNAMIC MECHANICAL ANALYZER AND ITS
APPLICATION TO CURE KINETICS STUDIES AND DYNAMIC MECHANICAL
PROPERTY ANALYSIS OF ORGANIC COATINGS

M. E. Koehler, A. F. Kah, C. M. Neag, T. F. Niemann
F. B. Malihi and T. Provder

Glidden Coatings and Resins
Division of SCM Corporation
16651 Sprague Road
Strongsville, Ohio   44136

INTRODUCTION

Efficient use of modern thermal analysis instrumentation
requires the availability of computer aided data collection and
analysis. This is true not only for the relatively complex
calculations involved in the kinetics analysis of curing systems,
but for simple routine calculations and engineering unit
conversions as well. Software provided by vendors for this purpose
is usually satisfactory for simple operations, but available
programs for kinetics analyses are often overly simplistic and
inflexible or unsuited for a particular application. This work
describes an automated data analysis system used in conjunction
with a DuPont Model 981 Dynamic Mechanical Analyzer (DMA) with a
Model 990 Programmer. This system collects data from the
instrument and transmits the data to a minicomputer system for
storage, analysis, reporting and plotting. A program was developed
to perform kinetics analysis of data from DMA and various other
instruments operated in the single dynamic temperature scan mode.

Data Acquisition System

Automated data analysis for the DMA was achieved by
interfacing the instrument to a microcomputer for data acquisition.
The microcomputer is responsible for all real time activities
involved in data collection. At the completion of the experiment,
data are transferred _via_ a serial line to the minicomputer for
storage, analysis, and report generation and plotting which may be

361

done at any time after the completion of the experiment. Details
of the mini–microcomputer system and its organization have been
reported elsewhere.[1,2,3]

Automated Instrument Analysis Process

There are four stages in an automated instrument analysis.
These are shown schematically in Figure 1. In the first stage,
the operator initiates the experiment by means of a dialog program
on the minicomputer. This dialog includes biographical
information about the sample as well as various instrument
parameters and conditions required for the collection and analysis
of the data. The minicomputer then transmits the parameters
entered through the dialog to the microcomputer for use by the
microcomputer in conducting the experiment and for incorporation
in the raw data file for use in the data analysis. The
microcomputer then turns on a "ready" light at the instrument
which is the signal to the operator that the system is ready for
automated data collection.

The second stage is data acquisition. This stage is entered
when the operator starts the instrument and pushes the computer
start button to initiate data collection. The data are collected
at the predetermined rate on the Y, Y' and T channels of the 990
console. The third stage is data transmission during which the
microcomputer transmits the entire data set for the sample to the
minicomputer where it is stored on disk until the operator
initiates the fourth stage which is data reduction. The data
reduction takes place in the minicomputer and is programmed in
FORTRAN. Reports and plots are generated at this time.

Data Analysis Methods

The previously reported method of kinetics data analysis for
DSC data[3] is not readily adaptable to raw data obtained in the
integral form of fractional conversion or cure obtained from DMA,
TGA or spectroscopic methods such as FT–IR. The basic assumption
of the kinetics analysis for DMA is that the change in relative
modulus at a given time and temperature during the dynamic
temperature scan, divided by the change in relative modulus
exhibited by the fully cured system at the same temperature, is
proportional to the extent of cure at that point of the reaction.
This is then used as the fractional degree of cure in the
calculations. This normalization of raw DMA data to fractional
degree of cure $F(t,T)$ is defined as

1. Dialog

       Operator ⟷ Mini ⟶ Micro

2. Data Aquisition

       Micro ⟷ Instrument

3. Data Transmission

       Micro ⟶ Mini

4. Data Reduction

       Mini ⟷ Operator

Fig. 1. Steps in an Automated Instrument Operation.

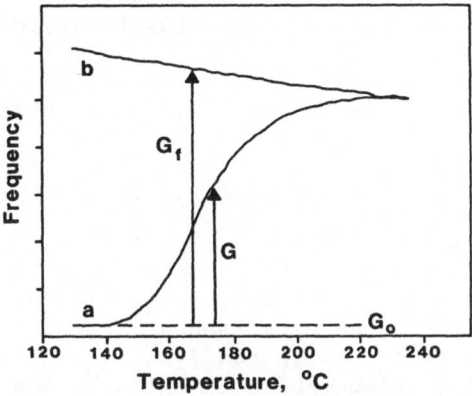

Fig. 2. Raw DMA Data. a. Curing System. b. Fully Cured System.

$$F(t,T) = \frac{G(t,T)-G_o(T)}{G_f(T)-G_o(T)} \tag{1}$$

where $G(t,T)$ is the frequency at a particular time and temperature, $G_o(T)$ is the baseline frequency representing the frequency of the uncured system over the cure region and $G_f(T)$ is the frequency of the fully cured system at temperature, T. This normalization is shown diagrammatically in Figure 2. For other instrumental methods such as TGA or FT-IR,[4] it is assumed that the measured property is proportional to the extent of the chemical reaction and can be normalized to fractional conversion. For DSC, the fractional conversion is taken to be proportional to the integrated heat of reaction at that point in the reaction, Figure 3.

Assuming that reagents are present in stoichiometric proportion and that there is only one slow step in the reaction mechanism, the general nth order rate expression can be written in terms of concentration in logarithmic form to yield the following expression.

$$\ln k(T) = \ln\left\{\frac{1}{C_o}\left(\frac{dC}{dt}\right)\bigg/\left(\frac{C_o-C}{C_o}\right)^n\right\} \tag{2}$$

where C_o is the initial concentration, C is the amount reacted at time, t, dC/dt is the rate of disappearance of reactants, n is the reaction order and $K(T)$ is the temperature dependent rate constant.

Rewriting equation (1) in terms of the time and temperature dependent fractional conversion, $F(t,T)$, we obtain

$$\ln k(T) = \ln\left\{\left(\frac{dF(t,T)}{dt}\right)\bigg/\left(1-F(t,T)\right)^n\right\} \tag{3}$$

The kinetic parameters n and E are determined by using a Nelder-Meade Simplex minimization routine.[4,5,6] In this method, successive segments of the experimental fractional conversion curve, designated as Y_a, to Y_b and Y_c to Y_d in Figure 4, are combined with the Gauss-Legendre numerical integration of the corresponding time and temperature portions of the curve, shown in equation (4),

$$I_{ab} = \int_a^b c^{-E/R(T_o+rt)}\, dt \tag{4}$$

where T is the initial temperature and r is the heating rate, to form an objective function, Ø defined as

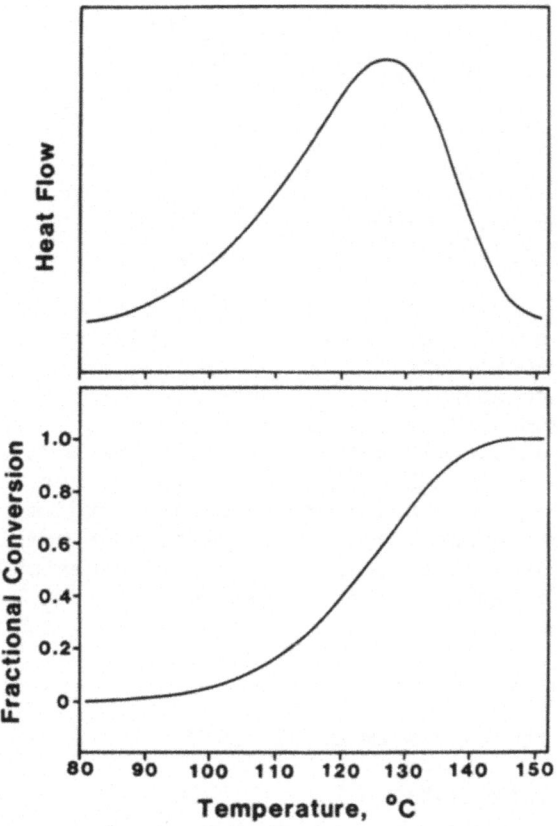

Fig. 3. Normalization of DSC Data. a. Heat Flow. b. Integrated
Heat Flow Normalized to Unity.

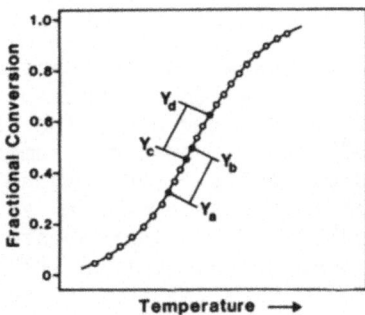

Fig. 4. Segmentation of Conversion Data for Cure Calculation.

$$\emptyset = \sum \left[\frac{\ln(1-Y_c) - \ln(1-Y_d)}{\ln(1-Y_a) - \ln(1-Y_b)} - \frac{I_{cd}}{I_{ab}} \right]^2 \qquad (5a)$$

for n equal to 1 and

$$\emptyset = \sum \left[\frac{(1-Y_d)^{1-n} - (1-Y_c)^{1-n}}{(1-Y_b)^{1-n} - (1-Y_a)^{1-n}} - \frac{I_{cd}}{I_{ab}} \right]^2 \qquad (5b)$$

for n not equal to 1. A detailed derivation of the objective function is shown in the Appendix. A segment is selected to contain about 20 percent of the data points over the cure region. The segments are stepped up the curve one point at a time to generate the sum used in the objective function. The parameters n and E are systematically varied to obtain the best fit between the experimental and the calculated conversion curves. The fit of the calculated conversion on the experimental curve improves as the objective function approaches zero. After the completion of the Simplex fit, the natural logarithm of the Arrhenius frequency factor, ln A, defined by the expression

$$\ln A = \ln k(T) + \frac{E}{RT} \qquad (6)$$

is calculated for each segment, Y_a to Y_b, as

$$\ln A = \frac{\sum \left[\ln[\ln(1-Y_a) - \ln(1-Y_b)] - \ln I_{ab} \right]}{N} \qquad (7a)$$

for n equal to 1

$$\ln A = \frac{\sum \left[\ln \frac{(1-Y_a)^{1-n} - (1-Y_b)^{1-n}}{n-1} - \ln I_{ab} \right]}{N} \qquad (7b)$$

for n not equal to 1, and the average value for ln A is calculated as indicated.

Raw DMA data for a typical curing coating system is shown in Figure 2a. The corresponding scan of the fully cured system is represented in this figure by curve b. The raw DMA data is converted to 0 to 100% cure as shown diagrammatically in Figure 2 to yield a normalized composite curve, Figure 5. This normalized curve is then analyzed in the same manner by the data analysis program, KINET, regardless of the instrumental method by which it was generated. The operator interaction with KINET is shown in Figure 6. The operator enters the name of the file containing the data to be analyzed and selects the plots to be generated. Next, the operator enters the minimum and maximum percent conversions

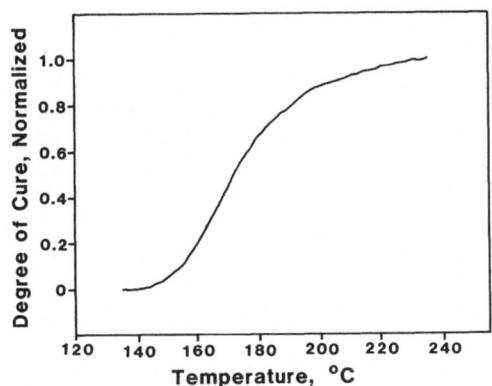

Fig. 5. DMA Data Normalized to Degree of Cure.

```
>RUN $KINET
    ENTER FILENAME <.DAT>                                  AIBN
    AIBN (INCHES), 15 DEG/MIN
    PLOTS: KINET,FUNCT,ORDER,ENERGY,ARRHEN,RESID ?   K,R
        LOWEST PERCENT REACTED  <0>
        HIGHEST PERCENT REACTED  <100>
        HEATING RATE (DEG.C./MIN.)                      15
    X-AXIS UNITS (SEC,MIN,HRS,INCHES)  <DEG.C>
    ENTER EDITING LIMITS IN DEGREES C.
        LOW  (LIMIT  60.00) <  5.0 % = 102.32 >   97
        HIGH (LIMIT 161.05) < 95.0 % = 141.89 >   149
        42 POINTS IN THE EDITED PORTION
    97.07 PERCENT OF CURVE IN EDITED REGION
    CHANGE DEFAULT VALUES FOR SIMPLEX (Y/N) ?       Y
    STARTING VALUES AND INTERVALS FOR SIMPLEX:
        ORDER OF REACTION     < 1.20, 0.12>       1,0
        ENERGY OF ACTIVATION <24.00, 2.40>
        NUMBER OF ITERATIONS < 500 >
        REQUIRED MINIMUM FIT <0.1000E-09>
        MIN AND MAX STEP FOR n                    0,0
        MIN AND MAX STEP FOR E
            COUNT =  50
                                               BEST          SECOND
        ORDER OF REACTION                     1.000000      1.000000
        ENERGY OF ACTIVATION                 30.596334     30.596334
        LN ARRHENIUS FACTOR                  34.359768
        FUNCTION VALUES                      0.4491E+00    0.4491E+00
        NUMBER OF ITERATIONS                 119
    REPORT: PRINT OR SAVE  < NO REPORT > ?         S
    REPORT FILE KNETIC.LST CREATED
        PLOT FILES CREATED
            KINET.T1 AND T2
            RESID.T1
```

Fig. 6. Operator Interaction with Program KINET.

represented by the data. Default conversion limits are 0 to 100%.
The heating rate of the linear temperature scan is entered and the
portion of the curve to be included in the analysis is selected by
entering the low and high temperature limits of the desired
region. The default limits of analysis are 5 to 95% conversion.
Default starting conditions for the Simplex may be taken, or any
of the parameters may be entered by the operator. It is possible
to constrain either the reaction order or the energy of activation
to a particular value, or range of values as shown in the example.
The program periodically displays the number of iterations
completed to indicate the progress of the calculation. Upon
completion, the best values of the reaction order and energy of
activation are displayed along with the Arrhenius frequency
factor. A report may be printed and the selected plots generated.

    A sample report from KINET is shown in Figure 7. Included in
the report are the results as well as the conditions for the
calculation which were selected by the operator. Figure 8 shows a
plot of the input data with the result of the calculated best fit.
Other plots which may be generated are the difference between the
calculated curve and the data, Figure 9, the variation of the
Arrhenius frequency factor over the data, Figure 10, and plots of
the objective function, reaction order and energy of activation as
a function of the number of iterations in the simplex calculation
which may be used as a diagnostic tool.

MATERIALS

    The samples used in this study included 2,2'-azo-bis-(isobuty-
ronitrile) (AIBN) and a commercial electrocoat system. AIBN samples
were recrystalized twice from di-n-butyl phthalate (Eastman Chem.
Products, Inc.) before use in DSC thermal decomposition studies.

EXPERIMENTAL

DSC

    A DuPont Model 910 DSC cell base was used in conjunction with a
DuPont Model 990 thermal Analyzer for studies of the thermal decomposition of AIBN in di-n-butyl phthalate. Sample sizes were limited
to 4-6 mg. Changes in sample enthalpies were recorded as a function
of time at a heating rate of 5°C per minute except as otherwise
noted under a dry nitrogen purge (0.1 std. ft./hr.). Samples were
cooled to approximately -90°C and scanned through 200°C.

DMA

    The DuPont Model 981 Dynamic Mechanical Analyzer was used in

```
DATA FILE AIBN.DAT
12-APR-83

AIBN (INCHES), 15 DEG/MIN

                                          BEST        SECOND

ORDER OF REACTION                        1.0000       1.0000
ENERGY OF ACTIVATION                    30.5963      30.5963
LN ARRHENIUS FACTOR (AVERAGE)           34.3598
   LN(A) RANGE, 34.28 TO 34.46
FUNCTION TO BE MINIMIZED              0.4491E+00   0.4491E+00
NUMBER OF SIMPLEX ITERATIONS              119

INITIAL GUESSES FOR SIMPLEX:             START         STEP

   ORDER OF REACTION                     1.00          0.00
   ENERGY OF ACTIVATION                 24.00          2.40
   MAXIMUM ITERATIONS                    500
   REQUIRED MINIMUM                   0.10E-09

   TEMPERATURE LIMITS             60.00 TO   161.05
   EDITING LIMITS                 97.00 TO   149.00
   NUMBER OF RAW DATA POINTS                 65
   POINTS IN EDITED PORTION                  42
   PERCENT WITHIN EDITING LIMITS           97.07

   HEATING RATE, DEG.C./MIN.                15.00
   X AXIS UNITS = DEGREES

   Y AXIS LIMITS                  -0.01 TO     4.77
   PERCENT                         0.00 TO   100.00
```

Fig. 7. Sample Report from Program KINET,

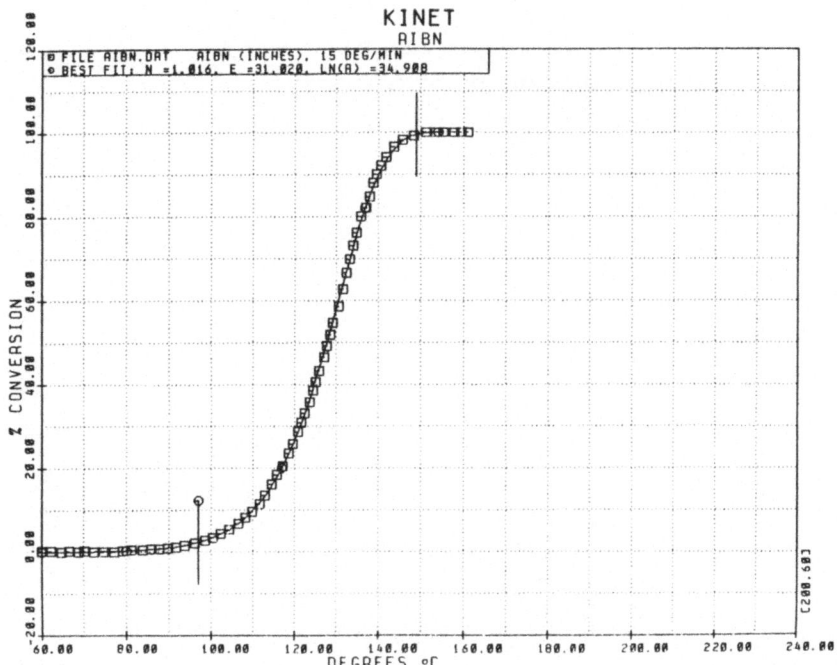

Fig. 8. Plot of Experimental and Calculated Conversion,

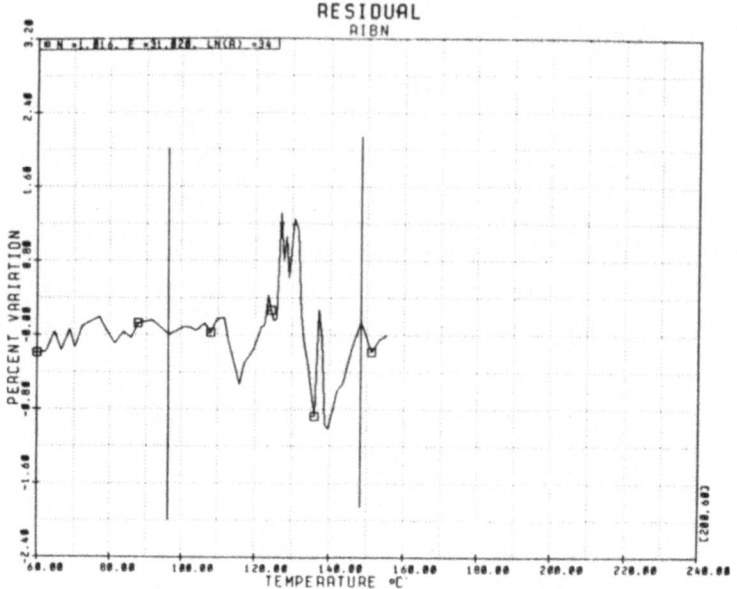

Fig. 9. Plot of Difference Between Experimental and Calculated
Conversion.

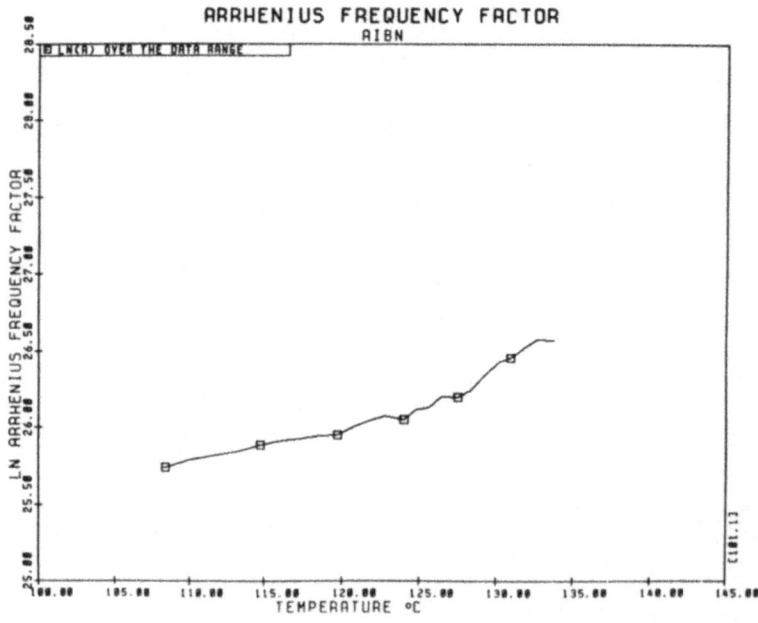

Fig. 10. Plot of Variation of Calculated Arrhenius Frequency
Factor with Temperature.

conjunction with the DuPont Model 990 Thermal analyzer. Woven
fiberglass braids (Potter Industries, Hasbrouck, N.J.) 1" x 0.5",
were used as inert substrates for the uncured electrocoat systems.
Electrocoat samples were applied uniformly to fiberglass braid
pre-mounted horizontally in the instrument clamps. Relative
modulus and energy disipation of all samples were recorded as a
function of time at a heating rate of $5^{\circ}C$ per minute under a dry
nitrogen purge of 5 l/min. Samples were cooled to approximately
$-80^{\circ}C$ and scanned through $225^{\circ}C$, cooled and rerun through the same
temperature range.

RESULTS AND DISCUSSION

     In order to test the program and methodology with a known and
well characterized sample, DSC data for the decomposition of
2,2'-azobisisobutylnitrile (AIBN) was analyzed by our program
THERM,[3] which calculates the kinetics parameters n, E and ln A by
a non-linear regression method. The data was integrated to yield
conversion curves suitable for analysis by KINET. The results of
both programs for the same experiments at various heating rates
are shown in Table 1. The f-test for analysis variance shown no
significant difference between the results of THERM and KINET for
this data. Table 2 shows a summary of the results for nine
separate experiments for the decomposition of AIBN at a scan rate
of $5^{\circ}C/min$. Again, no significant difference was found between
the results of the two methods for the same experiments. Also
shown in Table 2 is the DSC result for the decomposition of AIBN
performed by the ASTM E-698 method as well as a published value by
Tobolsky[7] using classical methods. Both ASTM E-698 and the
Tobolsky method assume a reaction order of one.

Applications

     Kinetic parameters determined by DMA are used to predict cure
behavior and oven bake schedules for coatings systems.[8] The
actual temperature-time profile of the coated part is of the
utmost importance. Figure 11 shows the experimental surface
temperature-time profiles for three different substrates. It
readily is seen that the heavier wheel stock substrate takes
longer to heat up and to cool down than do standard guage test
panels. The effect of these different temperature profiles on
cure for a typical electrocoat system is shown in Figure 12. If
0.95 fractional cure is taken to be sufficient for this material,
it is seen that a 20 minute bake schedule at $200^{\circ}C$ represents
substantial overbake for the light guage test panels, but is
insufficient to adequately cure the coating on the heavier guage
wheel stock. It can also be seen that there is a small degree of
cure which takes place on the wheel stock sample following removal
from the oven.

Table 1.   Results of Kinetic Analysis of AIBN Decomposition by
           Programs KINET and THERM

HEATING RATE $^\circ$C/min	KINET			THERM		
	n	E_o	ln A	n	E_o	ln A
2	1.00	30.9	35.6	1.08	30.8	35.5
5	0.97	28.0	31.6	1.00	29.1	33.1
10	0.93	26.0	29.0	0.97	25.9	35.6
15	0.92	26.5	29.1	1.01	27.0	30.3
20	1.03	30.4	34.7	1.05	30.2	34.4
MEAN	0.97	28.4	32.1	1.01	28.6	33.8
S.D.	0.04	2.2	3.0	0.03	2.1	2.2

F-Test (ANOVA) shows no significant difference between THERM and
KINET

Table 2.   Comparison of Results for Kinetics of Decomposition of
           AIBN

	n		E_o		ln A	
	mean	s.d	mean	s.d	mean	s.d
KINET*	1.021	0.036	29.04	2.16	33.20	2.78
THERM*	1.025	0.040	28.62	1.81	32.48	2.41
ASTM E-698 (OZAWA)	(1.00)		24.2		30.5	
TOBOLSKY	(1.00)		30.8		35.0	

*9 runs at 5°C/min

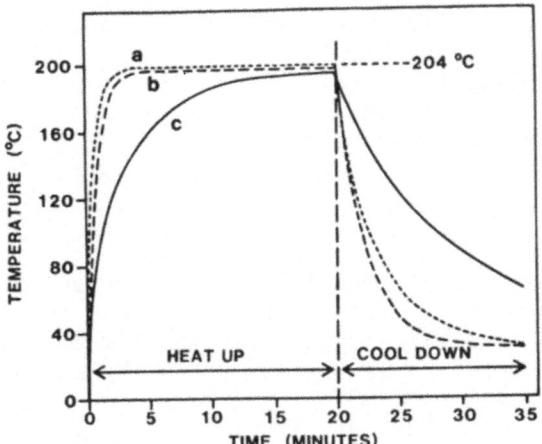

Fig. 11. Surfact Temperature Profiles of Test Panels. a. Uncoated
Standard Panel, b. Coated Standard Panel. c. Wheel Stock
Panel.

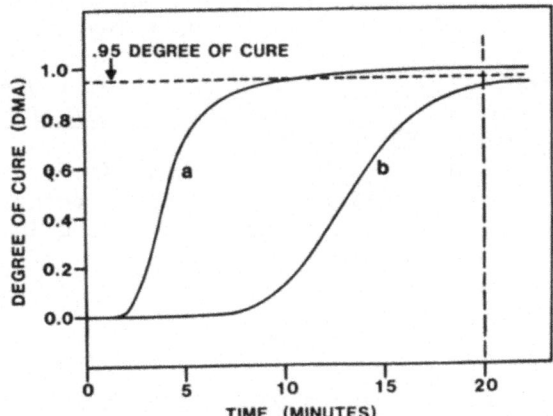

Fig. 12. Calculated DMA Degree of Cure for a Commercial Electrocoat
System for Temperature Profiles Shown in Fig. 11,
a. Standard Coated Panel. b. Wheel Stock Panel.

CONCLUSIONS

The ability to predict and to model cure behavior in terms of mechanical properties for a variety of coatings systems can be of significant value in the development and marketing of these coatings. The ability to monitor the effects of formulational changes through the individual kinetics parameters also gives insight into the cure mechanism of the coating.

REFERENCES

1. T. F. Niemann, M. E. Koehler and T. Provder, Microcomputers used as Laboratory Instrument Controllers and Intelligent Interfaces to a Minicomputer Timesharing System, in: "Personal Computers in Chemistry", P. Lykos ed., John Wiley and Sons, New York (1981).
2. A. F. Kah, M. E. Koehler, T. F. Niemann, T. Provder and R. R. Eley, An Automated Ferranti-Shirley Viscometer, in: ACS Symposium Series, vol, 197, "Computer Applications in Applied Polymer Science", T. Provder ed., American Chemical Society, Washington, D.C. (1982).
3. A. F. Kah, M. E. Koehler, T. H. Grentzer, T. F. Niemann and T. Provder, An Automated Thermal Analysis System for Reaction Kinetics, in: ACS Symposium Series, vol. 197, "Computer Applications in Applied Polymer Science", T. Provder ed., American Chemical Society, Washington D. C. (1982).
4. P. M. Olsson, L. S. Smith, Technometrics, 17:45 (1975).
5. J. M. Smith, "Mathematical Modeling and Digital Simulation for Engineers and Scientists," John Wiley and Sons, New York (1977).
6. P. M. Olsson, J. Quality Technology, 6:53 (1974).
7. J. P. Van Hook and A. V. Tobolsky, Journal of the American Chemical Society, 80:781 (1958).
8. T. Provder, C. M. Neag, G. Carlson, C. Kuo and R. M. Holsworth, "Cure Reaction Kinetics Characterization of Some Model Organic Coatings Systems by FT-IR and Thermal Mechanical Analysis", this volume.

APPENDIX

For the case of $n \neq 1$, equation (2) can be integrated over the conversion segment Y_c to Y_d in Figure 4 as follows:

$$\int_{Y_c}^{Y_d} \frac{dF(t,T)}{(1-F(t,T))^n} = \int_{t_c}^{t_d} A \exp(-E/RT)dt \qquad (A1)$$

$$(1-Y_d)^{1-n} - (1-Y_c)^{1-n} = (n-1)AI_{cd} \qquad \text{(A2)}$$

where

$$I_{cd} = \int_{t_c}^{t_d} \exp(-E/RT)dt \qquad \text{(A3)}$$

For the conversion segment Y_a to Y_b in Figure 4, the above procedure yields the following equation:

$$(1-Y_b)^{1-n} - (1-Y_a)^{1-n} = (n-1)AI_{ab} \qquad \text{(A4)}$$

Division of eq (A2) by (A4) results in an equation which has two unknowns, n and E.

$$\frac{(1-Y_d)^{1-n} - (1-Y_c)^{1-n}}{(1-Y_b)^{1-n} - (1-Y_a)^{1-n}} = \frac{I_{cd}}{I_{ab}} \qquad \text{(A5)}$$

The objective function is formed by summing the segments over the entire conversion curve as shown in equation (5a).

For the case of n=1, analogous equations to equations (A2) and (A4) are obtained:

$$\ln(1-Y_c) - \ln(1-Y_d) = AI_{cd} \qquad \text{(A6)}$$

$$\ln(1-Y_a) - \ln(1-Y_b) = AI_{ab} \qquad \text{(A7)}$$

In a similar manner, dividing equation (A6) by (A7) results in an equation with two unknowns, n and E.

$$\frac{\ln(1-Y_c) - \ln(1-Y_d)}{\ln(1-Y_a) - \ln(1-Y_b)} = \frac{I_{cd}}{I_{ab}} \qquad \text{(A8)}$$

Again, the objective function is formed by summing the segments over the entire conversion curve as shown in equation 5b.

CURE REACTION KINETICS CHARACTERIZATION OF SOME MODEL ORGANIC

COATINGS SYSTEMS BY FT-IR AND THERMAL MECHANICAL ANALYSIS

T. Provder, C. M. Neag, G. Carlson, C. Kuo and
R. M. Holsworth
Glidden Coatings and Resins
Division of SCM Corporation
16651 Sprague Road
Strongsville, Ohio 44136

INTRODUCTION

During the last several years the coatings industry has met the challenges of government regulations concerning volatile organic compound emissions and the increasing costs of petroleum based solvents by developing a variety of new coatings technologies. This has led to a need for improved methods of materials characterization. Techniques for assessing degree of cure and fractional conversion based upon fundamental reaction kinetics are included in these needs.

Mathematical modeling of the cure process coupled with the automation of various thermal analytical instruments and Fourier Transform Infrared Spectroscopy (FT-IR) have made possible the determination of quantitative cure and chemical reaction kinetics from a single dynamic scan of the reaction process. This paper describes the application of FT-IR, differential scanning calorimetry (DSC) and dynamic mechanical analysis (DMA) in determining cure and reaction kinetics in some model organic coatings systems.

Cure Kinetics Methodology

By assuming that the chemical and physical changes occurring during a reaction are proportional to the extent of reaction, $F(t,T)$, a general nth order rate expression can be used to characterize reaction kinetics information obtained from FT-IR, DSC and DMA. In the equation

$$dF(t,T)/dt = A \exp(-E/RT) [1-F(t,T)]^n \qquad (1)$$

n is the order of reaction, E(kcal/mole) is the activation

377

energy and $A(\text{sec}^{-1})$ is the Arrhenius Frequency Factor. Methods
for generating fractional conversion curves from DSC data and
degree of cure curves from DMA data have been described previously[1-3]; the methodology for DMA cure is summarized in Figure 7.

In the single dynamic temperature scan FT-IR method,
absorbance bands for reactive functionalities which have decreased
or increased during a reaction are compared to absorption bands
corresponding to non-reactive, and therefore unchanging, functionalities. Changes in the ratio of these values during a reaction
are illustrated in Figure 1a. Normalization of the curve in
Figure 1a according to equation (2) leads to the conversion curve
in Figure 1b.

$$F(t,T) = \frac{A(t,T) - A_o}{A_f - A_o} \qquad\qquad (2)$$

where A_o is the absorbance of the functionality of interest prior
to the start of the reaction, $A(t,T)$ is the absorbance at a
specific time and temperature, and A_f is the absorbance at the end
of the reaction. Evaluation of DSC data to obtain the kinetics
parameters n, E and lnA involved a multiple regression method[1-3]
applied to equation (1). Evaluation of FT-IR and DMA data to
obtain the kinetics parameters n, E and lnA involved a simplex
minimization method applied to fractional extent of conversion or
degree of cure curves, respectively[4].

MATERIALS

The samples used in this study include a modified blocked
isocyanate, isophorone diisocyanate (IPDI), (B-1370, Huls AG,
Gelsenkichen-Buer, W. Ger.) and both catalyzed and uncatalyzed
polyester gel coat resins. Gel coat cure reactions were initiated
with the addition of 1.0 wt. percent of methyl ethyl ketone
peroxide (MEKP, Lucidol Div., Pennwalt Corp., Buffalo, N.Y.) and
catalyzed with the addition of 0.15 wt. percent cobalt octoate
solution (12% cobalt in mineral spirits, Mooney Chemicals,
Cleveland, OH.).

EXPERIMENTAL

FT-IR

Spectra were obtained at a resolution of 4 cm^{-1} with a Digilab
FTS-15E Fourier Transform spectrophotometer. Thirty scans were
co-added to produce one spectrum. A reference spectrum of the
NaCl salt crystal was collected before the material to be analyzed

was applied. A thin film of reactants in solution was then cast
onto the salt crystal, the solvent allowed to evaporate, and the
coated crystal placed into the instrument's heated cell mount
(Model #018-5322, Foxboro/Analabs, North Haven, CT.). The sample
was heated at a linear heating rate of 5°C per minute to approximately 220°C. The instrument cell temperature was controlled by a
Dupont Model 900 Differential Thermal Analyzer and monitored with
a thermocouple mounted on the cell wall. Excellent agreement was
observed between the monitored cell wall temperature and the salt
crystal surface temperature. Following completion of the scan,
data were plotted as absorbance spectra.

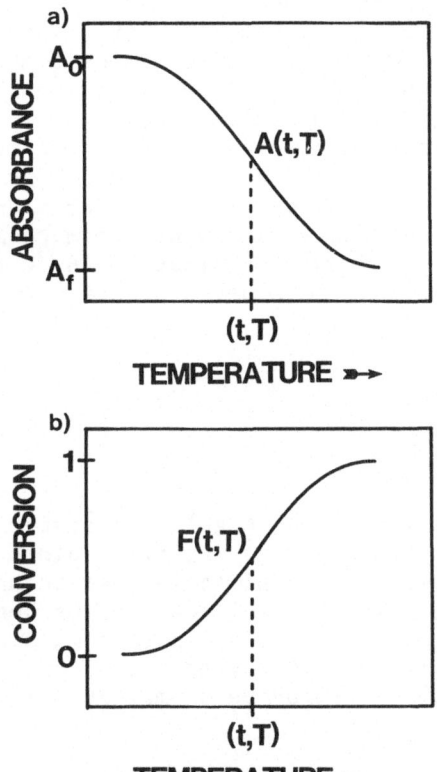

Fig. 1. Generalized dynamic FT-IR spectra for (a) absorbance
        versus time-temperature and (b) fractional conversion
        versus time-temperature.

Absorbance bands corresponding to reactive functionalities were identified visually or by spectral subtraction. Increases or decreases in the peak heights were normalized against unreactive, and therefore unchanging, functionalities.

DSC

A Dupont 990/910 DSC cell base was used for studies of cure kinetics. Gel coat sample sizes were limited to between four and six milligrams. Changes in sample enthalpies were recorded as a function of time at a heating rate of 5°C per minute under a dry N_2 purge (0.1 std. ft^3/hr.). Samples were scanned from room temperature to 175°C, cooled, and rerun from approximately -80°C through 200°C.

DMA

A Dupont 990/981 Dynamic Mechanical analyzer system was used in assessing changes in the mechanical behavior of the gel coats during cure. Woven fiberglass braids (Potter Industries, Hasbrouck, N.J.), 1" x 0.5", were used as inert substrates for the uncured gel coat resins. After thorough mixing with MEKP, samples were applied uniformly to fiber-glass braid pre-mounted horizontally in the instrument clamps. Relative modulus and energy dissipation of all samples were recorded as a function of time at a heating rate of 5°C per minute under a dry nitrogen purge of 5ℓ/minute. Samples were scanned from room temperature to 160°C, cooled, and rerun from approximately -80°C through 175°C. Experimental details unique to kinetic studies by DMA have been described elsewhere[4-5].

RESULTS AND DISCUSSION

Cure Of Polyester Gel Coat

The cure reaction in polyester gel coat systems makes them particularly useful model coatings materials since the cure kinetics can be studied by several analytical techniques. The initiation of cure in polyester gel coat resins begins with the decomposition of hydroperoxide (MEKP is predominately hydroperoxide) and the subsequent copolymerization of styrene with the unsaturated groups, either fumarate or maleate, incorporated into the resin backbone[6].

Because of the high degree of unsaturation present in the polyester resin, the fully cured system was highly crosslinked. Assuming typical polymerization kinetics apply to these systems, elementary kinetics analysis[7] indicates that the reaction order

for the cure reaction should be 1.5, first order in monomer and
half order in initiator.

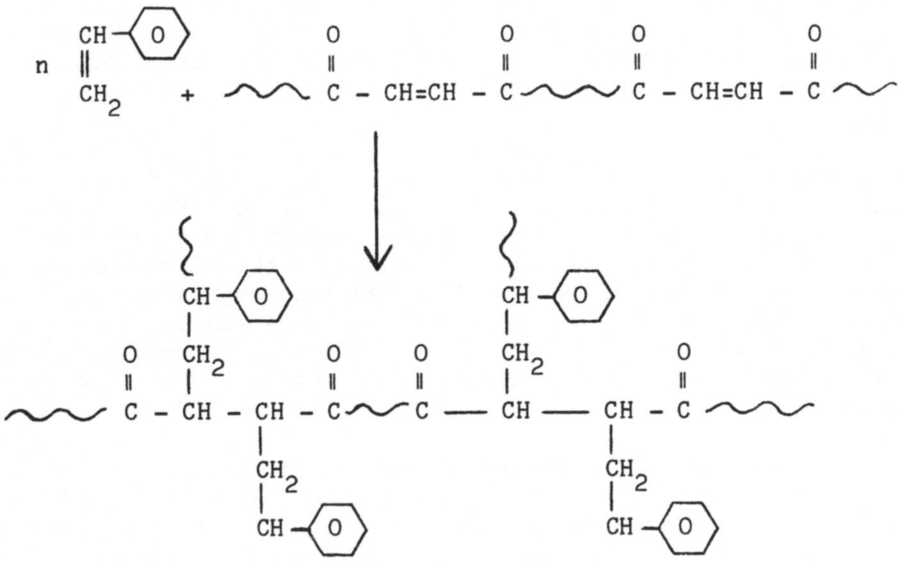

Scheme 1.

Since ambient cure is desirable, cure of polyester gel coat
systems is commonly catalyzed with the addition of a cobalt
compound to facilitate hydroperoxide decomposition and, consequently, lower temperature cure.

$$ROOH \xrightarrow{\Delta} RO\cdot + OH$$

The heat of reaction evolved by the curing system can be
monitored by DSC to obtain information concerning cure kinetics
and FT-IR can be used to directly obtain chemical kinetics
information by observing the disappearance of starting materials,
for example, styrene. Degree of cure kinetics can be obtained
from DMA by measuring the increase in relative modulus as
crosslinking proceeds. By combining information from the
aforementioned techniques, a detailed picture of the cure process
can be generated.

DSC

Although attempts were made to evaluate cure kinetics in both
catalyzed and uncatalyzed resins, only thermograms of the uncatalyzed system yielded results that were readily interpretable in

terms of the nth order rate equation. Thermograms of the catalyzed resins were characteristically bimodal and may be a consequence of side reactions unrelated to the curing process, e.g.
thermal decomposition of excess MEKP at elevated temperatures or
other minor decomposition reactions. The uncatalyzed resin, on
the other hand, produced monomodal curves over the entire cure
range (Figure 2).

Cure begins near 75°C and ends just prior to 150°C in the uncatalyzed resin. The temperature range (72°C to 148°C, Figure 2)
selected for the evaluation of reaction kinetics was based on
FT-IR spectra. These results indicated that the disappearance of
styrene, and presumably its copolymerization, begins and ends
between these temperatures. Evaluation of the kinetic parameters
for gel coat cure yielded a reaction order of 1.56, close to the
anticipated value of 1.50, an activation energy of 25.1 kcal/mole
and an lnA of 28.4.

The Arrhenius plot generated from cure kinetics parameters
(Figure 3) for this system essentially is linear through the cure
region. The excellent fit obtained with the linear least squares
regression over the temperature range of the cure reaction
confirms the validity of the nth order kinetic model used to
describe the cure of the uncatalyzed gel coat resins.

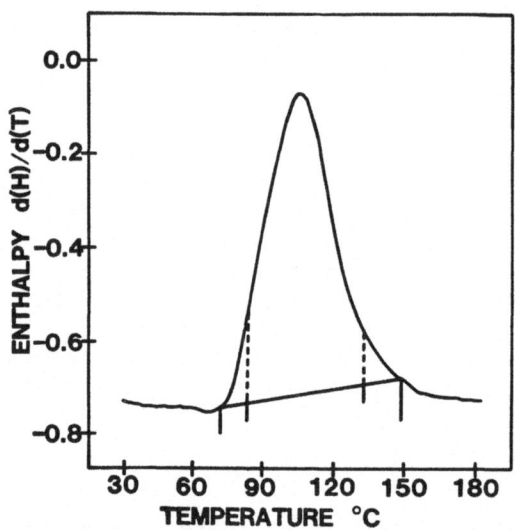

Fig. 2. DSC scan of curing uncatalyzed gel coat resin showing the
        reaction exotherm (72-148°C) and the temperature range
        (84-133°C) selected for kinetics evaluation.

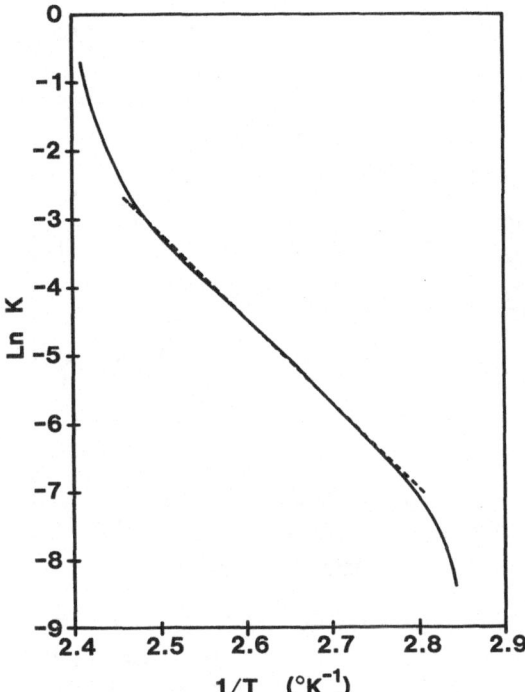

Fig. 3. Arrhenius plot for kinetics parameters determined from DSC
and a linear regression line over the temperature range
selected for kinetics evaluation.

FT-IR

Evaluation of gel coat cure kinetics were based on the
disappearance of an absorbance band at 779 cm^{-1}. This band is
associated with the out of plane C-H bending vibrations of the
styrene vinyl group and proved to be a sensitive gauge of the
ongoing chemical changes during the reaction. Figure 4 shows the
spectra of the uncured and cured material, emphasizing the
disappearance of the 779 cm^{-1} absorbance band during the reaction.
The absorbance due to the out of plane bending of the phenyl ring
at 731 cm^{-1} essentially remained unchanged during the cure
reaction and was used as a reference peak to correct for any
changes in sample thickness during heating.

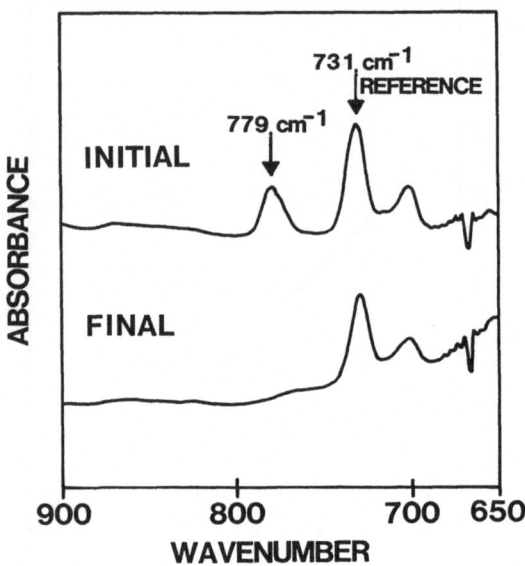

Fig. 4. Initial (uncured) and final (cured) FT-IR absorbance
        spectra of an uncatalyzed gel coat specimen. 779^{-1} =
        styrene vinyl group absorbance; 731 cm^{-1} = phenyl ring
        (reference) absorbance.

The normalized absorbance at the 779 cm^{-1} band obtained for
each spectrum during cure, e.g. like that for the uncatalyzed gel
coat resin in Figure 5, was converted into conversion versus
temperature curves shown in Figure 6. The temperature range for
the cure in the uncatalyzed resins was from $70^\circ C$ to approximately
$145^\circ C$. The kinetics parameters determined for the cure of the
uncatalyzed system by FT-IR, n = 1.54, E = 25.67 and lnA = 29.15
are in remarkable agreement with those obtained from DSC (Table
1). A similar analysis of the cure reaction of the catalyzed gel
coat resin using FT-IR generated the second conversion curve shown
in Figure 6. The similarity between these curves implies that the
cobalt catalyst is not altering the cure mechanism and supports
the hypothesis that the bimodal DSC thermograms maybe due to side
reactions unrelated to the actual cure chemistry. Analysis of the
reaction kinetics monitored by FT-IR yielded a reaction order of
1.57, and activation energy of 23.11 kcal/mole and a value of
30.10 for lnA. As expected, the primary effect of the catalyst
was a reduction in the activation energy of the cure reaction.

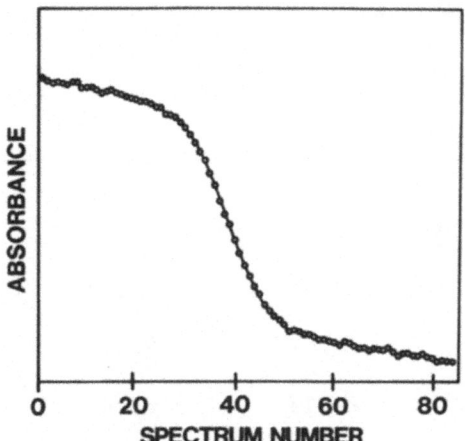

Fig. 5. Normalized FT-IR absorbance peak vs. spectrum number (temperature) for the disappearance of the styrene vinyl functionality (779 cm^{-1}) during gel coat cure.

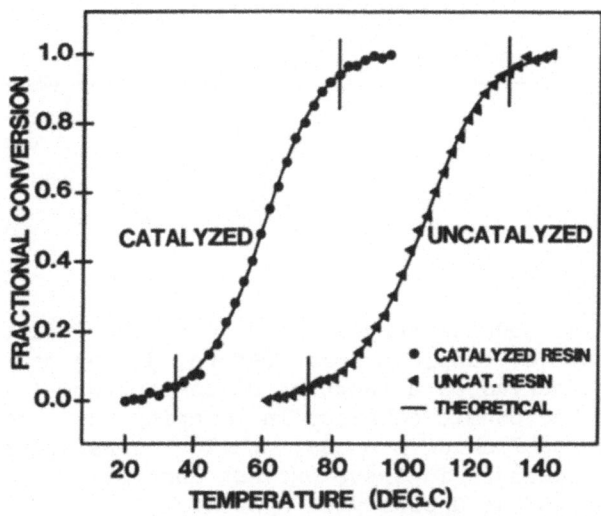

Fig. 6. Fractional conversion curves obtained from FT-IR data. (●) catalyzed; (◄) uncatalyzed; (——) calculated conversion curve showing temperature range selected for kinetics analysis.

Table 1.   Experimentally Determined Kinetics Parameters For The
           Cure of Catalyzed and Uncatalyzed Gel Coat Resins

	FT-IR	DSC	DMA
Uncatalyzed			
n	1.54	1.56	2.1
E	25.67	25.1	28.7
lnA	29.15	28.4	33.0
Catalyzed			
n	1.57	–	0.7
E	23.11	–	11.7
lnA	30.10	–	10.9

DMA

    The relative modulus curves shown in Figure 7 illustrate the
typical response of sample modulus before, during and after cure.
Degree of cure curves as a function of temperature (Figure 8) show
the effect of the cobalt catalyst in lowering the temperature
range of the crosslinking process. While the relative rate of
cure appears unchanged in these curves, cure starts and ends 75°C
to 80°C higher in the uncatalyzed system.

    Combining the experimentally determined conversion curves for
DSC and FT-IR along with the degree of cure curve from DMA for the
uncatalyzed system (Figure 9) shows clear agreement between DSC
and FT-IR results. The onset of the curing reaction measured by
DMA, however, starts about 10°C higher than the onset of the
chemical reaction as measured by either DSC or FT-IR.

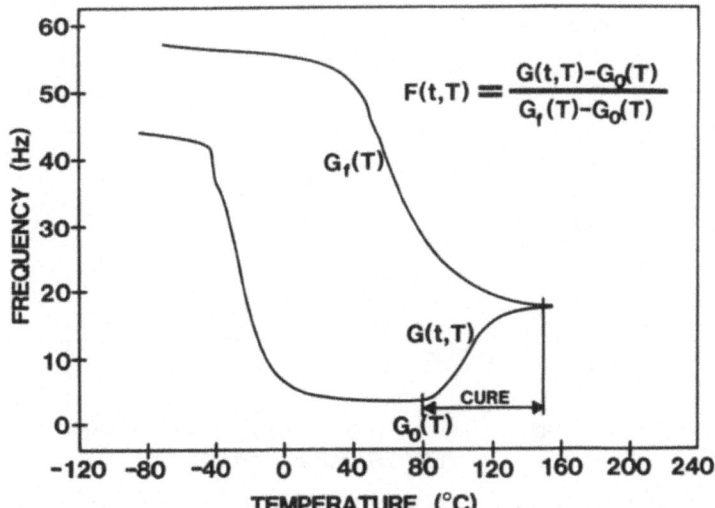

Fig. 7. Dynamic relative modulus (DMA) for the uncured (thermoplastic), curing, and fully cured (thermoset) gel coat.
$G_o(T)$ – thermoplastic modulus, $G(t,T)$ – modulus at a given
time and temperature during cure; $G_f(T)$ – thermoset modulus; $F(t,T)$ degree of cure at given time and temperature;
T_o = temperature at onset of cure; T_f = temperature at end
of cure.

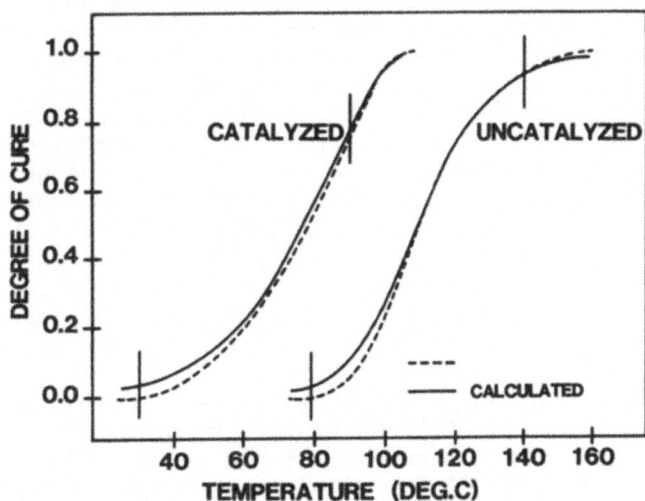

Fig. 8. Experimentally determined fractional degree of cure from
DMA data ((---), uncatalyzed and catalyzed) and calculated
conversion curves (—) showing temperature range selected
for kinetics evaluation.

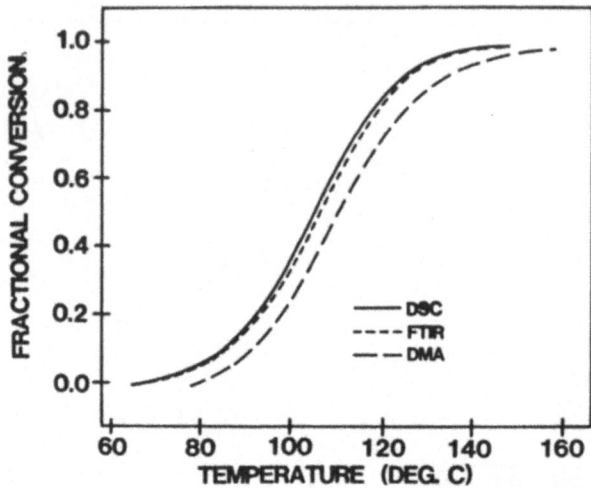

Fig. 9. Calculated fractional conversion curves (5°C/min heating
    rate). (——) DSC; (————) FTIR; (— —) DMA.

    This small but significant difference is expected since the
DMA directly measures changes in the physical properties (relative
modulus) of the curing resin as a function of temperature and time
while the DSC and FT-IR are directly measuring chemical changes
(i.e. heat of reaction and changes in reactive functionalities
respectively) during cure. Furthermore, the modulus increase
during chemical cure should start sometime after the chemical
curing reaction begins, e.g. crosslinking in epoxy resins usually
begins after some 50-60% conversion of the unreacted material. In
fact, it is likely that the onset of crosslinking observed in DMA
is directly related to the gel point of the cure reaction. If
this premise is correct, then the gel point occurs at about 10
percent chemical conversion of the polyester gel coat resins. In
DMA the onset of crosslinking is a function of molecular weight
between crosslink sites, backbone rigidity, functional group
reactivity and the viscosity of the reacting system.

    The kinetic parameters determined from the analysis of DMA
data yields expected differences and re-emphasizes the differences
in the properties monitored by the DMA as compared to DSC and
FT-IR. A comparison of the kinetics parameters determined by DMA
for the uncatalyzed and catalyzed resins reveals a drop in reaction order from 2.1 to 0.7, a decrease in activation energy from
28.7 kcal/mole to 11.7 kcal/mole and a reduction of lnA from a
value of 33.0(sec^{-1}) to a value of 10.9(sec^{-1}). Calculated degree
of cure curves fit reasonably well with experimentally determined

degree of cure curves (Figure 8), particularly in the middle and
upper regions of the curves. Variation between the two curves at
the onset of the crosslinking process may reflect some minor
inadequacies of the nth order rate expression for modeling the
chemorheological processes monitored by DMA. Nonetheless, the
results from DMA kinetics analysis clearly reflect relative
quantitative differences in the crosslinking process as a
consequence of the addition of catalyst.

Kinetics of Isocyanate Deblocking

Neat blocked isocyanates (IPDI) were used in this study to
avoid competing reactions between the isocyanate functionality and
hydroxy groups which are present in typical crosslinking reactions. Thermal dissociation of blocked isocyanates is thought to
proceed by one of two mechanisms[8]. In the first (Scheme 2),
thermal dissociation of the blocked isocyanate is followed by
crosslinking with available functional groups.

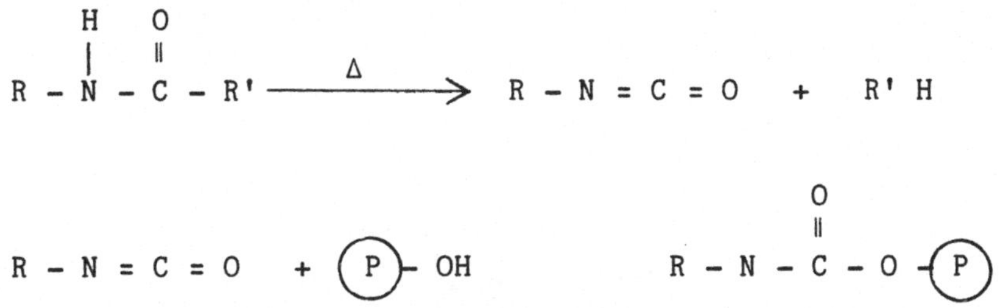

$$
\begin{array}{c} H \quad\; O \\ | \quad\;\; || \\ R-N-C-R' \end{array} \xrightarrow{\;\Delta\;} R-N=C=O \quad + \quad R'H
$$

$$
R-N=C=O \quad + \quad \textcircled{P}-OH \qquad R-N-\overset{\overset{\displaystyle O}{||}}{C}-O-\textcircled{P}
$$

Scheme 2

and in the second (Scheme 3), deblocking of the blocked isocyanate
occurs through a concerted mechanism in which the functionality of
the resin participates in the reaction. Since this study was
carried out in the absence of hydroxyl functionality, the thermal
dissociation reaction studied proceeded through the first mechanism in all cases.

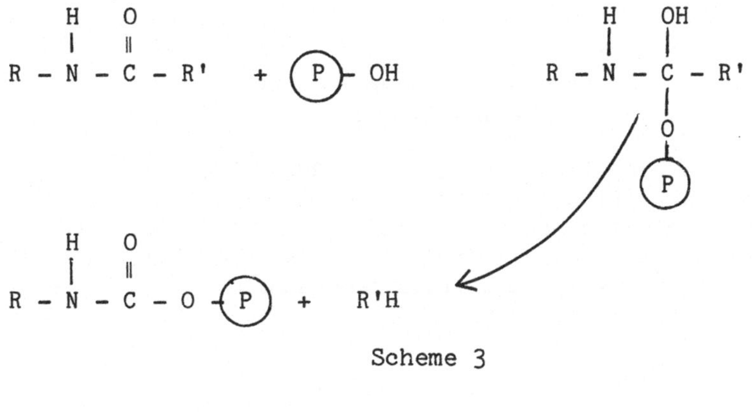

$$
\begin{array}{c} H \quad\; O \\ | \quad\;\; || \\ R-N-C-R' \end{array} + \textcircled{P}-OH \qquad \begin{array}{c} H \quad\;\; OH \\ | \qquad | \\ R-N-C-R' \\ | \\ O \\ | \\ \textcircled{P} \end{array}
$$

$$
\begin{array}{c} H \quad\; O \\ | \quad\;\; || \\ R-N-C-O-\textcircled{P} \end{array} + \quad R'H
$$

Scheme 3

Kinetics analyses were carried out only by FT-IR. Sample
volatility prevented kinetics analysis with DSC and only preliminary results of DMA kinetic studies are available, although
these results imply that the crosslinking process in some systems
cannot begin until the deblocking reaction is completed. Typical
initial and final spectra collected for the deblocking reaction of
the blocked IPDI are shown in Figure 10. Two functionalities, the
blocked isocyanate carbonyl at 1738 cm^{-1} and the blocked isocyanate N-H at 1503 cm^{-1}, disappear during the course of the
reaction. An absorbance corresponding to the isocyanate functionality at 2256 cm^{-1} subsequently appears during the reaction,
increasing in magnitude until the deblocking reaction is completed. All peaks were normalized against an aliphatic absorbance at
1446 cm^{-1} since this peak essentially remained constant throughout
the reaction.

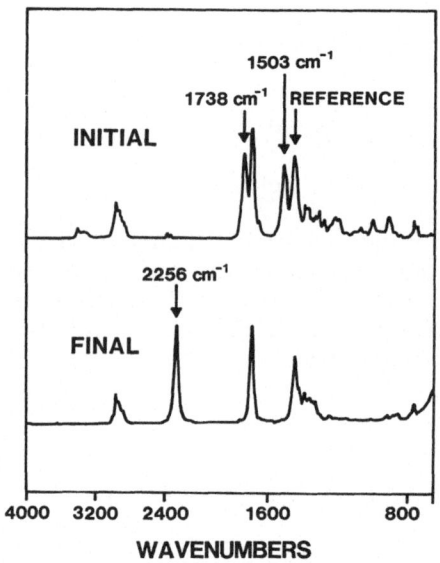

Fig. 10. FT-IR absorption spectra for blocked (initial) and
        deblocked (final) isophorone diisocyanate. 2256 cm^{-1} =
        unblocked isocyanate carbonyl; 1738 cm^{-1} = blocked
        isocyanate carbonyl; 1503 cm^{-1} = blocked isocyanate N-H;
        1446 cm^{-1} = aliphatic reference peak.

Normalized absorbance curves for the absorbances at 1738 cm^{-1}
and 2256 cm^{-1} (Figure 11) reveal that the deblocking reaction
begins near 120°C and ends at approximately 200°C. Figure 12
shows percent conversion as a function of temperature for the
experimental data and the conversion curves calculated from the
experimentally determined kinetic parameters. The excellent fit
between the experimental and calculated curves shown in Figure 12
demonstrates that the kinetics parameters determined from the
appearance of isocyanate functionality accurately describe the
thermal dissociation of the blocked IPDI.

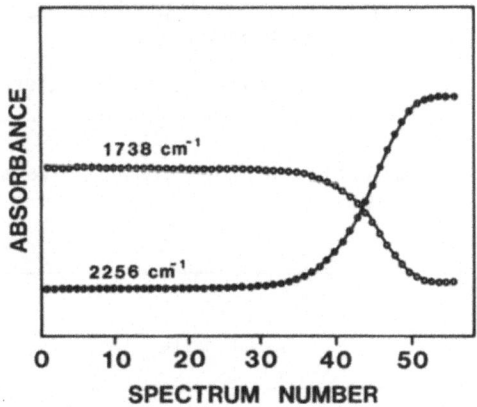

Fig. 11. Normalized FT-IR absorbance bands for the blocked
         isocyanate carbonyl (1738 cm^{-1}) and the unblocked
         isocyanate functionality (2256 cm^{-1}).

Reaction kinetics parameters were determined as averages from
three replicate runs. The most reliable kinetics parameters
(Table 2) were obtained with the isocyanate peak (2256 cm^{-1}) since
it was isolated from overlapping side peaks and had a readily
defined baseline. In fact, the kinetics parameters obtained with
the isocyanate peak for the thermal dissociation reaction agree
well with those of Kordomenos et. al.[8] as shown in Table 2.

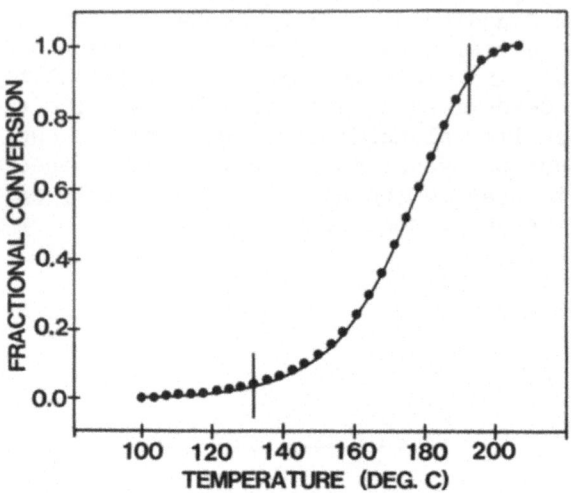

Fig. 12. Fractional conversion curve for the thermal dissociation
         reaction of the blocked isocyanate.

Table 2. Kinetics Parameters For The Thermal Dissociation of
         Blocked Isocyanates

Peak cm^{-1}	n	E (kcal/mole)	lnA (sec^{-1})
2256	0.93	23.82	21.54
1738	0.83	25.54	23.23
1503	0.78	25.56	25.55
Lit.*	1.00**	23.90	23.20

*-Kordomenos[8]
**assumed

CONCLUSIONS

    Single Dynamic temperature scan FT-IR and DSC are rapid and
effective means for directly obtaining quantitative chemical
kinetics of reacting or curing systems. Kinetics parameters
obtained from FT-IR for the thermal dissociation of blocked

isophorone diisocyanate are in excellent agreement with thermo-
gravimetrically obtained kinetics parameters for the same reaction.   Kinetics parameters determined for the cure of uncatalyzed gel coat polyester systems by dynamic temperature scan FT-IR
and DSC are in remarkable agreement. Degree of cure kinetics for
gel coat cure obtained by single dynamic temperature scan DMA seem
to indicate that the gel point for this reaction occurs at about
10% chemical conversion. FT-IR results indicate that the
competing side reactions observed during the cure of the catalyzed
gel coat resin by DSC were unrelated to the actual cure chemistry.
The combined kinetics information obtained from DSC, FT-IR and DMA
yield a detailed picture of the curing process in unsaturated
polyester gel coat systems. The concerted use of these techniques
provide a comprehensive means for assessing cure and reaction
kinetics in polymeric systems as well as providing insights into
the physics and chemistry involved in the cure mechanism.

REFERENCES

1. A. F. Kah, M. E. Koehler, T. H. Grentzer, T. F. Niemann and T.
    Provder, "An Automated Thermal Analysis System For Reaction
    Kinetics", Preprints, Org. Coatings Plastics Div., Amer. Chem.
    Soc., 45, 480 (1981).
2. T. H. Grentzer, R. M. Holsworth and T. Provder, "The
    Application Of The Dynamic Mechanical Analyzer To Organic
    Coatings", Preprints, Org. Coatings Plastics Div., Amer. Chem.
    Soc., 44, 515 (1981).
3. T. Provder, F. B. Malihi, C. M. Neag, R. M. Holsworth, and M.
    E. Koehler, "Simulation Of Non-isothermal Cure Of Coatings
    From DMA And DSC Reaction Kinetics", Preprints, Org. Coat.
    Plastics Div., Amer. Chem. Soc., 47, 493 (1982).
4. M. E. Koehler, A. F. Kah, C. M. Neag, T. F. Niemann, F. B.
    Malihi, and T. Provder, "Computer Automation Of The Dynamic
    Mechanical Analyzer And Its Application To Cure Kinetics
    Studies And Dynamic Mechanical Property Analysis Of Organic
    Coatings", this volume.
5. C. M. Neag, R. M. Holsworth, T. Provder and F. B. Malihi, "A
    Study Of Cure Kinetics By Dynamic Mechanical Analysis",
    Preprints, IUPAC Symposium, 855, 1982.
6. K. J. Saunders, Organic Polymer Chemistry, Chapman and Hall,
    Ltd., London, 1973, Chap. 10.
7. G. Odian, Principles of Polymerization, McGraw Hill, 1970, pp.
    172-178.
8. P. I. Kordomenos, A. H. Dervan, and J. Kresta, "Kinetics Of
    Thermal Dissociation Of Blocked Isocyanate Crosslinkers", J.
    Coatings Tech., 53, No. 673, 35 (1982).

CONTRIBUTORS

Ali-Asghar Alamolhoda, Chemistry Department, The University of
    Akron, Akron, OH 44325

Bernd K. Appelt, General Technology Division, IBM Corporation,
    Endicott, NY 13760

J. Bock, Corporate Research - Science Laboratories, Exxon Research
    and Engineering Company, Linden, New Jersey 07036

K.-H. Breuer, Mineralogisch-Petrographisches Institut, Universität
    Heidelberg, Germany

J. W. Bulock, Monsanto Company, 800 N. Lindbergh Blvd., St. Louis,
    MO 63167

G. Carlson, Glidden Coatings and Resins, Division of SCM Corpora-
    tion, 16651 Sprague Road, Strongsville, Ohio 44136

Jen Chiu, Polymer Products Department, Experimental Station, E. I.
    du Pont de Nemours & Company, Inc., Wilmington, Delaware
    19898

Pamela J. Cook, Baker Laboratory, Department of Chemical Engi-
    neering, Cornell University, Ithaca, NY 14850

Bill Davidson, Sciex, 55 Glen Cameron Road, Thornhill, Ontario
    L3T 1P2

Edward Donoghue, Department of Mathematics, Amherst College,
    Amherst, MA 01002

Susan M. Dyszel, U. S. Customs Service, 1301 Constitution Avenue,
    N.W., Washington, D.C. 20229

Charles M. Earnest, The Perkin-Elmer Corporation, Main Avenue - M/S
    131, Norwalk, CT 06856

395

John P. Elder, I.M.M.R., University of Kentucky, Lexington, KY
    40512

Thomas S. Ellis, Department of Polymer Science and Engineering,
    University of Massachusetts at Amherst, Amherst, MA 01003

W. Eysel, Mineralogisch-Petrographisches Institut, Universität
    Heidelberg, Germany

P. H. Foss, Institute of Materials Science and Department of
    Chemical Engineering, University of Connecticut, Storrs,
    Connecticut 06268

Paul D. Garn, Chemistry Department, The University of Akron, Akron,
    OH 44325

P. S. Gill, Clinical and Instrument Systems Division, Photo Products
    Department, E. I. du Pont de Nemours & Co., Wilmington, DE

Sharon Gorman, Departments of Chemistry and Physics, University of
    Southern Mississippi, Box 5043 Southern Station, Hattiesburg,
    MS 39406

Charles Gramelt, Research and Development, Owens-Corning Fiberglas
    Corporation, Granville, Ohio

A. R. Greenberg, Department of Mechanical Engineering, University
    of Colorado, Boulder, CO 80309

Anselm C. Griffin, Departments of Chemistry and Physics, University
    of Southern Mississippi, Box 5043 Southern Station, Hatties-
    burg, MS 39406

Joeseph Hakl, Sandoz AG, 4002 Basel, Switzerland

R. M. Holsworth, Glidden Coatings and Resins, Division of SCM
    Corporation, 16651 Sprague Road, Strongsville, Ohio 44136

William E. Hughes, Departments of Physics and Astronomy, University
    of Southern Mississippi, Box 5043 Southern Station, Hatties-
    burg, MS 39406

Valdis Ivansons, E. I. Du Pont De Nemours & Company, Central
    Research & Development Department, Experimental Station,
    Wilmington, DE 19898

Robert C. Johnson, E. I. Du Pont De Nemours & Company, Central
    Research & Development Department, Experimental Station,
    Wilmington, DE 19898

A. F. Kah, Glidden Coatings and Resins, Division of SCM Corporation, 16651 Sprague Road, Strongsville, Ohio 44136

Frank E. Karasz, Department of Polymer Science and Engineering, University of Massachusetts at Amherst, Amherst, MA 01003

Masao Kimura, Materials Research Laboratory, Polymer Science and Engineering Department, University of Massachusetts, Amherst, Massachusetts 01003

M. E. Koehler, Glidden Coatings and Resins, Division of SCM Corporation, 16651 Sprague Road, Strongsville, Ohio 44136

C. Kuo, Glidden Coatings and Resins, Division of SCM Corporation, 16651 Sprague Road, Strongsville, Ohio 44136

R. P. Kusy, Dental Research Center, University of North Carolina, Chapel Hill, NC 27514

Pierre Le Parlouër, Setaram, 101-103, rue de Sèze, 69451 Lyon Cedex 6 - France

J. D. Lear, Clinical and Instrument Systems Division, Photo Products Department, E. I. du Pont de Nemours & Co., Wilmington, DE

Paul H. Lindenmeyer, Dynamic Materials, Inc., 165 Lee Street, Seattle, WA 98109

F. B. Malihi, Glidden Coatings and Resins, Division of SCM Corporation, 16651 Sprague Road, Strongsville, Ohio 44136

L. Mandelkern, Department of Chemistry and Institute of Molecular Biophysics, Florida State University, Tallahassee, Florida 32306

P. J. M. Mathieu, Department of Chemistry and Institute of Molecular Biophysics, Florida State University, Tallahassee, Florida 32306

J. J. Maurer, Corporate Research - Science Laboratories, Exxon Research and Engineering Company, Linden, New Jersey 07036

Donald L. Melchior, Department of Biochemistry, University of Massachusetts Medical School, Worcester, Massachusetts 01605

Thomas E. Munns, Polymeric Composites Laboratory, Department of Chemical Engineering, University of Washington, Seattle, Washington

Takayuki Murayama, Monsanto Triangle Park Development Center, Post
      Office Box 12274, Research Triangle Park, North Carolina
      27709

C. M. Neag, Glidden Coatings and Resins, Division of SCM Corpor-
      ation, 16651 Sprague Road, Strongsville, Ohio 44136

T. F. Niemann, Glidden Coatings and Resins, Division of SCM Corpor-
      ation, 16651 Sprague Road, Strongsville, Ohio 44136

Genia Paul, Mettler Instrument Corporation, Hightstown, NJ 08520

Roger S. Porter, Materials Research Laboratory, Polymer Science and
      Engineering Department, University of Massachusetts, Amherst,
      Massachusetts 01003

R. Bruce Prime, IBM Corporation, San Jose, CA

T. Provder, Glidden Coatings and Resins, Division of SCM Corpor-
      ation, 16651 Sprague Road, Strongsville, Ohio 44136

Krishnan Rajeshwar, Department of Electrical Engineering, Colorado
      State University, Fort Collins, CO 80523

D. N. Schulz, Corporate Research - Science Laboratories, Exxon
      Research and Engineering Company, Linden, New Jersey 07036

James C. Seferis, Polymeric Composites Laboratory, Department of
      Chemical Engineering, University of Washington, Seattle,
      Washington

M. T. Shaw, Institute of Materials Science and Department of
      Chemical Engineering, University of Connecticut, Storrs,
      Connecticut 06268

S. H. Shaw, Institute of Materials Science, University of Connecti-
      cut, Storrs, CT 06268

Bori Shushan, Sciex, 55 Glen Cameron Road, Thornhill, Ontario
      L3T 1P2

D. B. Siano, Corporate Research - Science Laboratories, Exxon
      Research and Engineering Company, Linden, New Jersey 07036

G. M. Stack, Department of Chemistry and Institute of Molecular
      Biophysics, Florida State University, Tallahassee, Florida
      32306

E. P. Tam, Monsanto Company, 800 N. Lindbergh Blvd., St. Louis, MO
      63167

R. A. Weiss, Institute of Materials Science, University of Connecticut, Storrs, CT 06268

Lecon Woo, Material Development, Travenol Laboratories, Inc., Round Lake, IL 60073

H. K. Yuen, Monsanto Company, 800 N. Lindbergh Blvd., St. Louis, MO 63167